T0382950

Nanobiotechnology & Nanobiosciences

Volume **1**

Pan Stanford Series on **NANOBIOTECHNOLOGY**

Nanobiotechnology & Nanobiosciences

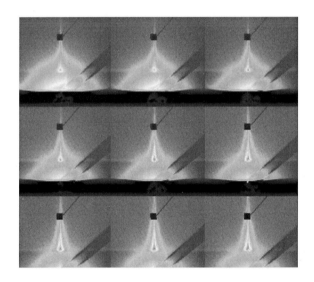

CLAUDIO NICOLINI
University of Genoa, Italy

PAN STANFORD PUBLISHING

Published by

Pan Stanford Publishing Pte. Ltd.
5 Toh Tuck Link
Singapore 596224

Distributed by

World Scientific Publishing Co. Pte. Ltd.
5 Toh Tuck Link, Singapore 596224
USA office: 27 Warren Street, Suite 401-402, Hackensack, NJ 07601
UK office: 57 Shelton Street, Covent Garden, London WC2H 9HE

British Library Cataloguing-in-Publication Data
A catalogue record for this book is available from the British Library.

Pan Stanford Series on Nanobiotechnology — Vol. 1
NANOBIOTECHNOLOGY AND NANOBIOSCIENCES

Copyright © 2009 by Pan Stanford Publishing Pte. Ltd.

All rights reserved. This book, or parts thereof, may not be reproduced in any form or by any means, electronic or mechanical, including photocopying, recording or any information storage and retrieval system now known or to be invented, without written permission from the Publisher.

For photocopying of material in this volume, please pay a copying fee through the Copyright Clearance Center, Inc., 222 Rosewood Drive, Danvers, MA 01923, USA. In this case permission to photocopy is not required from the publisher.

ISBN-13 978-981-4241-38-0
ISBN-10 981-4241-38-5

Printed in Singapore by Mainland Press Pte Ltd

To My Mother Camilla

Preface

This volume introduces in a coherent and comprehensive fashion the Stanford Series on Nanobiobiotechnology by defining and reviewing the major sectors of Nanobiotechnology and Nanobiosciences with respect to the most recent developments. Nanobiotechnology indeed appears capable of yielding a scientific and industrial revolution along the routes correctly foreseen by the numerous programs on Nanotechnology launched over the last decade by numerous Councils and Governments worldwide, beginning in the late 1995 by the Science and Technology Council in Italy and by the President Clinton in USA and ending this year with President Putin in Russian Federation.

The aims and scope of the Series and of this Volume are to cover the basic principles and main applications of Nanobiotechnology as an emerging field at the frontiers of Biotechnology and Nanotechnology. The publishing policy of the Series consists of at least one volume per year up to two per year with the title and authors chosen among leading scientists active in the field and will have the form of full manuscript single or multiauthored, and of a coherent collection of chapters-reviews edited by one or more leading scientists to cover a given topics. A Scientific Committee formed by Wolfgang Knoll (Max Planck Institute Mainz), Joshua LaBaer (Harvard University, Boston), Michael Kirpichnikov (Moscow University) and Christian Riekel (ESRF, Grenoble) has been established by the Publisher to act as an Editorial Board to discuss with myself as Series Editor for the purpose of sourcing and peer-reviewing manuscripts and providing advice on the series of future high quality Volumes.

I am particularly grateful to all numerous members of the Nanoworld Institute and Fondazione EL.B.A. active in this field cited in the list of publications contained in the Bibliography. Our research activities here reported were supported by several multinational companies, by the Italian Ministry of University and Scientific - Technological Research through an annual allocation granted to the Fondazione EL.B.A. and through numerous FIRB and FISS research contracts on Organic and Biological Nanosciences and Nanotechnology to both the Fondazione Elba and the CIRSDNNOB-Nanoworld Institute of the University of Genoa, including a very recent one on Functional Proteomics jointly with Harvard University.

Claudio Nicolini
(www.claudionicolini.it)

Genoa, 30 March 2008

Contents

Preface	vii
1. Nanoscale Materials	1
1.1 Produced Via LB Technology	2
1.1.1 Cells	2
1.1.2 Proteins	5
1.1.2.1 Light sensitives	6
1.1.2.2 Metal-containings	8
1.1.2.3 Others	12
1.1.3 Genes and Oligonucleotides	14
1.1.4 Lipids and Archaea	16
1.2 Produced Via Organic Chemistry	17
1.2.1. Conductive polymers	18
1.2.1.1 Amphipilic conjugated polymers	22
1.2.2 Carbon nanotubes and their nanocomposites	25
1.2.2.1 Interactions between conjugated polymers and single-WN	32
1.2.2.2 SWNT for hydrogen storage	33
1.2.3 Nanoparticles	34
1.3 Produced Via LB Nanostructuring	37
1.3.1 Langmuir-Blodgett	39
1.3.1.1 Protective plate	41
1.3.1.2 Heat-proof and long range stability	43
1.4 Produced Via APA Nanostructuring	48
1.4.1 Focus ion beam	49
2. Nanoscale Probes	54
2.1 Surface Potential	54
2.2 Atomic Force Microscopy	58
2.2.1 AFM spectroscopy	64
2.2.2 Scanning tunneling microscopy	65

2.3 Nuclear Magnetic Resonance 67
  2.3.1 Circular dichroism 69
2.4 Brewster-Angle Microscopy 70
  2.4.1 Ellipsometry 71
2.5 Electrochemistry 72
2.6 Infrared Spectroscopy 74
2.7 Nanogravimetry 77
  2.7.1 Quality factor 79
2.8 Biomolecular Microarrays 81
  2.8.1 Gene expression via DNASER 82
  2.8.2 Protein expression via Nucleic Acid Programmable Array ... 86
2.9 Biophysical Informatics 90
  2.9.1 Bioinformatics 91
  2.9.2 Biophysical molecular modelling 93
    2.9.2.1 Three-dimensional structure of octopus rhodpsin 94
    2.9.2.2 Three-dimensional structure of cytochrome P450scc . 96
    2.9.2.3 Protein crystallization 97
    2.9.2.4 Nanobiodevice implementation 100
2.10 Mass Spectrometry 101
  2.10.1 Mass spectrometry of label-free NAPPA 105
2.11 Synchrotron Radiation 106
  2.11.1 Diffraction 109
  2.11.2 Grazing Incidence Small Angle X-ray Scattering 111

3. Nanoscale Applications in Health and Science   118
  3.1 Nanobiocrystallography 118
    3.1.1 Radiation resistance 127
    3.1.2 New protein structures 128
    3.1.3 Three-dimensional engineering 132
    3.1.4 Basics of crystal formation 135
      3.1.4.1 Cytochrome P450scc 136
      3.1.4.2 Lysozyme 145
  3.2 Nanomedicine 148
    3.2.1 Carbon nanotubes biocompatibility and drug delivery 148
    3.2.2 Photosensitization of titanium dental implants 152
    3.2.3 Biopolymer sequencing and drug screening chip 155
  3.3 Nanogenomics 158
    3.3.1 Human T lymphocytes cell cycle 158
    3.3.2 Organ transplants 167
    3.3.3 Osteogenesis 173
  3.4 Nanoproteomics 174
    3.4.1 Cell cycle 175

		3.4.2 Cell transformation and differentiation 180

- 3.5 Nanomechanics and Nanooptics 183
    - 3.5.1 Nanocontacts for addressing single-molecules 183
    - 3.5.2 Nanofocussing 191
    - 3.5.3. Optical tweezers 193
    - 3.5.4 Magnetism 195
- 3.6 Cell Nanobioscience 197
    - 3.6.1 Nucleosome core 197
        - 3.6.1.1 DNA deformation 201
        - 3.6.1.2 Water and ions 201
    - 3.6.2 Protein stability to heat and radiation 202
        - 3.6.2.1 Bioinformatic analysis 204
        - 3.6.2.2 Structural comparisons 207
        - 3.6.2.3 Structural comparisons of homologous thermophilic/mesophilic pairs 207
        - 3.6.2.4 Water comparisons of homologous thermophilic/mesophilic pairs 209
        - 3.6.2.5 Detailed comparison of mesophilic versus thermophilic thioredoxin 212

4. Nanoscale Applications in Industry and Energy Compatible with Environment   219
    - 4.1 Nanobioelectronics 220
        - 4.1.1 Nanosensors 220
            - 4.1.1.1 Protein-based nanosensors 220
            - 4.1.1.2 Organic nanosensors 237
        - 4.1.2 Passive elements 246
            - 4.1.2.1 Resistors 247
            - 4.1.2.2 Capacitors 250
            - 4.1.2.3 Wires 254
        - 4.1.3 Active elements 255
            - 4.1.3.1 Schottky diode 256
            - 4.1.3.2 Led 257
            - 4.1.3.3 Optical filtering and holography   261
            - 4.1.3.4 Displays 265
            - 4.1.3.5 Monoelectronic transistors 267
        - 4.1.4 Quantum dots and quantum computing 271
    - 4.2 Nanoenergetics Compatible with Environment 275
        - 4.2.1 Photovoltaic cells 275
            - 4.2.1.1 Reaction centers-based 278
            - 4.2.1.2 Purple-membrane based 279
        - 4.2.2 Batteries 282

>   4.2.2.1 Lithium ion batteries elements 285
>   4.2.2.2 The cathode 285
>   4.2.2.3 The anode 288
>   4.2.2.4 The electrolyte 291
>   4.2.3 Hydrogen storage and fuel cells 296
>   4.3 Nanobiocatalysis 301
>   4.3.1 Bioreactors 306
>   4.3.1.1 From lab scale to industrial scale 309
>   4.3.2 Bioactuators 312

Bibliography   313

Index   363

Chapter 1

Nanoscale Materials

This chapter overviews the present status of new materials by organic and biological nanotechnology and their applications with respect to the development of organic and biological nanotechnology defining Nanobiotechnology and Nanobiosciences capable to yield a significant scientific and technological progress.

Particular emphasis is placed on what has been accomplished in our laboratory in the last eight years, whereby the details on the supramolecular layer engineering and its application to industrial nanotechnology can be found in recent complete reviews (Nicolini *et al.*, 2001; 2005; Nicolini and Pechkova, 2006). Material technology has changed our lives within only a few decades. Sand, the starting material, has been turned into a versatile "high-tech" product. Techniques have been developed for the production of silicon wafers and for the modification of this raw material into the final "high-tech" product - miniature electronic logic functions. Research now focuses on the molecular manufacturing (Nicolini, 1996c), from the microstructures to the nanostructures for a new generation of nanomaterials. The rapid increase in our knowledge of the function of both biological and organic materials has also directed interest into this area. The functionality, efficiency, and flexibility of biomaterials (see paragraph 1.1) are impressive; they are beyond the capabilities of synthetic chemistry already quite powerful (see paragraph 1.2).

The development of gene-technological methods has opened the way to the controlled modifications of proteins, which appear already in position to allow the production of newly designed biomaterials. Several molecules of native organic and biological origin being investigated in the last years are here summarized to exemplify potentially useful nanomaterials. They can be divided into two major classes, namely those produced via biotechnology and those produced via chemical synthesis.

1.1 Produced Via LB Technology

1.1.1 *Cells*

Several well-established cell culture systems, like mammalian CHO-K1, H9c2 or HeLa cells or microorganisms, are utilized in different context for various applications (Adami *et al.*, 1992, 1995a; Hafemann *et al.*, 1988; Garibaldi *et al.*, 2006, Spera and Nicolini, 2007).

Chinese hamster ovary fibroblasts (CHO clone K1) for Nanoproteomics, as supplied by American Type Culture Collection, Rockville, MD, is a stable hypodiploid cell line derived by spontaneous transformation from a fibroblast culture (Misteli, 2001). Cells were cultured in F12 medium supplemented with 10% fetal calf serum and 0.2% gentamycin at 37 °C in 5% CO_2 atmosphere. To reverse transform the cells they were treated with a solution 10^{-3} M dibutyryl cyclic 3',5' monophosphate adenosine sodium salt (Bt_2cAMP, Sigma Chemical Co., St. Louis, MO) that was added to the normal culture medium for six hours before the analysis. It has been shown that single cells of CHO-K1 in the native state grow equally well on plastic surfaces or in suspension (Figure 1.1). In the presence of reverse transformation conditions, however, excellent growth is still achieved on the plastic surface but no growth whatever occurs in suspension (Hsie and Puck, 1971) as monitored by a phase contrast microscope (Wilovert, Wesco).

Primary cultures of rat hepatocytes for Nanosensors were also used for experiments with biosensing potentiometric systems because they are easy to obtain and, like all hepatocytes, they maintain, in the first hours *in vitro*, their metabolic skills practically unchanged with respect to the *in vivo* situation (Nicolini *et al.*, 1995a). Hepatocytes were isolated from liver of Sprague-Dawley albino rats (200–250 g) by *in situ* collagenase perfusion according to Williams (Williams, 1977).

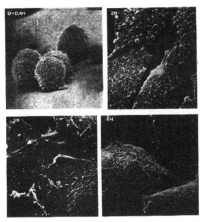

Figure 1.1 Chinese hamster ovary fibroblasts (CHO clone K1). Electronic micrographs of CHO-K1 cells at t equal 0 hour (mitosis), 2 hours (G1 phase), 3 hours (S phase) and 6 (G2 phase) hours after mitosis. (Nicolini and Rigo, Biofisiche e Tecnologie Biomediche, Zanichelli Editore, 1992).

Cardiac muscle cells for Drug Delivery via Carbon Nanotubes are a rat heart cell line H9c2 (2-1), obtained from American Type Culture Collection (Rockville, MD) at passage 16. Cells at passages from 22 to 24 were used for the experiments. H9c2 cells were cultured in Dulbecco's modified Eagle's medium with 4 mM L-glutamine adjusted to contain 1.5 g/l sodium bicarbonate, 4.5 g/l glucose, 1.0 mM sodium pyruvate, 10% FBS, with 50 units penicillin/ml, 100 mg streptomycin/ml at 37°C in 5% CO_2, and the medium was changed every 2–3 days. Subconfluent cells were detached with trypsin and seeded 24 hours before treatment at a density of 25000–30000 cells/cm^2 in 100 mm petri dishes. Cells were seeded in Petri dishes and let to grow for 24 hours before treatments. While untreated control cells were administered the complete medium without nanotubes, sonicated in parallel with Single Wall NanoTube (SWNT) suspension, treated cells were administered complete medium with suspended 0.2 mg/ml SWNT for serial time steps. A positive control was set to induce cell damage and death by treating H9c2 with 200 μM and 1 mM hydrogen peroxide. The experiments were performed in triplicate. Cardiomyocytes treated or not with SWNT were evaluated at relevant time points by light microscopy to assess cell proliferation and viability. Digital photomicrographs were recorded and the number of viable cells computed. Myocytes were incubated for 24, 48 and 72 hours at standard culture conditions to determine the viability following treatments. The number of viable cells was determined with trypan blue exclusion. In brief, cell monolayers were rinsed twice with PBS and resuspended with trypsin and EDTA. The cells were immediately stained with 0.4% trypan blue, and the number of viable cells was determined using a hemocytometer under a light microscope. In order to assessment apoptosis the cells were labeled with annexin V-FITC and propidium iodide, and fluorescent positive cells were detected by flow cytometry. Cellular fluorescence was determined by a flow cytometry apparatus (FACS-SCAN, Becton-Dickinson, Franklin Lakes, NJ, USA). Flow cytometric analysis was performed on a minimum of 1×10^4 unfixed cells per sample. Cardiomyocytes were trypsinized with trypsin-EDTA, resuspended in PBS, and loaded with 10 µM propidium iodide (PI; Sigma-Aldrich) and 1 µM annexin V-FITC (Sigma Aldrich) at 4°C for 10 min. Apoptosis was identified as cells with low forward scatter (FSC) on side scatter (SSC)/FSC dot plots, PI dim staining on FSC/PI dot plots, and annexin V-positive and PI-negative staining on annexin V/PI dot plots. Fluorescence probes were excited with an 488 nm argon ion laser. The emission fluorescence was monitored at 525 nm for annexin V-FITC and 620 nm for PI. Data were analyzed with the Cell Quest software package (Becton-Dickinson).

Human T lymphocytes for Nanogenomics, as previously shown (Nicolini *et al.*, 2006a) are obtained by a sample of heparinized peripheral blood specimens diluted 1:1 with sterile phosohate buffer saline (PBS, pH 7.2). The sample was centrifuged on a Ficoll-Hypaque gradient (specific gravity 1.080) at 1500 rpm for 40 min at room temperature without brake (Abraham *et al.*, 1980). The lymphocytes rich interface was collected and washed twice with PBS, and the number of cells counted in presence of trypan blue. The 2×10^6 cells were seeded into each well containing 2ml of RPMI 1640 culture medium supplemented with 10% heat-inactivated fetal bovine serum, penicillin 100 unit/mL and streptomycin 50 µg/mL. The cells has then been incubated at 37 °C in a humid chamber with 5% of CO_2 for 24, 48, 72 hours. The cells have been then collected and counted and then it has been proceeded to RNA extraction.

1.1.2 *Proteins*

Recombinant protein expression and purification is the single most important prerequisite for the effective engineering of nanostructured protein-based materials.

Typically the cDNA encoding the mature form of the protein being immobilized is subcloned into the proper expression vector, highly expressed in *Escherichia coli* and the expressed protein is typically purified by affinity chromatography (Pernecky and Coon, 1996; Ghisellini *et al.*, 2004; Wada *et al.*, 1991; Amann *et al.*, 1988) and evaluated spectrophotometrically using the appropriate extinction coefficient at the appropriate wavelength.

In the case of overall cell proteins extraction, as required in Mass Spectrometry at the nanoscale, a protein extraction kit (Subcellular Proteome Extraction Kit, Calbiochem) is typically chosen for the total protein extraction from cells (Spera and Nicolini, 2007). It is designed for extraction of cellular proteins from adherent and suspension-grown cells according to their subcellular localization. For the sequential extraction of the cell content, the kit takes advantage of the differential solubility of certain subcellular compartments in special reagent mixtures. Upon extraction of culture cells, four partial proteomes of the cells are obtained:
- cytosolic proteins,
- membrane and membrane organelle proteins,
- nuclear proteins,
- cytoskeleton proteins

The kit contains four extraction buffers, a protease inhibitor cocktail to prevent protein degradation and a nuclease (Merck) to achieve an efficient removal of contaminating nucleic acids. The sub-cellular extraction was performed according to the protocol for freshly prepared adherent culture cells. We performed extraction from 10^6 CHO-K1 cells in logarithmic phase at 80% confluence (Spera and Nicolini, 2007). To monitor the extraction procedure, morphological changes of the cells were examined by a phase contrast microscope (Wilovert, Wesco). The protein fractions were stored at -80°C until the Mass Spectrometry analysis. The amount of total proteins recovered from each extraction step has been evaluated by Bradford assay (Tjio and Puck, 1958), using bovine serum albumin as standard.

1.1.2.1 *Light sensitives*

Photosynthetic reaction centers from *Rhodobacter sphaeroides* and bacteriorhodopsin (bR) from purple membrane (PM) have been used for their unique optoelectronic properties and for their capability of providing light-induced proton and electron pumping.

Once assembled they display extremely high thermal and temporal stability (Nicolini 1997a, 1998a,b, Nicolini *et al.*, 1999). These features make them two of the most promising biological molecules for developing devices (Nicolini 1996a, Nicolini *et al.*, 2005). RC proteins can conduct electrons in only one direction in presence of a suitable light source (photoconduction) or of an external electric field (conduction). Both these conductive processes are due to the tunnelling of a single electron into the structure of the protein.

In the case of bR the situation is different (Figure 1.2). BR is the main part of purple membranes (about 80%) and is already close packed in it. It is difficult to extract bR in the form of individual molecules, for they are very unstable (Shen *et al.*, 1993). The modification of the retinal protein bacteriorhodopsin gave birth to the generation of new versatile media for optical processing (Zeisel and Hampp, 1992). Gene technology as the key technique in this new direction of nanomaterials research may be the counterpart to photolithography in semiconductor technology. The design of complex molecular functions is a goal extremely difficult to achieve with the classical approach of chemistry and supramolecular chemistry, whereas nature developed a large variety of such "nanomaterials". It is indeed now possible to begin the construction of such complex nanomaterials *ab initio*, by starting out from the molecules that we discover in nature and try to modify their properties and adapt them towards the demands of technical applications, for both metalloproteins and light-sensitive proteins. Modification of the genetic code facilitates manipulations of organisms that can produce these new "high-tech" nanomaterials with conventional biotechnological methods.

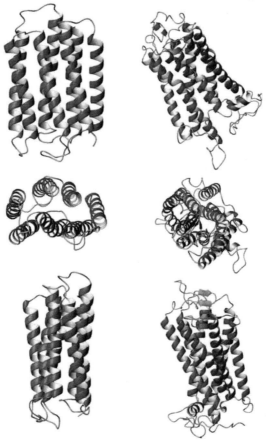

Figure 1.2. PDB image of (left) Bacteriorhodopsin (PDB id 2BRD, Grigorieff, *et al.*, 1996) and (right) Octopus Rhodopsin (PDB id 2AUL, Sivozhelezov and Nicolini 2006), top) front view, middle) top view, bottom) side view.

Therefore, the price of such products will be affordable. An intensive screening for functional biopolymers with technically relevant functions and their variation and identification will create a new class of "biomimetic" materials. The experimental studies done with bacteriorhodopsin have shown that this new approach leads to competitive materials in selected areas of opto-electronics and optical filtering; a similar route is presently being pursued with metallo-proteins also here summarized, as it will be shown see later.

1.1.2.2 *Metal-containings*

Expression and purification of a typical metal-protein, *i.e.* recombinant native P450scc, occur in *E. coli* JM109cells, transformed with pTrc99A-P450scc expression plasmid, being grown and induced as described in Wada *et al.* (1991) with few modifications.

The δ-aminolevulinic acid (1 mM), a precursor of heme biosynthesis, was added at the same time as isopropyl-1-thio-β-D-galactopyranoside (1 mM) and the cells were grown for 72 h at 28 °C by shaking at 150 rpm. Typically expressed cytochrome P450scc is purified by three different chromatographic steps: DEAE cellulose, hydroxyapatite and Adx-sepharose 4B columns. The sample is solubilized in 10 mM K phosphate buffer (pH 7.4) containing 0.1 mM EDTA, 0.2 sodium cholate and 20% glycerol and stored at -80°C. The properties of two P450scc cytochromes, namely *wild-type* versus recombinant, were systematically characterized in structural and functional terms (Nicolini *et al.*, 2001), in order to probe if conformational variations are associated with the two different processing pathways used for protein maturation (co-translational and post-translational). Initially sequencing was carried out using Pulsed-liquid Phase Sequencer Mod. 477A (Applied Biosystems) in order to determine the primary structure of both proteins. As recombinant type with two signal sequences (secretory and mitochondrial ones) and *wild-type* ("wt") with mitochondrial sequence were examined according to the same routine, the obtained results show identity of the primary structure of the N-terminus of both proteins (Ghisellini *et al.*, 2002, 2004). The first aminoacid was determined to be Ile, which agrees with theoretical suggestion about the position of the digestion site of the signal protease. These data give evidence that the recombinant protein is the mature one and goes trough the entire processing pathway. The fact that the recombinant construction has two functioning digestion sites for signal proteases seems to be very promising in view of biotechnological applications. The identity of their primary structure, in the presence of significant structural (Figure 1.3) and functional alterations, is consistent with the hypothesis that the processing pathways have a role in determining the function and the structure of proteins. Proteins are either *wild-type*, namely isolated from the corresponding cell tissues, or *recombinant*, namely obtained from clones properly genetic engineered with quite high yield and extreme purity and homogeneity.

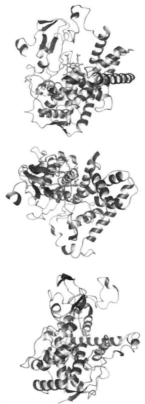

Figure 1.3. PDB image of cytochrome P450scc, top) front view, middle) top view, bottom) side view (Sivozhelezov *et al.*, 2006).

In the latter, the cDNA of the cytochrome P450scc obtained from genomic library of the bovine surrenal cortex (precursor gene) was cloned "down-stream" the secretory signal of *Kluviromyces lactis* in the ORF of GAL UAS promoter in the vector derived from pYeDP 1/10.

The yield of the product detected in the culture broth after induction of GAL UAS promoter was quite satisfactory.

The signal-appended cytochrome P450scc was translocated across the cells compartments and processed to yield authentic, haem-assembled cytochrome P450scc and was exported from the yeast cells. The eventual processing of the P450scc precursor was somehow unstable but unusual with respect to the presence of mitochondrial signal and the known reaction specificity of signal peptidase. The secreted protein was tested spectrophotometrically, electrophoretically, immunologically and by visual transformation of the culture broth into red color. Thus, the mitochondrial membrane protein (that carries the signal of post-translational transfer) was efficiently processed to mature haemo-protein and secreted from *S. cereviseae* by way of co-translational transfer. On the basis of the HPLC data it was concluded that the purity level of this protein as secreted was extremely high (approximately 98%). The production rate of correctly matured exported cytochrome P450scc was 10.2 mg per liter of culture. The obtained mature haem-bound cytochrome P450scc was indistinguishable in many of its characteristics from the native counterpart or was even better, with the only exception that its production was quite unstable. Because of this instability we have then utilized homologous expression within the *E. coli* bacteria and isolated the same cytochrome P450scc with similarly high degree of purity and yield, but with the needed reproducibility and stability (Ghisellini *et al.*, 2002, 2004). The problem of obtaining metalloproteins from microorganisms is crucial for development of applications in nanobiotechnology. As shown above, in comparison with a method of extraction from natural source this approach is cheaper and more advantageous for obtaining high quality protein of interest in large amounts. Cytochrome P450scc in nature is localized in inner membrane of mitochondria of the bovine adrenal cortex and plays a crucial role in the steroid metabolism. This protein is globular, contains the heme group and two active sites: "oxygen pocket" and "substrate pocket". The NADPH, adrenodoxine and adrenodoxine reductase are necessary for the functioning of the protein as a "mini chain of electron transfer". The hem-assembled P450scc in some of the characteristics is superior to its natural analogue. It can have as substrates cholesterol (5-cholesten-3-

βol), styrene (99% pure grade), clozapine (8-chloro-11-(4'-methyl)piperazine-5-dibenzo[*b,e*]-1,4-diazepine) and several other chemicals. Cytochrome P450scc native recombinant *(*product of CYPA11A gene*)* was cloned in *E. coli* system expression: cDNA gene of P450scc mature form was sub-cloned in the pTrc99A vector to obtain bacterial expression. The cDNA gene of the mature protein was obtained by deleting the N-terminal mitochondrial targeting sequence coding the first 39 aminoacid residues (Wada *et al.,* 1991; Amann, *et al.,* 1988).

Figure 1.4. PDB image of crystal structure of the catalytic subunit of human protein kinase CK2 (PDB ID 1NA7), top) front view, middle) top view, bottom) side view (Pechkova *et al.,* 2003).

The cDNA encoding for another mature form of cytochrome called P4502B4 was also subcloned for applications in Nanobiotechnology into the pGEX-KN expression vector and highly expressed in *Escherichia coli* by isopropylβ-d-thiogalactoside (IPTG) induction. The expressed protein, fused to glutathione-S-transferase (Figure 1.4) was purified by glutathione sepharose 4B affinity chromatography (Pernecky and Coon

1996). The purified cytochrome P4502B4 recombinant was in a 10 mM K-phosphate buffer (pH 7.4) containing 0.1 mM EDTA, 0.005% Tween 21 and 20% glycerol.

Cytochrome P4501A2 cDNA clone was inserted into the expression vector and used to transform *Escherichia coli*. Extraction and purification of cytochrome P4501A2 recombinant was performed according to Fischer *et al.* (1992). The purified cytochrome P4501A2 was solubilized in 100 mM K phosphate buffer (pH 7.25) containing 1 mM EDTA, 1 mM DTT, and 20% glycerol. Protein concentration was determined using the BCA assay (Pierce) or by the Bradford method, using bovine serum albumin as a standard. 10% SDS-polyacrylamide gel electrophoresis was performed as described by Laemmli (1970). Spectra were recorded using a JASCO 7800 Spectrophotometer (Japan) at room temperature. The purified cytochromes (P4501A2, P4502B4, P450scc) were found to be almost completely in the low spin iron configuration (Guryev *et al.*, 1996, 1997). The heme content for all iso-forms was measured according to Omura and Sato (1964) using an extinction coefficient of 91 mM^{-1} cm^{-1} for the absorbance difference between 450 and 490 nm. The concentration of the recombinant cytochrome is 1.2 mg/ml, and the pure grade of each sample was determined taking into consideration the absorbance values at A_{280} and the absorbance maximum at A_{417} (Soret peak). The absorbance A_{417}/A_{280} ratios for cytochromes P4501A2, P4502B4, and P450scc were 0.5, 0.8, and 0.9 respectively.

1.1.2.3 *Others*

Other different proteins, both water-soluble and membrane-bound, have been utilized and extensively characterized by a wide variety of biophysical probes in our efforts to develop new protein-based nanotechnology (Nicolini, 1995; Nicolini and Pechkova, 2006).

In the case of most proteins, the best way to examine the 3D structure is to make X-ray analysis. However, some proteins are not suitable for this technique, as the size of the crystal is too small, or the protein is very difficult to produce and cost too much to make large (submillimeter) 3D crystals. Instead, for high-resolution nuclear magnetic resonance (NMR) only minute amount of protein is necessary to make 3D atomic structure in solution. Thus NMR is very suitable for this task. As model systems for *soluble proteins* we have explored several proteins, namely glutathione-S-transferase, three metalloproteins, namely cytochromes P450scc (Figure 1.3), P4502B4 and P4501A2, lysozymes, human kinase CK2 (Pechkova *et al.*, 2003), thermophilic and mesophilic thioredoxin, an ubiquitous protein with many functions, such as in thiol-dependent redox reactions.

Table 1.1 Key proteins of interest to Nanobiotechnology present in RCSB PDB Data Bank.

Protein	Source	Method	Reference
P450scc	Bovine	Ab-initio	Sivozhelezov *et al.*, 2006b
Lysozyme LB	Egg	X-ray	Pechkova and Nicolini, 2006
Thioredoxin	*Bacillus acidocaldaricus*	NMR	Bartolucci *et al.*, 1997
Thioredoxin	*Escherichia coli*	NMR	Jeng *et al.*, 1994
Kinase CK2α	Human	X-ray	Pechkova *et al.*, 2003
IF2β	*Sulfolobus*	NMR	Vasile *et al.*, 2008
Bacteriorhodopsin	*Halobacterium halobium*	X-ray	Chou *et al.*, 1992
Rhodopsin	Octopus	Ab-initio	Sivozhelezov and Nicolini 2006

As model system for *membrane-bound proteins* we have chosen photosynthetic reaction centers from *Rhodobacter sphaeroides* and from *Rhodopseudomonas viridis*, bovine rhodopsin, octopus rhodopsin (Sivozhelezov and Nicolini 2006) and bacteriorhodopsin (bR). In the system of *Halobacterium halobium*, bacteriorhodopsin is found as a 2D crystalline lattice in purple membranes. This purple membrane protein is usually 1000 nm in diameter and 50 Å in thickness and makes part of the overall cell membrane of *Halobacteria* (Chou *et al.*, 1992).

Bacteriorhodopsin is the main protein in the biological function of these bacteria. BR converts light energy into chemical energy by transferring protons from the inside to the outside of the cell membrane, thus functioning as a light-driven proton pump. As a result, ATP is regenerated from ADP by an enzyme called ATP-synthase. Another class of proteins of wide interest is the ribosomal proteins such as Initiation Factor 2 from *Sulfolobus solfataricus* (Vasile *et al.*, 2008), and Bone

Morphogenetic Proteins representing a family of 13 proteins capable to stimulate bone formation. Nearly all the chosen proteins have been well characterized in solution by X-ray crystallography or solution NMR and their three-dimensional structure is now known at the atomic resolution for all proteins of our interests in Nanobiotechnology and Nanobiosciences as discussed later (Table 1.1).

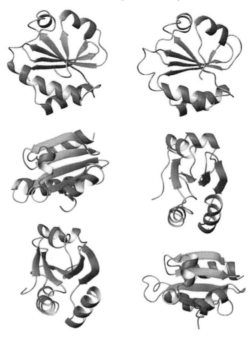

Figure 1.5. PDB image of Thioredoxin from *Escherichia coli* (PDB ID 1XOB, Jeng *et al.*, 1994) and from *Bacillus acidocaldaricus* (PDB ID 1QUW Nicastro *et al.*, 2000), top) front view, middle) top view, bottom) side view.

1.1.3 *Genes and Oligonucleotides*

To analyze either a fluorescently labelled DNA sample of unknown sequence (Jacobs and Fodor, 1994) or gene expression (Butte, 2002) in human tissues or cells, chip comprising an array of short oligonucleotides or of genes as hybridization probe constitutes the core technology of nanogenomics (Nicolini, 2006).

DNA microarray (see chapter 3) allows the study at the nanoscale of an immense amount of genes (over 10,000) with only one experiment and therefore can draw a picture of a whole genome.

The expression of each gene is analyzed by the hybridization of each gene spot with cDNA prepared by reverse transcription of total RNA isolated from cell or tissue being characterized. The CyScriptRT enzyme, together to RNA and to dCTP Cy3 and dCTP Cy5 nucleotides, is employed in a synthesis reaction lead to 42 °C for 1 hour and ½ in order to obtain the selective fluorescence labelling by retrotranscription (Figure 1.6).

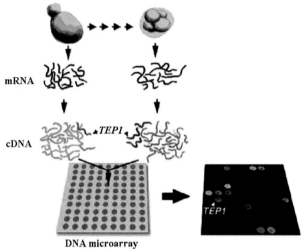

Figure 1.6. The cDNA coming from control cell and from cell to be tested are labeled with two different fluorescent probes, respectively with green (Cy3) and red (Cy5) emission, allowing thereby to monitor differentially in the same experiment the test genes and the control genes.

The synthesis product of this reaction, marked cDNA, has been then subordinate to a purification step. Total RNA was extracted and amplified using T-7 *in vitro* transcription. The cDNA marked samples have been purified employing a purification kit and chromatographic columns supplied by Amersham Biosciences. The cDNA obtained has been precipitated and resuspended in bi-distilled water to quantify, by the employing of the spectrophotometer, the samples and to verify the labelling. The cDNA marked samples have been subsequently lyophilized and resuspended in an opportune volume (120 μl) of hybridization buffer (Salt-Based ibridation). For the array hybridization 1 μg of cDNA marked with Cy3 and 1 μg of cDNA marked with Cy5 they have been mixed in a tube and resuspended in the hybridization buffer together to the control sample (Arabidopsis control). The sample thus obtained has been denatured and spotted on the array. The matrices have been then put in a hybridization chamber at 42 °C for 20 hours. To eliminate the aspecific binding the array has been washed with SSC buffers of decreasing concentration. On the other side in genome sequencing (Jacobs and Fodor, 1994) conventional solid-phase oligonucleotide synthesis was time ago involved to step-wise assembly 5'-dimethoxytrityl (DMT)-protected nucleoside monomers in the 3' to 5' direction. In a typical coupling procedure, the 5'-hydroxyl of an immobilized nucleoside is deprotected with mild acid, the liberated hydroxyl group is phosphatylated with a DMT-protected deoxynucleoside 3'-phosphoramidite, and the resulting phosphate is oxidized to a phosphotriester (Caruthers, 1985). The process is repeated until the desired oligonucleotide has been prepared. This technique was adapted to include light-directed parallel chemical synthesis by replacing the 5'-protecting group DMT with a substituted nitroveratryl derivative, and incorporating a nitroveratryloxycarbonyl (Nvoc), protected hydroxyl linker into the synthesis substrate (Fodor *et al.*, 1993). Hydroxyl groups are selectively photo-deprotected as described previously, and arrays of oligonucleotides assembled using standard peptide chemistry. It is worth to notice that using the orthogonal-stripe method (Jacobs and Fodor, 1994), it was possible time ago to assemble all 65566 possible octanucleotides (4^8) in only 32 chemical steps ($4 \times n$; where n = 8) (Pease *et al.*, 1994).

1.1.4 *Lipids and Archaea*

Archaea, a philogenetically coherent separate group of microorganisms, which differs from *Eubacteria* and *Eukaria*, comprises a variety of extremophilic bacteria living under extreme conditions such as high temperature, acidic or alkaline pH, and saturated solutions (Kandler, 1992).

Later we will see how nanotechnology may allow to mimic and even enhance their thermophilic properties in mesophilic proteins (Sivozhelezov *et al.*, 2007; Nicolini and Pechkova, 2006a). In this context we will discuss only the archaea membrane lipids useful to construct nanodevices for their unusual properties and structure (Gambacorta *et al.*, 1994), based on isoprenoid chains of different lengths with ether linkage to glycerol or to more complex polyols. In particular, membrane lipids of thermophilic archaea, such as *Sulfolobus solfataricus* (optimal growth conditions are 87 °C and pH 3), possess bipolar architecture (De Rosa *et al.*, 1983, 1986; Luzzati *et al.*, 1987) characterized by the presence of two polar heads and hydrophobic isoprenoid moiety of practically double average length with respect to that of classical ester lipids. The lipids play a key role in stabilization of the membrane under extreme conditions. Intense X-Ray scattering studies of the lipids from *S. solfataricus* carried out time ago by Gulik *et al.* (1985) pointed to a variety of phases observed when the temperature or water content in the sample was changed. In this respect, it was interesting to compare the structure of artificial multilayer systems created from these bipolar lipids with that of conventional amphiphilic molecules (Nicolini, 1996a).

1.2 Produced Via Organic Chemistry

Among the numerous organic compounds produced via chemical synthesis one of the first ones used in Nanobiotechnology was *fullerene*, with a structure consisting of C_{60} molecules investigated by X-ray diffraction (Meiney *et al.* 1991), electron diffraction (Krätschmer *et al.*, 1990) and scanning tunnelling microscopy (STM) (Wilson *et al.*, 1991) techniques. STM images of C_{60} samples show close packing of spherical molecules with lattice spacing of about 1.1 nm, in accordance with X-ray diffraction studies yielding a face-centred cubic (FCC) lattice spacing of 1.404–1.411 nm.

In the last decade are conductive polymers, carbon nanotubes, nanoparticles and their nanocomposites in combination with biopolymers to yield the most interesting properties in a wide range of applications.

The major drawback of the polymeric materials is the multi-step synthesis required for the functionalization and the stringent process requirements of the condensation polymerization. It was therefore our efforts in the last few years (Nicolini *et al.*, 2005) to find shorter synthetic routes to process the polymers with predictable absorption wavelengths of light. Several types of polymer derivatives have been synthesized in our laboratory:

- poly(p-phenylenevinylene) (PPV), namely poly(2-methoxy-5-(2'-ethyl)hexyloxy-p-phenylenevinylene) (MEHPPV), whereby the Gilch route has been modified in order to increase the processability for specific device application;
- poly(phenylene vinylene) (PPVs) are main-chain conjugated polymers, which have very interesting electrical and photoconjugated properties, and that make them suitable for applications in opto-electronics and microelectronics devices, such the PV cells;
- inorganic nanoparticles, such as CdS, PbS and TiO_2, and organic systems, namely conducting polymer, dyes, fullerene (C_{60}) nanocrystals, were prepared and organized in thin films to fabricate Donor (D) - Acceptor (A) supramolecular assemblies.

While the full details on the research efforts in nanotechnology-based organic materials can be found for the past in Nicolini *et al.* (2001a) and for the present in Nicolini *et al.* (2005), I will present in this subchapter only few key examples.

1.2.1. *Conductive polymers*

The structures of polymeric nanoscale materials are summarized in Figure 1.7. The conventional polymers, plastics, are traditionally used for their interesting chemicals, mechanicals and insulating properties, but not for theirs electronics properties.

The idea of using polymers also for their electronic semi-conductive properties in molecular electronic devices is relatively new (Su *et al.*, 1979; Nicolini, 1996a,b). Conjugated polymers behaving as insulators or semiconductors in their neutral form, can reach metallic conductivity as a result of the doping process (Salaneck and Brédas, 1994; Nicolini,

1996a). From 1977, the dream of combining the mechanical and processing properties of the polymers with the electronic and optical properties of the metals has been the driving force of the science and technology of the conjugated conducting polymers (Nicolini et al., 2001a, 2005; Nicolini, 1996a). Conjugated polymers play a fundamental role in transistors, integrated circuits and photovoltaic devices, Light Emitting Devices and solid-state laser (Nicolini, 1996a).

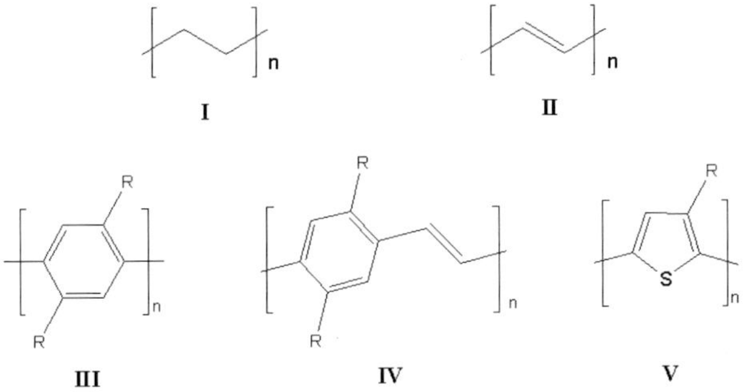

Figure 1.7. Structural formulae of some common polymers. The bonds with hydrogen atoms are not shown. (I) Polyethylene (PE); (II) *trans*-polyacetylene (PA); (III) poly(*para*-phenylenevinylene) (PPV); (IV) poly(*para*-phenylene) (PPP); (V) polythiophene (PT) (Reprinted with the permission from Nicolini *et al.*, New materials by organic nanotechnology and their applications, in Recent Research Development in Materials Chemistry 6, pp. 17–40, © 2005, Transworld Publishing).

A central topic in the physic of the π-conjugated polymers (and their parent oligomers) is the strong connections between electronic structure, geometrical structure and chemical structure. The latter can be termed lattice in parallel with the nomenclature of the condensed matter physics. In the '80 in the contest of transport and optical properties, was developed the polarons, bipolarons and solitons concepts (Nicolini, 1996a). The essential idea about the unusual nature of the species bearing charges and of the excited states of the conjugated systems has been intensively discussed in the last twenty years (see also Nicolini, 1996a). Recently the refinement of these ideas and the perceptions on the physical nature of the unique electronic properties of these conjugated polymers, as isolated molecules and as molecular solids leaded to the development of an higher sophisticated treatment level and consequently to a better understanding of some essential characteristics of the electronic structure of these polymers. In this contest the understanding of the nature of the π-conjugated systems is complicated by the electronic interactions, and by the strong interconnection and mutual influence of the electronic and geometrical structures (phonons).

The "electron-lattice interaction" is the strong influence that an extra electron or a hole or an excitation plays on the local geometry of the molecule (lattice, in the solid state physics terminology). Another fundamental point is that, despite the strong coupling between electrons or holes and the underlying lattice, the bearing charged species, or the neutral species in the excited states are surprisingly mobile (André et al., 1991). The geometrical structure of some discussed polymers is reported in Figure 1.7, where as a convention is reported only the monomeric repetition unit, or "unit cell".

The high band gap value in exclusively σ bonded polymer, $E_g(\sigma)$, makes this materials electrically insulating and generally not able to absorb the visible light. In the polyethylene, for instance, with monomeric repetitive unit defined by $-(CH_2-CH_2)-$, the optical band gap is about 8 eV. In the conjugated polymers exist a continuous network, often a simple chain, of carbon atoms in the hybridized state sp^2 or sp.

This chain of atoms with π-overposition of atomic orbital and with the periodical conditions imposed by the unit cell, leads to π-delocalized states along the polymeric backbone. As a result the π-bands forms the electronic structure border. In a monodimentional systems this π-states gives a *π-band gap*, $E_g(\pi) < E_g(\sigma)$, allowing to optical absorbance at lower photonic energies.

The essential properties of systems with delocalized π-electrons are the following:
- The electronic band gap E_g is relatively small (1–4 eV). This allows electronic excitation at low energies, and semiconducting behavior.
- The polymeric chain can be oxidized or reduced in a relatively simple way, generally by a charge transfer with the doping molecules.
- The mobility of the carriers is large enough to have high electrical conductivity in the doped state (oxidized or reduced).

- The charge bearing species are not free electron or holes, but quasi-particles, that can move freely through the materials, or along the uninterrupted polymeric chains at least.
- Due to the prevalently amorphous state of these polymers, the macroscopic electrical conductivity in the samples needs the hopping phenomena between the chains.

The geometrical structure is strongly dependent on the ionic states of the molecule and leads to unusual charge bearing species. These bearing species are self-localized (Table 1.2), in sense that the presence of an extra electronic charge leads to local variations in the atomic geometry (lattice), and at the same time, leading to localized variations in the electronic structure. The charge bearing species can be generated by optical absorbing of the neutral system, or by charge transfer doping.

Table 1.2. Self-localized excitations in the conjugated polymers. (Reprinted with the permission from Nicolini et al., New materials by organic nanotechnology and their applications, in Recent Research Development in Materials Chemistry 6, pp. 17–40, © 2005, Transworld Publishing).

State	Chemical term	Charge	Spin
Positive Soliton	Cation	+e	0
Negative Soliton	Anion	-e	0
Neutral Soliton	Neutral Radical	0	1/2
Positive Polaron	Radical Cation	+e	1/2
Negative Polaron	Radical Anion	-e	1/2
Positive Bipolaron	Di cation	+2e	0
Negative Bipolaron	Di anion	-2e	0
Exciton singlet	S_1	0	0
Exciton triplet	T_1	0	1

Reassuming, the band gap value (E_g) in the conjugated polymers is determined by the contribution of five terms, as shown in equation 1.1:

$$E_g = E^{bla} + E^{\theta} + E^{res} + E^{sub} + E^{int} \qquad (1.1)$$

where:
E^{bla}: bond length alternation - conjugation length.
E^{θ}: "rotational disorder" – mean deviation from co-planarity.
E^{res}: aromatic resonance – aromatic ring energy.

E^{sub}: effects of substituents – inductive and mesomeric electronic effects.

E^{int}: intermolecular coupling – interchain interactions in solid state.

This equation shows immediately, which are the structural variables that need to be dominated in the band gap control of the conjugated systems. As a consequence, the synthesis and assembling strategy devoted to design conjugated polymer with controlled band gap, must take into account the possibility of the control of energetic contribution of one or more parameters of Eq. 1.1 (Nicolini *et al.*, 2005).

The realization of electronic devices is composed by fundamental steps, such as: synthesis and project of supramolecular devices, development of architecture and chemical-physical assembling techniques. This now day represents one of the bigger scientific and technologic challenges of the new century. The researchers are called to radically change their own approach to the problems of the chemistry and physics of solid state. In research field of molecular control of materials, are being new emphasis the molecular self-assembling processes. This processes concerns the ability of single atoms or molecules to organize themselves in a rational way to give controlled macro and supramolecular structures. This new approach to the atomic/molecular world can, in first instance, appear non conventional, but in reality is typical of the biological world. In nature, in fact, every process follow a bottom up self-assembling mechanism, starting from single constituent atoms that self-assemble in more complexes structures with extraordinaire regularity.

1.2.1.1 *Amphipilic conjugated polymers*

Polythiophenes (PTs) usually do not form stable monolayers at the air-water interface because they are not sufficiently amphiphilic.

To obtain stable Langmuir films on the air-water interface, it is essential that the active species exhibit the necessary degree of amphiphilic character. It is well known that the low hydrophilic nature if thiophene ring makes very difficult the stabilization of the Langmuir films of the alkyl-substituted thiophenes on the interface and, as a consequence, the achievement of good transfer ratios to a solid substrate.

An alternative approach to synthesize an amphiphilic PT with high structural order was developed by Nicolini et al. (2005).

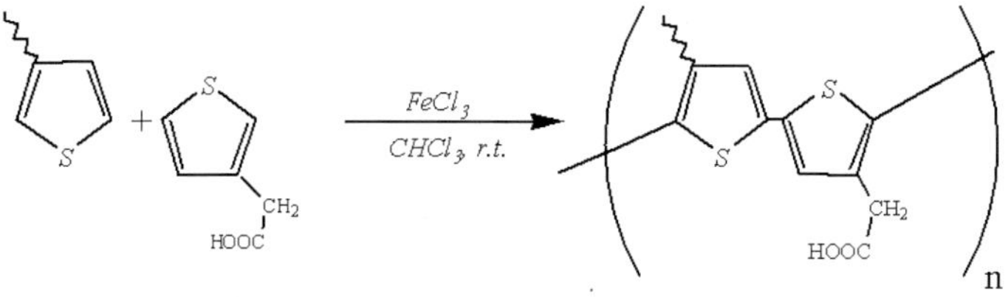

Figure 1.8. Polymerization of amphiphilic macromolecule starting from monomers with different affinity with the reaction solvent (chloroform), (Reprinted with the permission from Nicolini et al., New materials by organic nanotechnology and their applications, Recent Research Development in Materials Chemistry 6, pp. 17–40, © 2005, Transworld Publishing).

This method use the amphiphilic characters in the final polymer, and allows the formations of biological membrane like structure. This is possible using a simple chemical oxidative polymerization of 3-acetic acid thiophene and 3-hexylthiophene monomers as reported in the following scheme (Figure 1.8).

Table 1.3. – Comparison between the values of the peak absorption in the UV-vis spectra of polymers obtained with different synthesis methods (Reprinted with the permission from Nicolini et al., New materials by organic nanotechnology and their applications, in Recent Research Development in Materials Chemistry 6, pp. 17–40, © 2005, Transworld Publishing).

Chloroform solutions	Uv-Vis (λ_{max})
PAHT (FeCl$_3$ method) (Nicolini method)	442 nm
Polyhexylthiophene-PHAT (FeCl$_3$ method)	436 nm
Polyhexylthiophene-PHAT (McCullough method)	442 nm

Its represents maybe, the simplest, easy and cheap method to obtain this materials, in comparison with more expensive and sophisticated methods, as shown in Table 1.3 reporting the values of the wavelength absorbance maximum related to the band gap value of the polymer electronic structure.

The high structural order is due to the amphiphilic character of the polymer, which allows the coexistence of a fully conjugated and π-stacked polymer structural motif and a membrane, forming motif. In the

Figure 1.9, it is shown how the PHAT have a different morphology respect the polyalkylthiophenes and how it self assemble forming ordered homogeneous films on solid surface. The amphiphilic nature of the polymer allows it to assembly in solution and at the air-water interface forming a densely packed monolayer with highly ordered domains.

Using this self-assembling characteristic, it is possible use a bottom up approach in order to deposit on solid support with different dimensions and geometry polymeric films with variable thickness from monolayers to microns.

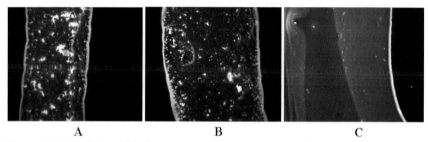

A   B   C

Figure 1.9. (a) Polyhexylthiophene on hydrophobic substrate; (b) polyhexylthiophene on hydrophilic substrate; (c) PHAT on hydrophilic substrate. The films are casted from chloroform solutions and the images are taken by fluorescence microscopy. (Reprinted with the permission from Nicolini *et al.*, New materials by organic nanotechnology and their applications, in Recent Research Development in Materials Chemistry 6, pp. 17–40, © 2005, Transworld Publishing).

A schematic representation, relative to the self-assembling process with Langmuir-Schaefer techniques using a polymer with amphiphilic characteristics, is reported on the following Figure 1.10. Fluorescence microcopy studies proves that at the air-water interface the hydrophilic side chains go into the water subphase while the alkyl chains stick up in the air, as depicted in Figure 1.10, forming dense and ordered films on hydrophobic substrate, while the LS deposition leads to non homogeneous monolayers on hydrophilic substrates with local domains.

The constant trends of increasing the density and complexity of semiconductor chip circuitry have frequently stressed the need for developing new revolutionary organic semiconductor technologies "bottom-up" (Reed et al., 1997; Collier et al., 1999; Reed, 1999; Nicolini, 1996a).

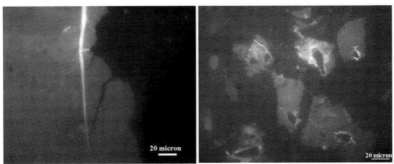

Figure 1.10. Images of PHAT monolayers deposited by LS technique on (left) hydrophobic substrate; (right) is hydrophilic substrate. The images are taken by fluorescence microscopy. (Reprinted with the permission from Nicolini et al., New materials by organic nanotechnology and their applications, in Recent Research Development in Materials Chemistry 6, pp. 17–40, © 2005, Transworld Publishing).

The several notable advances in the field of conducting polymer nanocomposite materials are indeed used for the photovoltaic and LED devices (see chapter 4). The defects such as solitons, polarons and bipolarons are the means to obtain the desired electrical conductivity in the conducting polymers. It was indeed proposed that non-linear excitations such as solitons and polarons play an important role in the transport of electrical charge in conducting polymers. Conducting polymers contain a variety of other defects such as cross-links, branch-point and conformational defects. Such defects may arise due to chemical linking of monomer units yielding an undesired linkage and/or breakage of chemical regularity. The cross-linking results in the rubbery nature of a conducting polymer. Some of the problems that have retarded practical applications of conjugated polymers are related to environmental stability, and loss of desirable mechanical properties on doping.

1.2.2 *Carbon nanotubes and their nanocomposites*

Carbon Nanotubes (NT) are new carbon allotropes (Figure 1.11), sharing similarities with graphite.

Their structure is a molecular scale wire shaped, called single-walled nanotubes (SWNT) or in concentric wires, called multi-walled nanotubes (MWNT), with dimensions of 1–1.5 nm in diameter and about up to many μm in length.

Single-wall carbon nanotubes (SWNT) are produced using catalytic metal particles in carbon arc vaporization, catalytic decomposition of organic vapors, plasma-enhanced chemical vapor deposition, and laser vaporization techniques (Iijima, 1991; Dillon *et al.*, 1997) SWNTs can be self-organized into ropes that consist of hundreds of aligned SWNTs on a two-dimensional triangular lattice, with an intertube spacing of van der Waals gap of approximately 3.2 Å (Odom *et al.*, 2002; Jost *et al.*, 2004). NT possess properties exploited in various fields as electronics or aerospace, and have recently received a great interest related to their use in biological systems (Baughman *et al.*, 2002, Penn *et al.*, 2003).

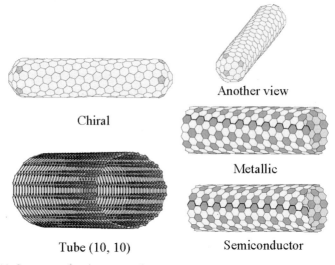

Figure 1.11. Structure of carbon nanotube.

The unique electrical and optical properties render nanotubes very sensitive to chemical or physical modifications of the surrounding environment. It follows that *in vivo* implants of bioelectronic sensors or delivery of molecules to cells could be improved with a targeted delivery with NT.

In recent years, two classes of organic materials like conducting polymers and carbon nanotubes have gained enormous interest for their attractive chemical-physical properties (Iijima, 1991). These characteristics have led to an interesting application of both the materials by the embedding of little quantity of carbon nanotubes, either single walled carbon nanotubes or multi walled carbon nanotubes, inside the polymer matrix for the synthesis of nanocomposites (Steuerman *et al.*, 2002). The fabrication of such nanocomposites materials can be obtained by means of either simple reaction carried out by easy steps of synthesis in the presence of a dispersion of carbon nanotubes or by assembling carbon nanotubes and just synthesized conducting polymers (Valentini *et al.*, 2004a; Bavastrello *et al.*, 2004; Erokhina *et al.*, 2002).

It is natural that the possibility of synthesizing materials by employing economic methods, can pave the path for the industrial applications of nanocomposite materials.

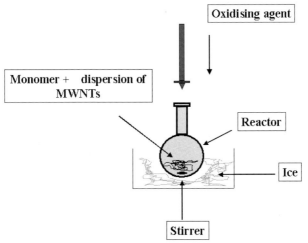

Figure 1.12. Schematic of reaction for the synthesis of nanocomposite materials by means of oxidative polymerisation. (Reprinted with the permission from Nicolini *et al.*, New materials by organic nanotechnology and their applications, in Recent Research Development in Materials Chemistry 6, pp. 17–40, © 2005, Transworld Publishing).

Among the conducting polymers utilizable for the synthesis of nanocomposite materials, the polyaniline and its derivatives can be chosen as good candidate since they have been deeply studied and have shown good electric properties, easy methods of synthesis and high environmental stability (Paul *et al.*, 1985; Ram *et al.*, 1999; Genies *et al.*, 1990; Nicolini *et al.*, 2001a).

The first method consists in the following steps. The monomers are dissolved into the medium of reaction, constituted by a dispersion of carbon nanotubes in an aqueous solution of hydrochloric acid. The dispersion is always obtained by means of ultrasonic equipment. Thus, by adding the oxidizing agent constituted by ammonium persulfate the reaction of polymerization is started. The growing polymeric chains embed the carbon nanotubes and form a matrix that contains them. The reaction is always continued for 12 hours at a temperature of 0–4°C, and before obtaining the final nanocomposite, the crude material undergoes an undoping process and several steps of purification carried out with different solvents. A schematic representing the reaction of synthesis of the nanocomposite materials is shown in Figure 1.12.

This method of synthesis implies the formation of non covalent bonds between the conducting polymer chains and the carbon nanotubes themselves. Anyway, in spite of the weak bonds obtained from the synthesis, this method represents a good way to dissolve non-functionalised carbon nanotubes into organic solvents. A specific example can be shown by the embedding of carbon nanotubes inside a polymeric matrix soluble in chloroform. The resulting synthesized nanocomposite will be then soluble inside the same solvent allowing carbon nanotubes in solution.

The second method consists in assembling carbon nanotubes previously deposited on a substrate and a consequent deposition of thin layer of the conducting polymer (Bavastrello *et al.*, 2004). In this method of fabrication carbon nanotubes are not embedded in the polymeric matrix but they are previously deposited on a silica substrate and then a layer of conducting polymers is spread on them. A good method for the preparation of a substrate of aligned carbon nanotubes is that known as *Plasma Enhanced Chemical Vapour Deposition.* For the deposition of the conducting polymers upon the mat of carbon nanotubes different techniques can be employed, according to the final task to be obtained. This depends on the thickness and the macromolecule order required.

Among them we can use in addition to the Langmuir-Schaefer, *spin coating and solution casting techniques.*

The first method of deposition, as shown in a previous paragraph, can be useful for the fabrication of thin films with a high grade of order, but the final thickness is maintained in nanometers scale since thicker layers would require long period of time.

The spin coating technique can be used to obtain thicker layers sacrificing something in the final order of the molecules constituting the film itself.

The solution casting technique is surely the less appropriate to obtain highly ordered films but is appreciate to get good thickness in only few depositions.

The nanocomposite materials usually show different properties with respect to the polyaniline derivative used in the synthesis. The nanocomposite materials based on carbon nanotubes and polyaniline derivatives can be successfully employed for the fabrication of different devices. An interesting application is for example the fabrication of sensors for acidity (Bavastrello *et al.*, 2004).

This is due to the fact that this class of conducting polymers is doped by means of protic acids and during the doping process it is able to change its conductivity of several orders of magnitude. This phenomenon, fundamental for the fabrication of sensor devices for acids, is deeply affected by the conducting polymers themselves, since the chemistry of polyanilines is generally more complex with respect to other CP.

This fact is due to their dependence on both the pH value and the oxidation states, described by three different forms known as leucoemeraldine base, the fully reduced form, emeraldine base, 50% oxidised form, and pernigraniline base, fully oxidised form.

The most important is the emeraldine base form and its protonation by means of H^+ ions generated from protic acids gives the emeraldine salt form, responsible of the strong increment of conducting properties (Epstein and MacDiarmid, 1991), and the mechanism of reaction is represented in Figure 1.13.

This process is reversible and it is possible for the presence of imine group basic sites located along the conducting polymer backbone. The remarkable fact that the chemical-physical properties of polyaniline and its derivatives are pH sensitive has led to the study of these materials as sensors (Lindino and Bulhoes, 1996).

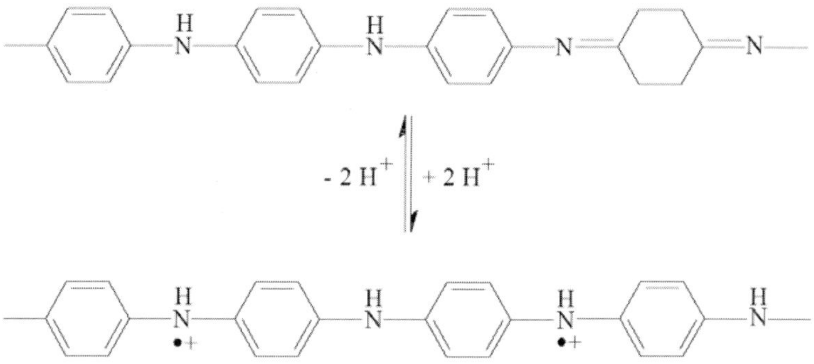

Figure 1.13. Mechanism of reaction for the doping process of polyaniline by means of protic acids of the emeraldine base form. (Reprinted with the permission from Nicolini et al., New materials by organic nanotechnology and their applications, in Recent Research Development in Materials Chemistry 6, pp. 17–40, © 2005, Transworld Publishing).

One of most important properties of nanocomposite materials is surely individuated in the possibility of enhancing the electrical properties with respect to the pure conducting polymers.

Recent studies carried out on a nanocomposite material based on poly(o-methoxyaniline) with multi walled carbon nanotubes, demonstrated that the insertion of carbon nanotubes embedded in the polymeric matrix provide better conducting properties with respect to the parent pure polymer.

For films of 30 monolayers, the nanocomposite material showed a specific conductivity of 2 S/cm, while for the pure conducting polymer a specific conductivity of 1.1×10^{-3} S/cm was found (Bavastrello et al., 2004a). Interestingly the conducting polymer in the doped form can be maintained for long periods of time till the material reacts with basic reagents and strongly changes its chemical-physical properties.

In other words, the reversibility of the process is not spontaneous. Anyway, studies carried out on the nanocomposite material obtained by embedding multi walled carbon nanotubes in the polymeric matrix, showed an anomalous behavior.

In fact, this nanocomposite tends to spontaneously release the doping agent.

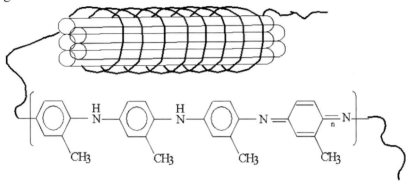

Figure 1.14: Representation of the conducting polymer chains wrapped around bundles of carbon nanotubes (Reprinted with the permission from Nicolini *et al.*, New materials by organic nanotechnology and their applications, in Recent Research Development in Materials Chemistry 6, pp. 17–40, © 2005, Transworld Publishing).

It means that the system is maintained in the doped only under a constant support of doping agent (Bavastrello *et al.*, 2004b). The possible explanation of this singular phenomenon is related to the chemical structure of the conducting polymer, where the aromatic rings constituting the polymeric chains bear two methyl groups. In Figure 1.14 is shown the representation of the conducting polymer chains wrapped around bundles of carbon nanotubes.

The methyl groups are then responsible of an increased sterical hindrance that impedes the conformational rearrangement that takes place during the doping process. It is also possible that the presence of carbon nanotubes inserted in the polymeric matrix accentuates this phenomenon (Bavastrello *et al.*, 2004).

Indeed the electrical behavior of the doped nanocomposite material along the time is such that the conductivity of the material, deposited as a thin film increased of about two orders of magnitude in 30 hours (Bavastrello *et al.*, 2004b).

1.2.2.1 *Interactions between conjugated polymers and single-WN*

The remarkable physical and chemical properties of single-walled carbon nanotubes (SWNTs) have made these materials attractive candidates for a host of structural and electronic tasks. However, a number of challenges must still be met before nanotubes can be exploited for most of these envisioned applications.

Those challenges include separating the tubes by conductivity types, scaling synthetic approaches to large-scale production, and developing chemical techniques for manipulating nanotubes as rational molecular materials. Progress on all, of these fronts has been proceeding rapidly over the past few years as shown by Gimzewski and co-workers (Schlittler *et al.*, 2001). In cooperation with California Nanosystem Institute (Steuerman *et al.*, 2002) we have recently carried out the chemical interactions between single walled carbon nanotubes (SWNTs) and two structurally similar polymers, poly{(*m*-phenylenevinylene)-*co*-[(2,5-dioctyloxy-*p*-phenylene)vinylene]}, or PmPV, and poly{(2,6-pyridinylenevinylene)-*co*-[(2,5-dioctyloxy-*p*-phenylene)vinylene]}, or PPyPV, are investigated. The fundamental difference between these two polymers is that PPyPV is a base and is readily protonated via the addition of HCl. Both polymers promote chloroform solubilization of SWNTs. We find that the SWNT/PPyPV interaction lowers the pKa of PPyPV. Optoelectronic devices, fabricated from single polymer-wrapped SWNT structures, reveal a photogating effect on charge transport, which can rectify or amplify current flow through the tubes. For PmPV wrapped tubes, the wavelength dependence of this effect correlates to the absorption spectrum of PmPV. For PPyPV, the wavelength dependence correlates with the absorption spectrum of protonated PPyPV, indicating that SWNTs assist in charge stabilization.

To further explore the interaction of conjugated polymers organization on SWNTs surface we have recently (Narizzano and Nicolini 2005) studied the interactions between poly{(2,6-pyridinylenevinylene)-co-[(2,5-dioctyloxy-p-phenylene)vinylene]}, or PPyPV, and SWNTs by UV-Vis absorption spectroscopy, suggesting a semi-quantitative mechanism with ropes formation of SW carbon nanotubes (Figure 1.15). The SWNTs appear to promote the polymer organization. The PPyPV is a Lewis base and can be doped by strong and weak Lewis acids. The basicity strength of the PPyPV depends on the polymer interchain interactions enhanced by the SWNTs presence, as the SWNT concentration is increased, and a Kb increment of PPyPV is observed.

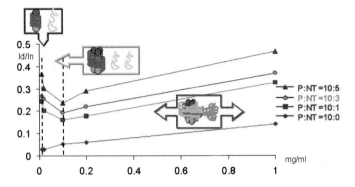

Figure 1.15. Semi-quantitative mechanism of the interaction process and ropes formation of SWNTs by a conjugate polymer (PPyPV), taking into account all the species in solution is proposed. Particular attention is paid to study the interactions as a function of nanotube and polymer concentrations ((Reprinted with the permission from Narizzano and Nicolini, Mechanism of conjugated polymer organization on SWNT, Macromolecular Rapid Communications 26, pp. 381–385 © 2005, Wiley-VCH Verlag GmbH & Vo. KGaA).

1.2.2.2 *SWNT for hydrogen storage*

Finally SWNTs appear to have many potential advantages for hydrogen storage (De Heer *et al.*, 1995) over currently available adsorbents. They have large theoretical surface areas that are on the order of those for high-surface-area activated carbons, but many open problems still remain. The key to this potential lies in the nanotube's unique structure, which in turn depends on the unique properties of its building material and the defects that can form in the network of carbon bonds.

Crystallized arrays of SWNTs have a very narrow pore-size distribution that has virtually all their surface area in the micropore region. In contrast, surface area in activated carbons is broadly distributed between macropores, mesopores, and micropores. The pore sizes in an array of tubes could be controlled by tuning the diameter of the SWNTs making up the array. Theoretical calculations by Ye *et al.* (1999) predicted that carbon nanotubes have very strong capillary forces for encapsulating both polar and nonpolar fluids. Dillon and coworkers (1997) used hydrogen adsorption on carbon soots containing small amounts of SWNTs with high hydrogen uptake.

With SWNT materials, hydrogen apparently adsorbs and desorbs over a narrower range of pressure, as shown by Ye *et al.* (1999), so storage systems could be designed to operate without wide pressure excursions. Similarly Darkrim and Levesque (1998) computed hydrogen adsorption by grand canonical Monte Carlo simulations and stressed that the results depended on the choice of intermolecular potentials between the hydrogen molecules and the carbon atoms. These studies prove the potential usefulness of carbon nanotube technology in this field and must be continued to search for new energy solutions compatible with the environment, a big challenge in front to the entire humanity in the near future.

1.2.3 *Nanoparticles*

Nanometer size semiconductor particles of cadmium sulphide inside Langmuir-Blodgett (LB) films of cadmium arachidate can be obtained by exposing films to an atmosphere of hydrogen sulphide according to the following reaction:

$$(CH_3(CH_2)_{18}COO)_2 Cd + H_2S \rightarrow 2(CH_3(CH_2)_{18}COOH) + CdS$$

where protons from hydrogen sulphide protonate the head groups of the arachidic acid, while sulphur binds to cadmium. Works on mercury, lead and cadmium sulphide (Zylberajch *et al.*, 1989, Erokhin *et al.*, 1995a, 1998; Facci *et al.*, 1994; Nicolini, 1996c) point out this approach as a general one for the formation of particles of sulphur salts of bivalent metals. Optical and electron diffraction investigations on the particles formed by this technique allowed to estimate that their sizes range between 5 nm and 20 nm.

Assembly of nanoparticles in organic conjugated heterostructures or superlattices is of great importance in current material science (Figure 1.16), as well as in molecular electronics (Gao *et al.*, 1997).

Figure 1.16. Atomic force microscopy of gold nanoparticles (Larosa *et al.*, in preparation).

Controlled thickness and order of these ultrathin structures by LB techniques play a crucial role in obtaining high quality molecular films (Nicolini, 1996b). CdS nanoparticles were formed (Smotkin *et al.*, 1988) by exposing cadmium arachidate to an atmosphere of H_2S gas, whereby carboxylic groups of arachidic acid were protonated, and nanometer size CdS was produced (Figure 1.17a).

H_2S gas was prepared by the reaction of FeS with diluted H_2SO_4 and samples are exposed into the chamber to the H_2S atmosphere for 12 h, that is, a time interval to complete the reaction.

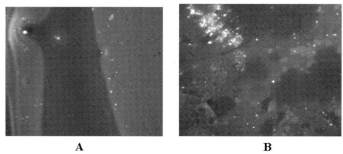

Figure 1.17. a). Images taken by fluorescence microscopy of 20 layers films of copolymer. (a) before exposure to H_2S; (b) after exposure to H_2S. (Reprinted with the permission from Narizzano *et al.*, A heterostructure composed of conjugated polymer and copper sulfide nanoparticles, The Journal of Physical Chemistry B 109, pp. 15798–15802, © 2005, American Chemical Society).

Distribution of nanoparticles observed by CCD camera, shown in Figure 1.17a, reveals their random growth resulting in the formation of isolated islands. The technique allowed us to obtain very good resolution of inorganic layer formation, 0.6 nm for each precursor bilayer.

The methodology was employed to prepare CdA (Facci *et al.*, 1995), PbS, CuS, HgS, *etc.* layers (Erokhin *et al.*, 2002). It was found that the technique does not yield continuous semiconductor monolayers and for films less than 25 nm thick low conductivity is observed (Figure 1.17b). The use of a conjugated matrix results in conductive connection of the semiconductor islands and, therefore, in the formation of a continuous conductive layer at film thickness less than 25 nm (Figure 1.17b). A conjugated polymer with amphiphilic properties that allowed the formation of a stable monolayer at the air-water interface was indeed needed in order to achieve structures with controlled thickness, by means of the LS technique (Narizzano *et al.*, 2005). This was obtained by chemical copolymerization of two different monomers: the 3-thiopheneacetic acid (3TAA) and 3-hexylthiophene (3HT). The former allows for reaction with copper ions, and consequently the formation of nanoparticles, and it has a hydrophilic character. The latter has a hydrophobic character and makes the final polymer soluble in common organic solvents. CuS nanoparticles were grown directly in the polymeric matrix using the carboxylic groups as nucleation centers. The reactions were monitored by quartz crystal microbalance (Figure 2.16 in chapter 2), Brewster angle (see Figure 2.11 in chapter 2), and fluorescence microscopy (Figure 1.17a). This new conjugated amphiphilic polymer acts as a growth matrix for semiconducting CuS nanoparticles in which they form islands/domains three-dimensional randomly distributed. Interestingly (Narizzano *et al.*, 2005) the electrical properties and the conductivity of the heterostructure strongly depend on the formation of nanoparticle islands in the conjugated polymer (Figure 1.17b). In thicker film large nanocrystal domains grow at the expense of smaller ones, reflecting a three-dimensional diffusion mechanism of individual particles from smaller to larger domains. This is driven by a decrease in the surface free energy as larger domains grow, since only few nanoparticles remain in energetically unfavorable sites.

In summary it is conservative to conclude that the nanoscale methodology via chemical synthesis represents a promising general-purpose tool for the design of new materials, products and processes for a wide variety of applications.

1.3 Produced Via LB Nanostructuring

The design, the engineering and the properties of nanostructured protein biofilms represent the nanomaterials at the higher level of supermolecular organization of both polymers and biopolymers. Each nanostructured biofilm exhibits numerous and interesting properties *in vivo* and *in vitro*, which deserve attention.

A large number of potential applications for organized protein monolayers had motivated considerable research activity in this field (Nicolini, 1997; Boussaad *et al.*, 1998; Kiselyova *et al.*, 1999), as shown also by numerous reviews (Nicolini *et al.*, 1995b, 2001a; Nicolini 1996a,b, 1998b; Vianello *et al.*, 2004; Hampp, 2000; Hampp and Brauchle, 2003).

Construction of specific interaction-directed, self-assembled protein films has been performed at the air-water interface. The Langmuir-Blodgett (LB) technique has been extensively used to order and immobilize natural proteins on solid surfaces (Nicolini, 1997; Tronin *et al.*, 1994, 1995; Facci *et al.*, 1993).

A surface pressure ranging between 15 and 25 mN/m was used to transfer the monolayers of the various proteins being immobilized from air-water interface to solid supports, and the subphase is buffered and contains high ionic strength.

The layer-by-layer (LBL) assembly processes based on electrostatic or other molecular forces represent instead a unique technique that presents an alternative approach to the formation of nanostructured architectures by adsorption of consecutively alternating polyelectrolytes. An application of the LBL technique is the fabrication of homogenous ultrathin film of conjugated polymers.

Molecular-level processing of conjugated polymers (*i.e.*, polypyrrole, polyaniline, poly(phenylene vinylene), poly(o-anisidine) by the LBL technique (Ferreira and Rubner, 1995; Ram *et al.*, 1999a; Decher 1996, 1997; Decher *et al.*, 1992) was indeed shown in the literature (Ram and Nicolini, 2000) (Figure 1.18).

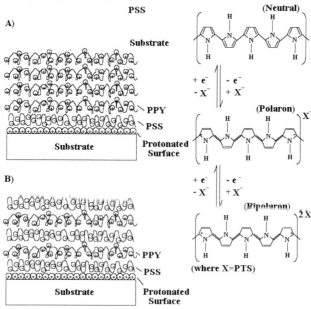

Figure 1.18. Layer-by-Layer deposition. (a) Schematic of *in situ* self-assembled layer-by-layer films of PPY with PSS. (b) Schematic of in-situ self-assembly of PPY on PSS surface as function of time (Reprinted with the permission from Ram and Nicolini, Thin conducting polymeric films and molecular electronics, in Recent Research Development in Physical Chemistry 4, pp. 219–258, © 2000, Transworld Publishing).

The self-assembly of charged polyelectrolytes (*i.e.*, proteins, nucleic acids, conducting polymers, zirconium phosphate, optical dyes, metal nanoparticles, aluminosilcates, and clay) by LBL can be considered an alternative approach to spin-coating and chemical vapor deposition techniques (Nicolini *et al.*, 2005).

None however appears for proteins as efficient as the Langmuir-Blodgett technique namely if utilizing our "protective plate" modification (Troitsky *et al.*, 1996a).

1.3.1 *Langmuir-Blodgett*

The fact that oil forms thin layers over the water surface is known from the ancient time. First statement that such films must be monomolecular in thickness was done by Lord Rayleigh (1879). Nevertheless systematic scientific study of such objects began from the works of Irving Langmuir and can be divided into several phases. During the first phase main attention was paid to the behavior of monolayers of amphiphilic molecules at the air/water interface.

The phase is connected with early works of Langmuir in 1920. As a result, monolayer formation process was characterized and several phase transitions were determined in two-dimensional system at the air/water interface. During the second phase it was shown that the layers could be transferred onto surfaces of solid substrates. The works were performed by Langmuir in collaboration with Blodgett, whose names began to be used to term the method itself (Blodgett, 1934, 1935; Blodgett and Langmuir, 1937). Their technique consisted in deposition of monolayers when the substrate moved vertically through the monolayer. A little bit later another deposition technique was suggested by Langmuir and Schaefer, where the substrate touched the monolayer horizontally (Langmuir and Schaefer, 1939). The method now is called in a literature as "horizontal lift" or Langmuir-Schaefer technique. It is interesting to note that the method was developed for deposition of protein layers (first attempt to work with protein layers was done in 1938 by Langmuir and Schaefer on pepsin and urease (Langmuir and Schaefer, 1938). After the initial interest in the subject at the beginning of the century resulting in a Nobel prize received by Langmuir in 1932, the activities in the field were not numerous involving only some academic interest for such two-dimensional systems. Third phase in the LB films investigations began with the works done in the group of Kuhn (Kuhn 1965; Drexhage and Kuhn, 1966). The works have demonstrated that it was possible to form complex structures with desired mutual orientation of functional groups of molecules by the method (Bücher *et al.*, 1967; Inacker *et al.*, 1976). The works on energy transfer in the films (Kuhn, 1981) created big resonance in the scientific world attracting a huge number of researchers to be involved in investigation of films. It is possible to consider that the fourth stage of the LB investigations began when the first international LB conference was organized (Nicolini, 1996a). It demonstrated that scientific forces of differing background began to be included into LB films investigations. During this stage the films became to be characterized by practically all experimental techniques available nowadays (Nicolini, 1996; Nicolini and Pechkova, 2006a). Application aspect of the films began also be taken into account (Nicolini *et al.*, 2001a; Nicolini and Pechkova, 2006a), with their development in an alternative technology for manifacturing.

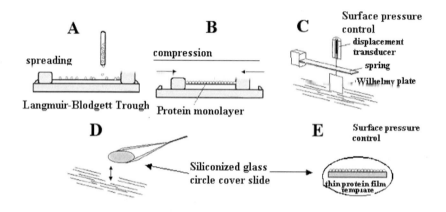

Figure 1.19. Schematic of the protein monolayer formation by the LB Trough (A-B), at the optimal surface pressure automatically controlled and monitored (C), and its manual horizontal transfer to a siliconized glass circle cover slide accordingly to the Langmuir Schaefer deposition (D-E). (Reprinted with the permission from Nicolini and Pechkova, Nanostructured biofilms and biocrystals, Journal of Nanoscience and Nanotechnology 6, pp. 2209–2236, © 2006, American Scientific Publishers, http://www.aspbs.com).

The floating monolayer can be transferred onto the surface of solid supports (Figure 1.19). Two main techniques are usually considered for the monolayer deposition, namely, Langmuir-Blodgett (or vertical lift) and Langmuir-Schaefer (or horizontal lift). In LB technique (Sánchez-González *et al.*, 2003), a specially prepared substrate is passed vertically through the monolayer. The monolayer is transferred onto the substrate surface during this passage. In those cases where it is important to have the monolayer electrically neutral, the deposition will not be performed when some charges in the monolayer molecule head groups are uncompensated and the electrostatic interaction of this charge with water molecules will be higher than the hydrophobic interactions of chains with the hydrophobized substrate surface. The Langmuir-Schaefer method of monolayer transfer from the air/water interface onto solid substrates is most frequently utilized for protein immobilization (Owaku *et al.*, 1989) and it is illustrated in Figure 2.2. It was developed in 1938 by Langmuir and Schaefer for deposition of protein layers. Prepared substrate horizontally touches the monolayer, and the layer transfers itself onto the substrate surface.

1.3.1.1 *Protective plate*

A modification of the original LB/LS technique, called "protective plate" method (Troitsky *et al.*, 1996a) and shown in Figure 1.20, appeared however mandatory to enhance the physical and functional properties of all class of proteins, mainly of enzymes.

The "protective plate" apparatus, described in Figure 1.20 was used to produce the multilayered nanostructures. The "protective plate" method allows us to combine the LB technique with adsorption of soluble compounds in such a way that during LB assembly formation the surface of the protein layer never crosses the air-water interface.

This feature eliminates the main reason for protein denaturation. Utilizing this apparatus, the deposition of LB monolayers and/or adsorption of dissolved compounds are carried out in different compartments in the required sequence.

To transfer the sample from one compartment to another, the deposited film is protected by a thin layer of water, which is held by capillary forces between the solid support and the protective plate.

In accordance with the required deposition sequence, the sample moves automatically between the compartments with different monolayers or dissolved compounds and the protective plate shifts up and down to either open or close the film surface.

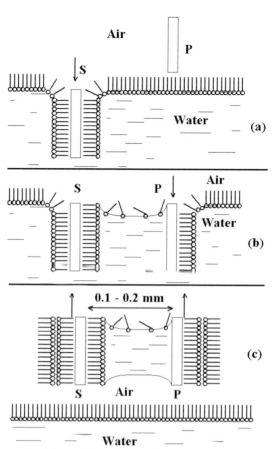

Figure 1.20. Schematic of "protective plate" method deposition. Principle of the proposed method. Monolayer is deposited (a), substrate is closed by plate (b), system substrate-plate is pulled out from aqueous subphase (c). (Reprinted with the permission from Troitsky *et al.*, Deposition of alternating LB monolayers with a new technique, Thin Solid Films 285, pp. 122–126, © 1996a, Elsevier).

The "protective method" technique (Troitsky *et al.*, 1996a, 2003) has several advantages not available with LB method:
a) protein monolayer formed at an air-water interface can be functionally protected by the thin water layer and then transferred to many different surfaces, including those prepared for electronic studies; protein multi-layers are obtained by repeating the monolayer transfer process as many times as desired;

b) the transferred film can be studied in different environments, such as air or solution (Guryev *et al.,* 1997).

The success of the technique draws on the property of amphiphilic molecules when spread and compressed at an air-water interface to form a compact monolayer, with hydrophobic and hydrophilic parts directed to air and water, respectively. When this technique is applied to non-amphiphilic water-soluble proteins, difficulties can be encountered which can be overcame by the "protective plate" method.

Simple water-soluble proteins at the air-water interface would tend indeed to expose their hydrophobic interior to the air, causing dramatic changes of their native structure, which may be detrimental to the function of the proteins or even cause the liganded cofactors to dissociate from the protein. In-plane order can be greatly enhanced by exploiting the combination of planar orientation and mobility, which this interface provides.

1.3.1.2 *Heat-proof and long range stability*

The physical properties of Langmuir-Blodgett (LB) films of proteins (Nicolini and Pechkova, 2006a; Tiede 1985; Hwang *et al.,* 1977) and lipid-protein complexes (Lvov *et al.,* 1991; Fromherz 1971; Phillips *et al.,* 1975; Heckl *et al.,* 1987; Kozarac *et al.,* 1988) were intensively studied and characterized by different techniques.

Numerous observations on the thermal stability of proteins organized in dense solid films, deposited by LB (Nicolini *et al.,* 1993; Facci *et al.,* 1994; Erokhin 1995a) or by self-assembling (Shen *et al.,* 1993), point to the role of decreased water content and molecular close packing (Nicolini *et al.,* 1993).

Bacteriorhodopsin (bR) and photosynthetic reaction centers from Rhodopseudomonas Viridis (RC) were studied in solution, LB film and self-assembled film. RC was extracted from the membranes by detergent (lauryldimethylamineoxide-LDAO); the solution contains the individual protein molecules surrounded by a detergent "belt" shielding the hydrophobic areas of the protein surface.

The initial solution of bR was instead the solution of sonicated membrane fragments. Described differences in the initial protein solution conditions, of course, differentiates the processes of film formation, both in the case of the LB technique and in the case of self-assembling.

Circular dichroism spectra of different samples of RC and bR after heating at different temperatures are presented in Figure 1.21 (Erokhin, et al., 1996a).

Comparison of the CD spectra of RC and BR in solution allows one to conclude that bR in solution is much more heat resistant with respect to RC. The other interesting point is that the temperature behaviors of the bR in LB and self-assembled films are absolutely identical.

Both of them demonstrate high thermal stability, and significant differences in the CD spectra appear only after heating up to 200 °C. The situation is absolutely different in the case of RC.

In fact, LB films show high thermal stability, and significant differences appear, as in case of bR, only after heating to 200 °C. Self-assembled films of RC, in contrast, are much more affected by thermal treatment.

Such differences in the secondary structure behavior with respect to temperature can be explained by suggesting that molecular close packing of proteins in the film is the main parameter responsible for the thermal stability.

In fact, as in the case of bR, we have close packing of molecules even in the solution (membrane fragments); there are practically no differences in the CD spectra of bR solution at least till 75 °C (denaturation takes place only for the sample heated to 90 °C).

RC in solution begins to be affected even at 50 °C and is completely denatured at 75 °C, for the solution contains separated molecules.

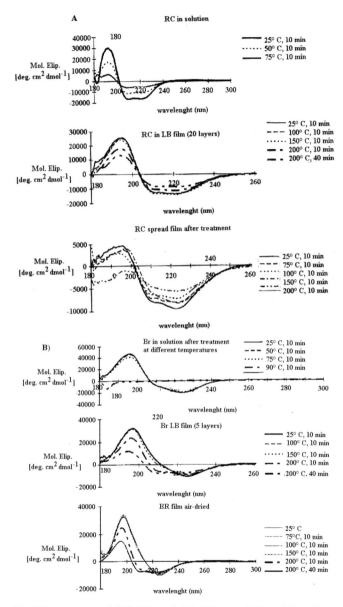

Figure 1.21. CD spectrum of RC (A) and bR (B) in solution, LB and spread films. (Reprinted with the permission from Erokhin *et al.,* On the role of molecular close packing on the protein thermal stability, Thin Solid Films 285, pp. 805–808, © 1996a, Elsevier).

Langmuir-Blodgett monolayers of photosynthetic reaction centers from *Rhodobacter sphaeroides* have been studied by scanning tunnelling microscopy (Facci et al., 1994a) (Figure 1.22).

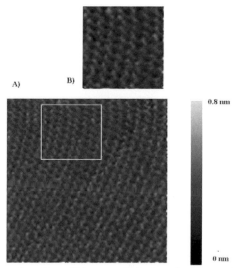

Figure 1.22. Heat and cooling of Photosynthetic Reaction Center films. STM image of an RC monolayer after heating at 150 °C for 10 min. (a) Image size 57.6 x 57.6 nm^2 (b) zoomed area (21.3 x 21.3 nm^2 from the outlined area. Images in the light and in the dark were identical. (Reprinted with the permission from Facci et al., Scanning tunnelling microscopy of a monolayer of reaction centers, Thin Solid Films 243, pp. 403–406, © 1994a, Elsevier).

Freshly deposited films were studied both in the dark and in the light. In the dark, images revealed molecular structure with 64Å and 30 Å periodicities, which correspond to protein and sub-unit sizes known from X-ray crystallography, while no periodic structure appeared in the light due to the tip action on the excited proteins.

STM voltage-current measurements showed the charge separation in single protein molecules in the film and their different behavior in the dark and light. Together with surface potential measurements at the macroscopic level, they indicated the preservation of reaction center activity in the monolayer. By fixing the protein layer with glutaraldehyde, it was possible to prevent the perturbing tip action and obtain a periodic molecular structure with 30 Å spacing even in the light.

After heating at 150° C, the unfixed film reorganized itself into a long-range ordered state with a hexagonal structure of 27 Å spacing but with no activity. It is important to notice that the heat-proof in LS protein film is systematically associated with long range stability at room temperature (Paddeu et al., 1996; Facci et al., 1998) (Figure 1.23).

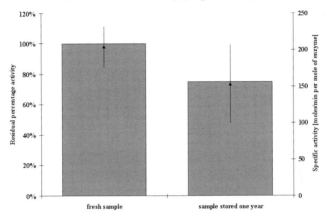

Figure 1.23. Long range stability of protein films. Enzymatic activity of fresh and of long term stored (one year) GST LB monolayer immobilised on silanized silicon surface. Each point represents the average value with the confidential interval. (Reprinted with the permission from Paddeu et al., Kinetics study of glutathione s-transferase Langmuir-Blodgett films, Thin Solid Films 284-285, pp. 854–858, © 1996, Elsevier).

In summary, LB organization of protein molecules in film not only preserved the structure and functionality of the molecules, but also resulted in the appearance of new useful properties, such as enhanced thermal stability coupled to long-range stability (Nicolini et al., 1993; Erokhin et al., 1995).

These are the properties that open the areas of application of protein films to many fields of potential industrial implications, namely nanobiocatalysis, nanosensors, nanoactuators and nanoelectronics (chapter 4).

1.4 Produced Via APA Nanostructuring

The process of ordered porous alumina fabrication and microstructuring (Nicolini and Pechkova, 2006a; Grasso *et al.*, 2006) is shown schematically in Figure 1.24.

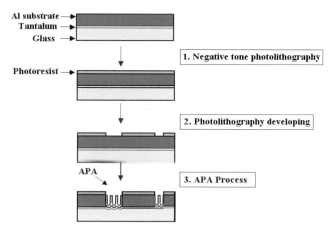

Figure 1.24. APA method. (Reprinted with the permission from Grasso *et al.*, Nanostructuring of a porous alumina matrix for a biomolecular microarray, Nanotechnology 17, pp. 795–798, © 2006, IOP Publishing Limited).

Briefly, ordered nanopore arrays are prepared by using a photolitographic technique that microstructures the nanopore arrays. A negative resist (SU-8, Micro-Chem) was used, that is a high contrast, epoxy based photoresist designed for micromachining and other microelectronic applications. SU-8 shows very high optical transparency above 360 nm, which makes it ideally suited for imaging near vertical sidewalls in very thick films. SU-8 is best suited for permanent applications since; when imaged, cured and left in place it gives hydrophobic properties to final nanostructured surfaces. An aluminium sheet (250 μm thick) was cleaned to obtain maximum process reliability followed by isopropyl alcohol cleaning and deionized water rinse. After the resist has been applied to the substrate, it was soft baked in order to evaporate the solvent and thicken the film. The resist used is optimized for near UV (350–400 nm) exposure and is virtually transparent and insensitive above 400nm. Hexagonally ordered pore domains were prepared by a self-organization process under specific anodization conditions. This nano-patterning technique leads to a sharp edge. The anisotropy of the process can be seen with FIB system. The sidewalls of the structures are very steep, and their roughness is determined by the quality of the mask (Figure 1.25). This second resist having hydrophobic properties increases specificity to biological sample linking.

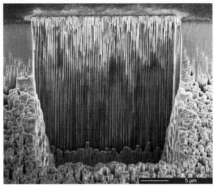

Figure 1.25. FIB system images of cross-sectional morphologies of the microarray spot, resulting at the end of photolithographic microstructuring technique and 2 step anodization process. FIB - FEI - measurement were done with gallium ions - 37 pA - both for cutting and imaging (Reprinted with the permission from Grasso *et al.*, Nanostructuring of a porous alumina matrix for a biomolecular microarray, Nanotechnology 17, pp. 795–798, © 2006, IOP Publishing Limited).

1.4.1 *Focus ion beam*

APA is a partially transmissive substrate, thus a luminous signal emitted into APA can be seen either at the top (direct vision) or at the bottom side (through the bottom layer). These advantages can be exploited in order to differentiate the instruments for sample analysis (laser scanner, fluorescence microscopy, DNA Array). APA can be obtained by removing the remaining aluminum substrate in a saturated $HgCl_2$ solution.

In alternative to a substrate in aluminum, a glass surface can be used, being possible to evaporate an aluminum layer with desired thickness directly on glass. The DNA solution is placed onto the hydrophilic spots of alumina of the arrays, while surface covered by the resist that surrounds the alumina spot shows hydrophobic properties. The DNA binding can be improved by means of a surface functionalization with Poly-L-Lysine.

The Poly-L-Lysine linkage exploits the phosphate groups onto alumina (Figure 1.26) and the final biomolecular nanostructured APA gave excellent results (Figure 1.27) with far reaching potential application in microarray technology for both DNA and protein.

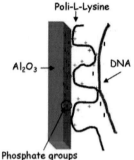

Figure 1.26. Gene-APA linkage via Poly-L-Lysine (Reprinted with the permission from Nicolini and Pechkova, Nanostructured biofilms and biocrystals, Journal of Nanoscience and Nanotechnology 6, pp. 2209–2236, © 2006, American Scientific Publishers, http://www.aspbs.com).

This photolithographic microstructuring technique for the ordered nanopore arrays fabrication is reported consists of a negative resist with hydrophobic properties increasing specificity to biomolecules linking.

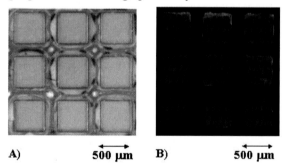

Figure 1.27. Nanostructured APA for gene microarrays probed by fluorescent labeling. (fluorescence microscope image, magnification 2X, filter set n° 15. (Reprinted with the permission from Grasso *et al.*, Nanostructuring of a porous alumina matrix for a biomolecular microarray, Nanotechnology 17, pp. 795–798, © 2006, IOP Publishing Limited).

Nanoporous alumina is formed by anodic process and yields straight holes with high aspect ratio: its use as substrates for DNA-microarray or protein-chip application offers several advantages over conventional supports, making them very attractive to use as supports for biological sample microarrays application. Oligonucleotide and antibody microarrays are currently intensively investigated for a broad range of applications in biomedical diagnostics, to simultaneously determine several parameters in individual samples from a limited amount of material (Templin *et al.*, 2002). On the other hand, the fabrication of nanochannel-array materials has attracted considerable scientific and commercial attention. Moreover, the application of ordered nanochannel arrays as two-dimensional photonic crystals has generated increasing interest in recent years.

Among the applications for these materials are the inhibition or enhancement of spontaneous emission and the fabrication of tunable optical filters and microcavities (Foresi *et al.*, 1997), waveguides (Yablonovitch 1987) and catalytic combustion (Suzuki *et al.*, 2003).

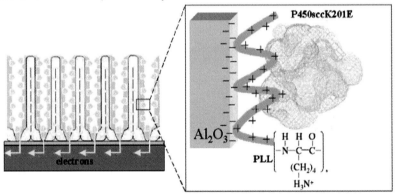

Figure 1.28. Schematic view illustrating the direct electron transfer between the cytochrome P450scc catalytic "core" and the APA modified working electrode. In the box is shown the specific interaction between the cytochrome P450scc negative surface (blue) and the positive charges of Poly-L-Lysine. (Reprinted with the permission from Stura *et al.*, Anodic porous alumina as mechanical stability enhancer for Ldl-cholesterol sensitive electrode, Biosensors and Bioelectronics 23, pp. 655–660, © 2007, Elsevier).

Anodic porous alumina, which has been studied extensively over the last five decades (See *et al.*, 1980; Thompson and Wood, 1983) has recently been reported to be a typical self-ordered nanochannel material (Masuda and Fukuda, 1995; Masuda *et al.*, 1997). Self-organization during pore growth, leading to a densely packed hexagonal pore structure, has been reported in oxalic, sulfuric, and phosphoric acid solution (Jessensky *et al.*, 1998).

The interpore distances of the regularly ordered pore arrangement have been extended to the large range of 50÷420 nm (Li *et al.*, 1998). Masuda and Satoh (1996) first reported a two-step fabrication method for straight nanoholes in a thin membrane of alumina. The structural characteristics of the ordered porous alumina make it not only a perfect template material for the fabrication of nanoscale structures, but also an outstanding candidate material for two-dimensional (2D) photonic crystals, which may show photonic bandgaps that are adjustable in the visible to ultra-violet spectral region.

This material offers several advantages over conventional supports: high surface area enlargement, improved microfluidic properties, easy and cheap manufacturability, flexibility in porous dimension and confinement effect as a mean for selectivity and maximization of light emission.

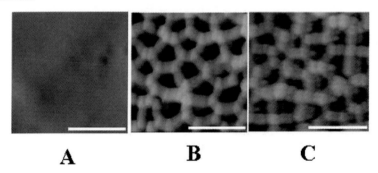

Figure 1.29. Topographic AFM image of rhodium-graphite s.p.e working electrode before, after the APA deposition and after functionalization. (a) Top view of rhodium-graphite s.p.e working electrode before the increase of its surface by APA membrane; (b) Top view of the same rhodium-graphite s.p.e after the APA deposition on its working electrode surface. Scale; (c) Top view of APA nanopores after the working electrode surface functionalization (physical adsorption) with PLL and cytochrome P450scc. Scale bars correspond to 1 μm. (Reprinted with the permission from Stura *et al.*, Anodic porous alumina as mechanical stability enhancer for Ldl-cholesterol sensitive electrode, Biosensors and Bioelectronics 23, pp. 655–660, © 2007, Elsevier).

For protein array applications APA was also used (Stura *et al.*, 2007) as Mechanical Stability Enhancer of Cytochromes Electrodes to improve the performances of the electrodes based on P450scc for LDL-cholesterol detection and measure, APA (Anodic Porous Alumina) was used. To optimize the adhesion of P450scc to APA, a layer of poly-L-lysine, a poly-cathion, was successfully implemented as intermediate organic structure (Figure 1.28).

This inorganic APA matrix has been used to functionalise the rhodium-graphite working electrode (Figure 1.28), and the corresponding pores can be tuned in diameter modifying the synthesis parameters in order to obtain cavities 275 nm wide and 160 µm deep (as demonstrated with AFM and SEM measurement (Stura *et al.*, 2007).

This allows the immobilization of P450scc macromolecules (Figure 1.29) preserving their electronic sensitivity to its native substrate, *i.e.* the cholesterol. Even if the sensitivity of the APA+P450scc system was slightly reduced with respect to the pure P450scc system, the readout was stable for a much longer period of time, and the measures remained reproducible inside a proper confidentiality band, as demonstrated with several cyclic voltammetry measures (Stura *et al.*, 2007).

Chapter 2

Nanoscale Probes

Over the last few decades numerous probes have been emerging at the nanoscale level which are being used in developing new basic scientific knowledge (see chapter 3), novel nanostructured materials (see chapter 1) and revolutionary applications to health (see chapter 3) and industry (see chapter 4). Several biophysical probes employed to characterize films and crystals, as well as single polymer or biopolymer earlier introduced, are here summarized stressing their unique contributions and features ranging from surface potential and AFM (Atomic Force Microscopy) to μGISAXS (micro Grazing Incidence Small Angle X-ray Scattering).

2.1 Surface Potential

As already mentioned in this book (see paragraph 1.3), LB technique allows to operate with biological objects, such as proteins, and to organize them in regular layers, which can be probed with surface potential technology. It is worth mentioning that LB films are of fundamental interest as they represent 2D systems, which can give origin to new types of phenomena not found in 3D objects.

As mentioned before, classic materials for the LB method are amphiphilic molecules, one side of which is a polar head-group and the other, a long hydrocarbon chain (Hann, 1990). In order to spread the molecules, they must be dissolved in a "strong" solvent at concentration that does not permit the formation of aggregates. Drops of the solution are placed on the water surface. The amount of the molecules in the drop must be rather small in order to have after spreading a monolayer where molecules are far from each other and do not interact. When such molecule is placed on the water surface, its polar head group interacts with water, while the hydrocarbon chain faces towards air, as it cannot be surrounded by water for entropy reasons. Floating molecules can be compressed by the barrier until condensed state is achieved and monitored with surface potential measurements (Figure 2.1). Usually the parameter under control during the compression is the surface pressure, which characterizes the decrease in the surface tension of the water surface due to the presence of the monolayer:

$$\pi = \sigma_{H_2O} - \sigma_{ml}$$

where σ_{H2O} is the surface tension of water without monolayer and σ_{ml} is the surface tension of the water surface covered by monolayer. One of important characteristics of the monolayers on the water surface is the dependence of surface pressure on the area occupied by single molecule in the monolayer. The dependence is usually called "compression isotherm" or "π-A isotherm" (the curve is measured at fixed temperature). This characteristic (Figure 2.1) is rather important since it allows to calculate the area per molecule in the monolayer and to reveal phase transitions in the monolayer structure.

The Langmuir-Blodgett (LB) technique was in recent time extended to quite more interesting biopolymers as proteins (Nicolini *et al.*, 1993; Nicolini 1997; Sanchez-González *et al.*, 2003; Owaku *et al.*, 1989). These molecules, being placed at the air/water interface, arrange themselves in such a way that their hydrophilic part penetrates water due to its electrostatic interactions with water molecules, which can be considered electric dipoles. The hydrophobic part (aliphatic chain) orients itself to air, because it cannot penetrate water for entropy reasons. Therefore, a few molecules placed at the water surface form a two-dimensional system at the air/water interface. Figure 2.1 shows the dependence of surface pressure upon area per molecule, obtained at constant temperature (Nicolini, 1997; Paternolli *et al.*, 2004; Nicolini and Pechkova, 2006a).

The compression of the interface yields a surface potential (Figure 2.2) for the cytochromes at the air-water interface (Nicolini *et al.*, 2001). Initially, the compression does not result in significant variations in surface pressure. Molecules at the air/water interface are rather far from each other and do not interact.

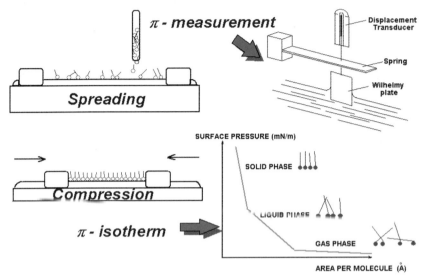

Figure 2.1 Typical π-A isotherm of cytochrome P450 monolayer obtained by the shown spreading and Wilhelmy plate on 10 mM K-phosphate, pH 7.4, buffer subphase.

This state is referred to as a two-dimensional gas. Further compression results in an increase in surface pressure. Molecules begin to interact. This state of the monolayer is referred as two-dimensional liquid, which can be separated in liquid-expanded and liquid-condensed phases. Continuation of the compression results in the appearance of a two-dimensional solid-state phase, characterized by a sharp increase in surface pressure, even with small decreases in area per molecule. Dense packing of molecules in the monolayer is reached. Further compression results in the collapse of the monolayer.

Surface potential measurements are typically made on the solid substrate using a home-made device (Erokhin *et al.*, 1995). This technique, which has been widely used for investigations of monolayers both at air/water interfaces (Tredgold and Smith, 1983) and, more recently, on solid substrates (Facci *et al.*, 1994a, Erokhin *et al.*, 1995), can measure the surface potential (with a sensitivity of 1 mV) that arises from dipole and charge distributions in one or more deposited monolayers (Figure 2.2); the resulting current is monitored in a circuit, one end of which is equipped with a vibrating electrode placed near (around 1 mm) the monolayer surface and which oscillates at its resonance frequency (282 Hz). Early reference (Erokhin *et al.*, 1995) illustrates the experimental setup and the measuring principle as implemented originally on the LB oriented RC layer. In the dark, LB provides a specific surface potential due to the preferential orientation of its molecules, which have a static dipole moment. On exposing to light, RC molecules begin to show an additional dipole behavior due to the charge separation inside them (electrons are displaced by about 3 nm) (Facci *et al.*, 1994a).

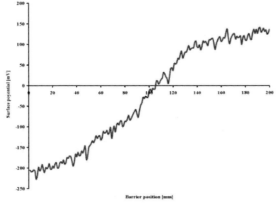

Figure 2.2. Typical surface potential versus barrier position of a monolayer of cytochrome P450 at the air-water interface. The subphase was a 10 mM K-phosphate, pH 7.4 (Reprinted with the permission from Paternolli *et al.*, Recombinant cytochrome P450 immobilization for biosensor applications, Langmuir 20, pp. 11706–11712 © 2004, American Chemical Society).

In that early paper by measuring the difference between the surface potential values of the RC layer in the dark and in the light, direct information is provided about the activity of the proteins in an LB film. Such an approach, moreover, seems to be very useful-especially for multiple investigations on the same monolayer as it couples the protein monolayer with a solid state electrode in a "soft" fashion (Facci *et al.*, 1994a). Non-oriented films do not show any preferential molecular orientation, and so it is impossible to monitor the protein activity using surface potential measurement. Only when the given protein, such as any cytochrome P450 displays a self-assembly significant surface potential appears evident (Figure 2.2) even when the surface pressure is zero (Figure 2.1).

2.2 Atomic Force Microscopy

AFM utilizes a sharp probe to scan across the surface of a sample with the laser being focused on the tip, and the beam being reflected to the split photodiode detector (Figure 2.3).

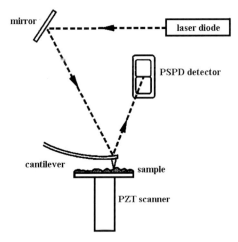

Figure 2.3. Atomic force microscopy configuration as described in the text.

As the cantilever is deflected over the sample, the photodiode monitors the changes of the laser beam. The changes are stored in the computer to produce a topographic image of the sample surface. There are three different modes in atomic force microscopy on protein film: *contact* mode, *tapping* mode, and *non-contact* mode.

In *contact* mode, the tip physically comes in contact with the sample. Atomic resolution can be reached in contact mode; however, there is a risk of damage to samples being soft.

During *tapping* mode, the cantilever is oscillated at or near its resonance frequency. The scanner, in a pendulum-type motion, enables the tip to "tap" the sample surface as the scanner comes to the bottom of its swing. There is less damage in tapping mode and higher lateral resolution; however, there is a slightly slower scan speed than in contact mode.

The cantilever is oscillated at a frequency slightly above the cantilever's resonance frequency during *non-contact* AFM. The tip oscillates above the adsorbed fluid layer on the surface. It does not come in contact with the sample surface. The sample is not damaged during non-contact mode; however, the scan speed is much slower, there is lower lateral resolution, and it may be used only with very hydrophobic samples. Our original homemade AFM instruments (Sartore *et al.*, 2000) operated in air or in water, at constant deflection with triangular-shaped, gold-coated Si_3N_4 microlevers. Originally the tips of the microlevers had a standard aspect ratio (about 1:1), and the levers had a nominal force constant of 0.03 N/m. The constant-force set point was about 0.1 nN, while the images acquired were 256 × 256 pixel maps. All images are standard top-view topographic maps, where the brightness is proportional to the quota of the features over the sample surface *i.e.*, light means mountain, dark means valley. Figure 2.4 shows an AFM of lysozyme LB film (Pechkova *et al.*, 2005a).

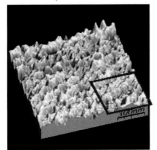 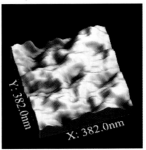

Figure 2.4. AFM image of LB lysozyme film. Images of the lysozyme template, obtained in tapping mode in a dry atmosphere (AFM: cantilever I type NSC14/Cr-Au MikroMash) (Reprinted with the permission from Pechkova *et al.*, µGISAXS and protein nanotemplate crystallization methods and instrumentation, Journal of Synchrotron Radiation 12, pp. 713–716, © 2005a, Blackwell Publishing).

A custom AFM instrument further optimised for protein crystal imaging in solution has been recently introduced (Figure 2.5) and tested on crystals and Langmuir-Blodgett films of two proteins having quite different molecular weight (Pechkova *et al.*, 2007b).

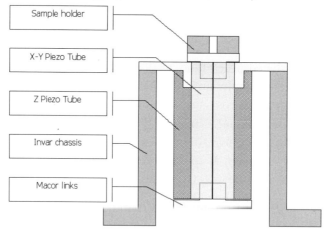

Figure 2.5. Piezo movers assembly in the AFM used for the experiments. Piezo tubes were purchased by Physik Intruments and perform a maximum 10x10 µm travel in XY and 2.5µm in Z direction (Reprinted with the permission from Pechkova *et al.*, Atomic force microscopy of protein films and crystals, Review of Scientific Instruments 78, pp. 093704_1–093704-7, © 2007b, American Institute of Physics).

AFM is a topography sensitive method, which in this last case is used with protein crystals in wet environments (Pechkova *et al.*, 2007b). In the "noncontact-tapping mode", the AFM derives topographic information from measurements of attractive forces. Images of the lysozyme crystal (Figure 2.6) have been grabbed in tapping mode with a cantilever I type NSC14/Cr-Au MikroMash in dry atmosphere utilizing our instrument based on in house SPMagic controller (Pechkova *et al.*, 2007b). The freeware WSxM© (http://www.nanotec.es) was utilized for the processing of the acquired images. The typical resonance frequency of the cantilever tip is between 110 [kHz] and 220 [kHz], and the proper positioning of the cantilever on the tip holder of AFM has been found at frequency of 92 [kHz] with intensity 0.6 Volt. The set point for loop control was at 0.2 Volt. The integral gain value during image grabbing has been set at 4.4 (I Gain) and the proportional gain value during image acquisition has been set at 8.03 (P gain). This approach allows to study the crystal periodicity and morphology in their mother liquid, preserving the native periodic protein crystal structure, typically destroyed with drying. Comfortingly it appears to distinguish the protein crystals from the salt crystals (Figure 2.6a), which under the optical microscope is frequently quite similar and often their difference is revealed only during X-ray analysis. AFM estimates of the given single proteins packing, order and morphology appear quite similar in the LB thin film and in the crystals, thereby allowing routine crystal measurements at high resolution. The AFM consists of a custom measuring head with a flexible SPM controller in house produced which can drive the head for contact, non-contact and spectroscopy modes, providing the user with an high degree of customization for the crystal measurement (Figure 2.6b).

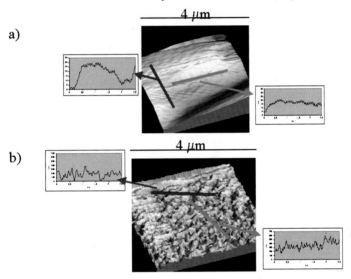

Figure 2.6. AFM images and profiles of (a) lysozyme crystal 4 μm × 4 μm and (b) salt crystal 4 μm × 4 μm (Reprinted with the permission from Pechkova *et al.*, Atomic force microscopy of protein films and crystals, Review of Scientific Instruments 78, pp. 093704_1–093704-7, © 2007b, American Institute of Physics).

The described AFM system for crystal was tested with standard samples (Figure 2.7a) in order to prevent any possible problem when working with more protein samples. We have imaged several portions of CD-ROMs, both burned and not, because they show regular geometries at standardized distances. Figure 2.7a shows a typical result of an unwritten sample. The image was acquired in tapping mode using a NSC-11 cantilever from MikroMasch. Another test was preformed on samples with a regular geometry and known dimensions. This test indeed can be used to calibrate the instrument in all scan directions. We have used several types of calibration gratings, namely TGX01, TGZ01, TGZ02 and TGZ03 from MikroMasch.

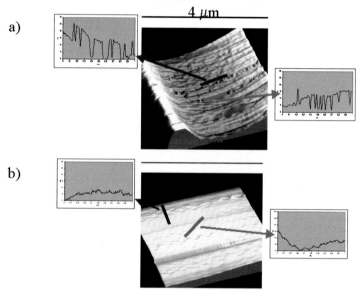

Figure 2.7. (a) AFM image of the standard sample: a typical result obtained with unwritten CD-ROM sample in tapping mode using a NSC-11 cantilever from MikroMasch; (b) AFM image of the standard sample: typical result obtained with TGZ01 grid in tapping mode using a NSC-18 cantilever from MikroMasch. The picture also shows a line profile (below on the left) taken along a step portion where a dust particle was imaged (zoomed image on the right) (Reprinted with the permission from Pechkova et al., Atomic force microscopy of protein films and crystals, Review of Scientific Instruments 78, pp. 093704_1–093704-7, © 2007b, American Institute of Physics).

The former grid shows a square pattern with 3μm pitch and is mostly used to calibrate X and Y directions. The latter ones show line step profiles with horizontal pitch of 3 μm and distinct vertical (Z) steps. Figure 2.7b shows a typical result obtained when imaging the TGZ01 grid in tapping mode. A NSC-18 cantilever from MikroMasch was used. The picture also shows a line profile taken along a step portion where a dust particle was imaged (zoomed in the same figure). The profile clearly indicates that the vertical dimensions are correctly calibrated, that no particular convolution or spherical effects are present and that small details can be reasonably acquired along the vertical axis.In conclusion, atomic force microscopy (AFM) has been frequently used to study protein, nucleic acid, and crystals *in situ*, in their mother liquors and as they grow.

From the sequential AFM images taken at brief intervals over many hours, or even days, the mechanisms and kinetics of the growth process was tentatively defined time ago (McPherson *et al.*, 2000; Wiechmann *et al.*, 2001). In few case three-dimensional microcrystals of integral membrane proteins as OmpC porin, air-dried slowly and imaged by AFM (Kim *et al.*, 2000), give some correspondence with X-ray diffraction even if only recently new insights emerge from a detailed unexpected study using laser irradiation on classical versus LB crystal (Pechkova *et al.*, in preparation). AFM studies allow also analyzing in details the molecularity of growth steps of tetragonal lysozyme crystals (Li *et al.*, 1999).

The incorporation of a wide range of impurities, ranging in size from molecules to microns or larger microcrystals, and even foreign particles were visually recorded. Thanks to these observations and measurements, a more complex understanding of the detailed character of macromolecular crystals has been emerging, one that reveals levels of complexity previously unsuspected. The new Atomic Force Microscopy configuration (Pechkova *et al.*, 2007b) allows acquiring "on real time" images at the atomic resolution also of very small micro crystals previously impossible.

For crystal characterization at subnanometric resolution precise control of operating conditions is indeed a critical aspect, in which the strong dependency of measurements on environmental factors, such as acoustic and mechanical noises, temperature, and humidity play an important role. This was shown initially using the tapping mode with the AFM measurements of protein in water giving a resolution ten times higher than in air (Pechkova and Nicolini, 2002a). Indeed, with the measurements in water, the disturb created by dust or other particles attached onto the surface by adhesion can be corrected, and the AFM image of P450scc micro crystal yielded the same geometric parameters of the P450scc protein obtained by homology modeling (Pechkova and Nicolini, 2002a; Sivozhelezov et al., 2006). The present AFM system for crystal (Pechkova et al., 2007b) has grown from an AFM for surface investigation installed in a chamber with controlled atmosphere developed time ago with hardware-software configuration based on neural network (Sartore et al., 2000; Salerno et al., 1999).

2.2.1 *AFM spectroscopy*

With AFM we can typically obtain from the force in action apparent in Figure 2.8 the intensity of the force versus the sample-tip distance.

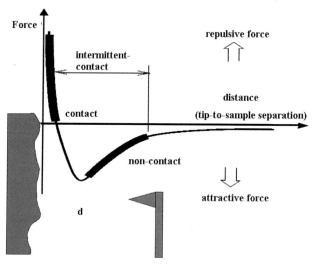

Figure 2.8. Forces in action.

Moving the tip along the Z-axis (*i.e.*, orthogonal to the sample) and recording the cantilever deflection gives then rise to the force-distance spectroscopy, which can be readily applied to the study of protein-protein interaction without the utilization of fluorescence labelling.

Doing the AFM analysis shown in Figure 2.8 with the proteins yields multi snap-off segments, where the two interacting proteins are fixed respectively on the sample and on the tip, namely gives rise to the AFM spectroscopy of proteins (Figure 2.9).

We have recently designed based on the above principles an AFM-based Label Free analysis of NAPPA microarrays (see paragraph 2.8.2) containing an opto-mechanical system for an AFM head capable to house multiple cantilevers, for differential measurements and driven by an appropriate electronics in order to detect the emerging signals with significant improvement in protein detection (Sartore *et al.*, in preparation).

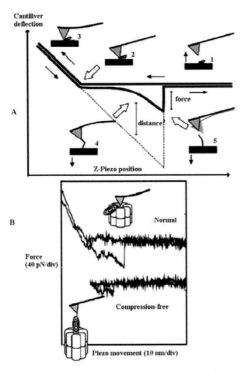

Figure 2.9. Spectroscopy of proteins, namely cantiliver deflection versus piezo Z position (a) and AFM force versus Piezo movement (b).

2.2.2 Scanning tunneling microscopy

Time ago Scanning Tunneling Microscopy (STM) was also used (Facci *et al.*, 1994) to characterize nanomaterials in terms of voltage-current (V-I) characteristic at single points.

All the images are typically obtained in air in constant current mode within a tunneling current range of 0.1–0.5 nA and voltage from -1.5 V to 1.5 V (MDT-AsseZ, Moscow-Padua).

STM analysis on different areas of several samples showed similar features, some of which are presented in Figure 2.10.

In all these cases it is possible to distinguish features shaped like wells in a corrugated matrix, sometimes isolated, sometimes preferentially arranged along lines.

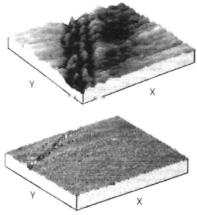

Figure 2.10. STM images of cadmium arachidate. (a) STM image of CdS film from one bilayer of cadmium arachidate on top of highly oriented pyrolytic graphite: constant current mode; image size 25.6 x 25.6 nm^2; maximum corrugation 1 nm; tunnelling parameters $V_t = 0.1$ V, $I_t = 1$ nA: scanning rate 12 Hz. (b) STM image of highly oriented pyrolytic graphite plate: constant current mode; image size 25.6 x 25.6 nm^2; maximum corrugation 1 nm; tunnelling parameters $V_t = 0.15$ V, $I_t = 1$ nA; scanning rate 12 Hz. (Reprinted with the permission from Facci *et al.*, Formation of ultrathin semiconductor films by Cds nanostructure aggregation, The Journal of Physical Chemistry 98, pp. 13323–13327, © 1994, American Chemical Society).

Their dimensions, ranging typically between 5–10 nm, match well with the sizes of CdS particles, formed by the above described procedure, estimated by analysis of optical absorption spectra and by X-ray and electron diffraction. This comparison allowed suggesting that the wells in the images could be connected with the CdS particles formed in LB film of arachidic acid (*i.e.*, particles could lay inside the wells). This fact is in agreement with general concepts coming from ellipsometry (Smotkin *et al.*, 1988) and X-ray analysis (Erokhin *et al.*, 1991) and with the suggestion that the initial 2D arrangement of cadmium atoms in the film can be preserved to a certain extent in the final film after the reaction.

2.3 Nuclear Magnetic Resonance

^{1}H NMR 500 MHz spectrometer has been used in this context to monitor the atomic structure of conductive polymers can be determined as well (Ram and Nicolini, 2000) and of proteins, either in solution as in the case of ribosomal proteins (Figure 2.11) (Vasile *et al.*, 2008) or derived from the dissolved corresponding crystals (Pulsinelli *et al.*, 2003; Pechkova and Nicolini, 2006).

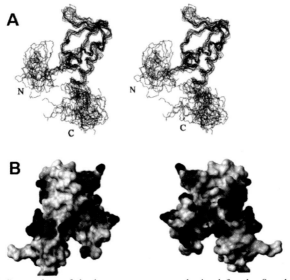

Figure 2.11. A) Stereoview of the best ten structure obtained for the β–subunit of aIF2 from *S. Solfataricus*. (PDB ID code 2NXU). B) Electrostatic surface plot generated with MOLMOL of *S. Solfataricus* aIF2β, acidic and basic residues are colored in red and blue respectively. The structures are rotated of 180° (Reprinted with the permission from Vasile *et al.*, Solution structure of the β-subunit of the translation initiaton factor Aif2 from Archaebacteria Sulfolobus solfataricus, Proteins Structure, Function and Bioinformatics 70, pp. 1112–1115, © 2008, John Wiley & Sons, Inc.).

Nuclear magnetic resonance (NMR) is a phenomenon that occurs when nuclei of certain atoms are immersed in a static magnetic field and exposed to a second oscillating magnetic field. Some nuclei experience this phenomenon, and others do not, depending upon whether they possess a spin. Nuclear magnetic resonance spectroscopy is the use of the NMR phenomenon to study physical, chemical, and biological properties of matter. Many of the dynamic NMR processes are exponential in nature. The simplest NMR experiment is the continuous wave (CW) experiment.

There are two ways of performing this experiment. In the first, a constant frequency, which is continuously switched on, probes the energy levels while the magnetic field is varied. The CW experiment can also be performed with a constant magnetic field and a frequency that varies. The signal of NMR spectroscopy results from the difference between the energy absorbed by the spins, and that depicts a transition from the lower energy state to the higher energy state and the energy emitted by the spins, which simultaneously make a transition from the higher to the lower energy state. The signal is thus proportional to the population difference between the states. NMR is a rather sensitive spectroscopic technique since it is capable of detecting these very small population differences. NMR samples are prepared by dissolving an analyte in a deuterium lock solvent. The concentration of the sample should be great enough to give a good signal-to-noise ratio in spectrum, yet minimize exchange effects found at high concentrations. The exact concentration of the sample in the lock solvent depends on the sensitivity of the spectrometer.

Recently also the atomic structure of a new conjugated amphiphilic polymer (Narizzano et al., 2005) was determined by ^{1}H-NMR spectroscopy in $CDCl_3$ solution using a Bruker AMX 500 instrument. The ^{1}H-NMR spectra revealed that the polymer contains 20% of 3HT. The molar fraction of HT was calculated by integrating the peak areas of methyl protons (~ 0.9 ppm), which are only present in the 3HT units, and of aromatic protons (~ 7.0 ppm). The normalized areas were found to be 0.52 for the methyl group, and 1.00 for the aromatic protons, respectively. The polymer exhibits good solubility in $CHCl_3$, maintaining the typical absorption peak of 3-alkyl-substituted polythiophenes, which are obtained via the $FeCl_3$ method at about 431 nm. This absorption, which is detected in the UV-visible spectrum of the non-doped sample in chloroform solution correspond to the typical π-π^* transition of the conjugated backbone.

2.3.1 Circular dichroism

The *Circular Dichroism* (CD) spectra at various temperatures is at times a very useful information on the secondary structure of polymers which can be to acquired quickly whenever needed, as in monitoring the increased stability of the protein as a result of their immobilization in the thin films in the range of 25–300 °C (Figure 2.12), a very useful information indeed to assess the unexpected and useful heat-proof property intervening in proteins whenever immobilized by LB technology.

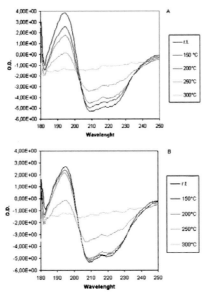

Figure 2.12. Circular Dichroism of protein monolayer. CD spectra as a function of the temperature. (A) Cytochrome P450 native recombinant film. (B) Mutant K201E. The data were collected in the far ultraviolet region (180-250 nm) with a wavelength step of 2 nm. Each spectrum was the result of an accumulation of three scans, and it was recorded at a rate of 50 nm/min with a time constant of 4 s. (Reprinted with the permission from Ghisellini *et al.*, P450scc mutant nanostructuring for optimal assembly, IEEE Transactions on Nanobioscience 3, pp. 121–128, © 2004, IEEE).

CD spectra are typically acquired using a spectropolarimeter to obtain the percentage of protein secondary structures in the film and in the solution, namely in terms of α-helix, β-pleated and random coil.

2.4 Brewster-Angle Microscopy

Brewster Angle Microscopy (BAM) allows to follow the Langmuir film formation and to display its morphology, as well as to correlate the 2D/3D transformations with the shape of the π-A isotherms recorded simultaneously.

The method is based on the fact that polarized light does not reflect from the interface when it reaches it at the Brewster angle, determined by the equation:

$$tg\varphi = n_2/n_1$$

where n_1 and n_2 are refractive indexes of the two media at the interface. The value of this angle for the air/water interface is 53.1°. Therefore, it is possible to adjust the analyzer position in such a way that it will bear a dark field when imaging the air/water interface. Spreading of the monolayer varies the Brewster conditions for both air/monolayer and monolayer/water interfaces, making visible the morphology of the monolayer. Langmuir film formation was imaged and analyzed by BAM. It was also possible to detect film dishomogeneities and breaks as well as the 2D-3D transformations, which occurred at different surface pressures. Such typical morphologies related to the investigation carried out on Langmuir films of both proteins and polymers are conformed in the images obtained in Paddeu *et al.* (1997) and in numerous papers reviewed by Nicolini and Pechkova (2006).

When semiconductor particles were grown directly on LS films of copolymer with copper ions by exposure to H_2S atmosphere, the morphology of the films, studied by Brewster angle microscopy, undergoes evident changes upon exposure to the H_2S atmosphere. In fact, the formation of CuS nanoparticles causes increased layer corrugation. As can be observed in Figure 2.13 (Narizzano *et al.*, 2005), the rather good homogeneity of the pristine film surface changes drastically after the reaction. Morphology of films was studied by Brewster angle microscopy (BAM-2, Nanofilm Technologie GmbH, Germany) at each step, *i.e.*, after deposition and after the particle formation process. Brewster angle microscopy images were acquired with an Instrument BAM-2 (Nanofilm Technologie GmbH, Göttingen, Germany) (Narizzano *et al.*, 2005). The standard laser of the BAM is a diode laser with wavelength of 690 nm. The high-power laser diode has 30 mW primary output.

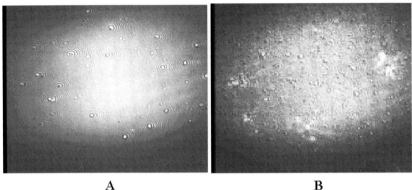

A   B

Figure 2.13. Brewster Angle Microscopic images taken for 20 layers films of semiconductor particles grown directly on LS films of copolymer with copper ions copolymer. (a) before exposure to H_2S; (b) after exposure to H_2S. The image size is 0.3 × 0.2 mm (Reprinted with the permission from Narizzano *et al.*, A heterostructure composed of conjugated polymer and copper sulfide nanoparticles, The Journal of Physical Chemistry B 109, pp. 15798–15802, © 2005, American Chemical Society).

2.4.1 *Ellipsometry*

Ellipsometry is a technique based on the measure of two parameters, namely the ratio of the vibration of the electric vector in the plane of incidence and perpendicular to it, and the difference of phases of these two vectors. The theory of the ellipsometry allows connecting these two parameters with the thickness and refractive index of a nanostructured layer (Drude, 1902). Usually, the layer is assumed to be not absorbing and isotropic. In principle, the assumption is not valid in the most of cases. Nevertheless, for the thickness estimation it seems to vary the value inside the experimental error.

The technique allows to determine the thickness of the monolayer during compression, revealing the reorganization of molecules at the water surface. It carries out the measurements of film thickness using a laser as a light source. Typically the mean values of the measurements at different angles are given for the thickness and the index of refraction (Tronin et al., 1994). According to the two-layer model the first lower layer accounts for the imperfections which always exist on the surface of most substrate in the form of traces of polishing, intrusions of the polisher, a thin oxide layer, *etc*.

2.5 Electrochemistry

Cyclic voltammetry offers important and sometimes unique approaches to the electroactive species. The cyclic voltammetry (CV) is the most versatile electroanalytical technique to study the electroactive species. CV consists of cycling the potential of an electrode, which is immersed in an unstirred solution and measuring the resulting current.

The detailed understanding of cyclic voltammetry technique can be gained by considering the Nernst's equation and the changes in concentration that occur in solution adjacent to the electrode during the electrolysis. The proper equilibrium ratio in reversible system at a given potential is determined by the Nernst equation. The electron transfer reaction between proteins such as cytochrome P450 in presence of cholesterol is an important system for investigating fundamentals regarding long-range electron transfer in biological systems; it has attracted considerable experimental and theoretical attention (Nicolini et al. (2001a). Cyclic voltammetry, as an efficient method to investigate electron transfer reactions between proteins in films, is being exploited increasingly for obtaining electrochemical information on proteins (Figure 2.14).

The electrochemical measurements were made by a potentiostat/galvanostat (either homemade or EG & G PARC, model 263A), which was supplied with its own software (M270). The working cell was homemade. A standard three-electrode configuration was used, where LB films of cytochrome P450scc were deposited on an ITO coated glass plate, which acted as a working electrode, having platinum as a counter electrode and Ag/AgCl as a reference electrode. The working electrode was cycled between initial and switch potentials of 500 and 500 mV, respectively, after holding the electrochemical system at the initial potential for 10 s. The scan rate used was 20 mV/s because the cathodic peak was most evident at this speed and, at the same time, the low background current was minimized.

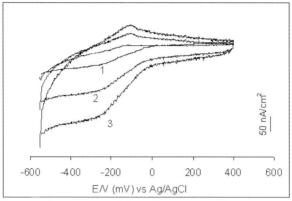

Figure 2.14. Current-voltage characteristics of cytochrome film in presence of increasing concentration of cholesterol. Cyclic voltammetry of cytochrome P450scc LB film with the addition of (1) 100, (2) 400, and (3) 750 μM cholesterol. (Reprinted with the permission from Paternolli et al., Recombinant cytochrome P450 immobilization for biosensor applications, Langmuir 20, pp. 11706–11712, © 2004, American Chemical Society).

All the measurements were repeated three times to verify the reproducibility. The electrical behavior of all films of either polymers, biopolymers or nanocomposites, is typically studied by I-V measurements. For istance, after the formation of nanoparticles in the polymeric matrix, the conductivity of LS films was changed, increasing by about 2 orders of magnitude. Conductivity shows a linear dependence on the film thickness (Narizzano and Nicolini, 2005). The relatively low value of the conductance depends on the limited number of polymeric layers and, as a consequence, on the low number of CuS nanoparticles as well as their random distribution. What is interesting is the dependence of the specific conductance (σ) of the polymer, with and without nanoparticles, as a function of the monolayer number (Narizzano and

Nicolini, 2005). As a consequence of the increase in the number of monolayers, we observed a net and linear increment of σ in the PAET/CuS heterostructure, while in the pristine polymer σ remains almost unchanged, going from 5 to 10 monolayers, with a slight increment for the 20 monolayer system. This could be due to the structural reorganization of the polymer that usually occurs as the number of monolayers increase, thus leading to a longer conjugation length. Nanoparticle formation is responsible for a monotonous increase. The noticeable change brought about by increasing the number of monolayers suggests that particles are able to move and to rearrange themselves within the films, leading to a structural reorganization of the PAET/CuS system (Narizzano and Nicolini, 2005).

2.6 Infrared Spectroscopy

Fourier transform infrared (FTIR) spectroscopy is a powerful analytical tool for characterizing and identifying organic molecules. Infrared radiation is defined as the electromagnetic radiation the frequency of which varies between 14300 and 20 cm^{-1} (0.7–500 μm). Within this region of the electromagnetic spectrum, chemical compounds absorb IR radiation provided there is a change in dipole moment during a normal molecular vibration, molecular orientation, and molecular rotation or from combination of difference in overtones of normal vibration.

The FTIR spectrum of an organic compound, such as cytochrome C (Figure 2.15) serves as its fingerprint and provides specific information about chemical bonds and molecular structure. Samples are run either as pure substance or in KBr pellets. A beam of infrared radiation is passed through the sample and the detector generates a plot of percent transmission of radiation versus the wave number or wavelength of the transmitted radiation. When the percent transmission is below 100, some of the light is absorbed by the sample. Each peak in the spectrum represents absorption of light energy, and is called an absorption band. It is a powerful analytical tool for characterizing and identifying organic molecules. The IR spectrum of an organic compound serves as its fingerprint and provides specific information about chemical bonding and molecular structure (Bramanti *et al.*, 1997; Pechkova *et al.*, 2007b). The FTIR involves the vibrations of molecular bonds. Molecules can bend or stretch at their bonds. They can also wiggle around in a wag, or twist. Molecules vibrate in certain modes, depending on the symmetry properties of the molecules' shape (Bramanti *et al.*, 1997; Pepe *et al.*, 1998).

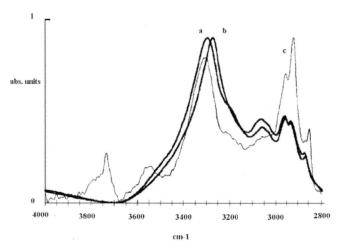

Figure 2.15. Fourier transform infrared (FTIR) spectra of a) cytochrome C film obtained by spreading the solution at pH 7.4, dried at 25 °C, b) cytochrome C film obtained by spreading the solution at pH 1, dried at 100 °C, c) LB film in the 4000-2800 cm^{-1} region. (Reprinted with the permission from Bramanti *et al.*, Qualitative and quantitative analysis of the secondary structure of cytochrome C Langmuir-Blodgett films, Biopolymers 42, pp. 227–237, © 1997, John Wiley & Sons, Inc.).

The energy it takes to excite a vibrational mode varies depending on the strength of the bond and the weight of the molecule. FTIR involves the conversion of energy to molecular vibrations. Infrared radiation (wave numbers of 4800–400 cm^{-1}) can be converted to vibrations in the molecule, which causes the molecule to go from a ground vibrational state to an excited vibrational state.

The amount of energy required to stretch a bond depends on many things, one of which is the strength of the bond and the masses of the bonded atoms. The higher wave numbers are achieved when the bonds are stronger and the atoms are found to be smaller. Two-dimensional

Fourier transform infrared spectroscopy has been recently applied to study the thermal stability of multilayer Langmuir-Schaefer (LS) films of lysozyme deposited on silicon substrates (Pechkova et al., 2007a).

The study has confirmed previous structural findings that the LS cytochrome films have a high thermal stability that is extended in a lysozyme multilayer up to 200 °C. 2D infrared analysis has been used here to identify the correlated molecular species during thermal denaturation (Figure 2.16).

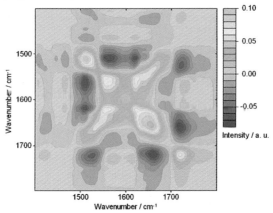

Figure 2.16. Synchronous 2D infrared correlation spectra in the 1800–1400 cm^{-1} interval of LS films during thermal treatment from 25 to 250°C. The analysis was done in transmission and in situ (Reprinted with the permission from Pechkova et al., Thermal stability of lysozyme Langmuir-Schaefer films by FTIR spectroscopy, Langmuir 23, pp. 1147–1151, © 2007a, American Chemical Society).

Asynchronous 2D spectra have shown that the two components of water, fully and not fully hydrogen bonded, in the high-wave number range (2800–3600 cm^{-1}) are negatively correlated with the amine stretching band at 3300 cm^{-1}. On the grounds of the 2D spectra the FTIR spectra have been deconvoluted using three main components, two for water and one for the amine.This analysis has shown that, at the first drying stage, up to 100 °C, only the water that is not fully hydrogen bonded is removed. Moreover, the amine intensity band does not change up to 200 °C, the temperature at which the structural stability of the multilayer lysozyme films ceases.

2.7 Nanogravimetry

By means of this technique it was possible to estimate the surface density and the area per molecule values by depositing the Langmuir-Blodgett film at different surface pressures, as well as the reliability of the deposition itself.

Gravimetric measurements were carried out with a Quartz Crystal Balance (QCM) utilising 10 MHz quartz resonators according to the procedure shown in literature. A layer of conducting polymer was deposited on both sides of the resonator and dried with nitrogen flux, and then the frequency shift due to the film deposition was registered. This frequency variation Δf was correlated to the mass change ΔM by the Sauerbray equation, written as:

$$\Delta M = K \Delta f$$

where constant K depends upon physical parameters of the utilized resonator, and defined as "the sensitivity of the instrument".

From the nanogravimetry curve (Figure 2.17), plotted as surface density as a function of the number of protein monolayers, the information on the uniformity and reproducibility of deposition can be obtained.

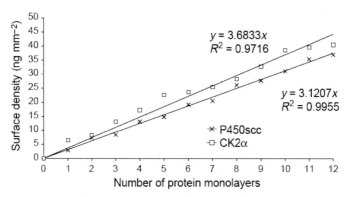

Figure 2.17. The dependence of the surface density of deposited protein (cytochrome P450scc and human kinase CK2α) LB thin film upon the number of the transferred monolayers. (Reprinted with the permission from Pechkova and Nicolini, Protein nanocrystallography: a new approach to structural proteomics, Trends in Biotechnology 22, pp. 117–122, © 2004a, Elsevier).

Moreover, knowing molecular weight of the protein molecule, it is possible to experimentally evaluate the area per one molecule in the obtained film and to compare it with that theoretically estimated for the closely packed system (Pechkova and Nicolini, 2002a).

The latter value can be easily calculated knowing the geometrical parameters of the protein from RCSB PDB data bank (Berman et al., 2002). In case the structure of the protein is not resolved yet, homologous protein parameters or geometrical features from molecular modelling can be used for calculation. It uses quartz resonators with a resonance frequency of about10 MHz to measure mass.

A simple circuit allows the quartz resonators to oscillate at their resonance frequency. The shift in the resonance frequency, induced on a quartz resonator by subsequent LB, deposition, is monitored by a frequency meter. The frequency shift Δf owing to the amount of mass Δm attached to the resonator surface is expressed in mass units by applying the Sauerbrey equation (Sauerbrey 1964):

$$\frac{\Delta f}{f_0} = -\frac{\Delta m}{A\rho l}$$

where f_0 is the resonance frequency of the quartz, ρ its density, and l and A the thickness and the area covered by the deposited monolayer respectively.

The knowledge of the area of the resonator covered by the deposited layers (measured by an optical microscope) allows us to convert these values to surface density units (ng mm^{-2}) (Erokhin et al., 1990) to calculate the packing degree (the ratio between the film surface density at the air-water interface and that after deposition onto the solid substrate).

Calibration of the quartz balance was performed according to Facci et al. (1993) and Lvov et al. (1990).

The protein film is deposited on both sides of the resonator, and afterwards dried by nitrogen flux; the frequency shift is registered after the covering (Facci et al., 1993; Antolini et al., 1995a; Paddeu et al., 1995a).

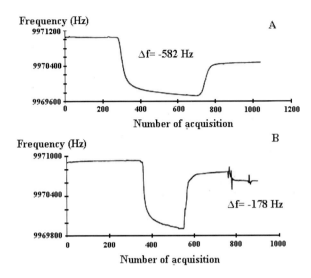

Figure 2.18. Example of specific reaction: a) quartz has been covered with BSA (by silanization) and the flowing solution contains antibodies specific to BSA (1 mg/ml); b) the same quartz has been contacted by flowing solution containing antibodies non-specific to BSA (GAM 1 mg/ml).

Results of the typical gravimetric study of deposited monolayers are presented in Figure 2.18.

2.7.1 *Quality factor*

In most situations the adsorbed film is not rigid and the Sauerbrey relation becomes invalid (Adami *et al.*, in preparation). Namely, a film being "soft" (viscoelastic) will not fully coupled to the oscillation of the crystal, hence the Sauerbrey relation will *underestimate* the mass at the surface.

A soft film thereby dampens the crystal's oscillation, where the damping, or dissipation, D of the crystal's oscillation (Figure 2.19), which reveals the film's softness (viscoelasticity) is defined as:

$$D = E_{lost} / E_{stored}$$

where E_{lost} is the energy lost (dissipated) during one oscillation cycle and E_{stored} is the total energy stored in the oscillator.

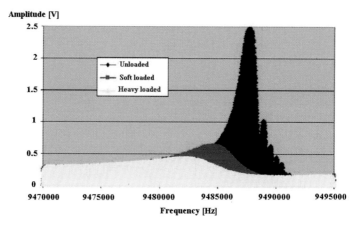

Figure 2.19. Quartz dumping.

The coupling of the crystal surface to a liquid drastically changes the frequency. When a quartz crystal oscillates in contact with a liquid, a shear motion on the surface generates motion in the liquid near the interface.

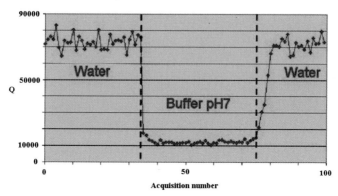

Figure 2.20. Quartz quality factor.

The oscillation surface generates plane-laminar flow in the liquid, which causes a decrease in the frequency, depending on the liquid density and viscosity: For a 10 MHz crystal with one face exposed to diluted solutions, near room temperature, Δf is about 2 KHz. The dissipation of the crystal is measured by recording the response of a freely oscillating crystal that has been vibrated at its resonance frequency. D is also equal to $1/Q$, the quality factor of the quartz crystal (Figure 2.20). A typical 10 MHz, AT-cut crystal working in a vacuum or gaseous environment has a dissipation factor in the range of 10^{-6}–10^{-4}. If a substance slips on the electrode, frictional energy is dissipated and D can be used to interfer the coefficient of friction of the adsorbed film; if the film is viscous, energy is also dissipated due to the oscillatory motion induced in the film and D can be used to interfer the internal friction in the film. An experimental set-up is presently being developed to measure D (Adami et al., in preparation), with the method being based on the principle that when the driving power to an oscillator is switched off at $t=0$, the amplitude of oscillation A decays as an exponential damped sinusoid:

$$A(t) = A_0 \, e^{\frac{-t}{\tau}} \sin(\omega t + \varphi) + \text{constant}$$

where τ is the decay time constant, φ is the phase and the constant is the dc offset.

The decay constant is related to D by:

$$D = \frac{1}{\pi f \tau}$$

2.8 Biomolecular Microarrays

Advance in genomics and proteomics have created a demand for miniaturize robot platforms for the high throughput (HT) study of proteins on a solid surface at high spatial density. Biomolecular arrays with a number of genes are currently available, chosen without a precise consideration of the particular target of the study, as described in the following subparagraph.

Two new methods nanotechnology-based to produce DNA-microarray and protein-chip (alternative to the one commercially available) were recently introduced, the first based on a novel biomolecular patterning on glass (Troitsky et al., 2002) and the second based on the nanostructuring of anodic porous alumina matrix for biomolecular microarray (Grasso et al., 2006; and chapter 1).

2.8.1 *Gene expression via DNASER*

A new matrix-integral part of the new DNA microarray instrumentation DNA analyzer (DNASER) here described was introduced based on a novel DNA patterning on the solid support surface (Troitsky *et al.*, 2002). Such patterning found the way to modify a glass surface for a precise positioning of small droplets of aqueous DNA solutions, without special robots (arrayers), within the boundaries of the modified regions. The physically heterogeneous surface consists of highly hydrophilic spots surrounded by a highly hydrophobic area leading to the surface patterning needed for a DNA microarray: a matrix of hydrophilic spots properly activated for immobilization of oligonucleotides has been fabricated on absolutely passive hydrophobic surface (Figure 2.21). The optimal efficiency of the above functionalitation technology of a glass-substrate in obtaining DNA microarray was confirmed by the Cy3-dCTP-labeled DNA sample, as shown by charge coupled device images of the DNASER previously described.

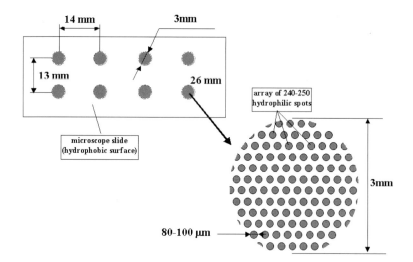

Figure 2.21. Samples with the arrays of hydrophilic spots prepared for the deposition of DNA solution. (Reprinted with the permission from Troitsky *et al.*, DNASER II. Novel surface patterning for biomolecular microarray, IEEE Transactions on Nanobioscience 1, pp. 73–77, © 2002, IEEE).

The novel bioinstrumentation named DNASER was introduced for the evaluation of gene expression via the real-time acquisition and elaboration of images from fluorescent DNA microarrays (Nicolini *et al.*, 2002). A white light beam illuminates the target sample to allow images grabbing on a high sensibility and wide-band charge-coupled device camera (ORCAII, Hamamatsu).

This high-performance device permits to acquire DNA microarrays images and to process them in order to recognize the DNA chip spots, to analyze their superficial distribution on the glass slide and to evaluate their geometric and intensity properties. Differently from conventional techniques, the spots analysis is fully automated and the DNASER does not require any additional information about the DNA microarray geometry.

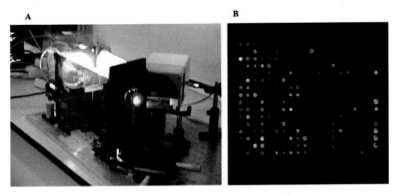

Figure 2.22. Nanogenomic diagnostics via DNASER. (A) The apparatus containing the optimally designed CCD camera (Nicolini *et al.*, 2002); (B) Fluorescent images of DNA solution spots with two color codes taken under the optimal drying procedures of the sample.

Using our DNASER (Nicolini *et al.*, 2002; Troitsky *et al.*, 2002), we confirmed the results by analyzing changes in gene expression after 24, 48 and 72 hours.

The validity of the DNASER measurements was confirmed by standard fluorescence microscopy equipped with CCD.

This experimental analysis proved that DNASER is appropriate for monitoring gene expression during the human lymphocytes cell cycle (Nicolini *et al.*, 2006a).

We calculated a final map of interactions among these 8 high-ranking genes in cell cycle of human T lymphocytes (Sivozhelevov *et al.*, 2006a; Chang *et al.*, 2004).

Typically the experimental datasets are derived from pangenomic microarrays as fully described elsewhere for the study of kidney transplant in humans (Braud *et al.*, 2008). Fifty-one individuals were included in the study: 8 patients tolerating a kidney graft (TOL) without any treatment and 18 patients with chronic rejection (CR) were evaluated against 8 healthy volunteers (HV) using a subset of the pangenomic (more than 35,000 genes) array displaying 6,865 genes (hence, "individual fullchip").

For every patient, 2 independent DNA amplifications were used. Data were expressed as mean values (log2) of the relative intensities [Cy3 (grafted patient) /Cy5 (pool of 169 kidney grafted recipients with stable graft function)].

This database emerges from our previous similar studies of original datasets called west-genopole based on different microarrays utilizing different gene nomenclature and obtained from 14 CR, 11 TOL and 6 HV patients (hence, "pool fullchip").

The data validity of the microarray was in this case assayed in details (Sivozhelezov *et al.*, 2008) by counting the fraction (%) of valid data *i.e.* data actually present in the microarray for each gene, separately for tolerance and rejection samples.

In the pool fullchip, the total numbers of the tolerance and rejection samples were 28 and 42 respectively. For example, for gene ABCA1-1A, the fraction is 39/42=93% among the "rejection" samples, and 18/28=64% among the "tolerance" samples.

Clustering analysis according to that parameter showed that about 74% "pro-rejection" genes and 71% "pro-tolerance" genes were classified in the top category, which we termed "reliable". Three more categories were revealed, termed "medium", "unreliable", and "very unreliable".

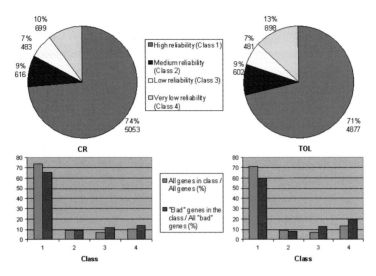

Figure 2.23. Top, percentages of classes of genes with belonging to the four categories with respect to reliabilities in the old "fullchip" raw dataset. Bottom, the same values (green) compared to fractions (%) of genes not adhering to the HGNC nomenclature ("bad") in each genes with respect to total "bad" genes. Left, CR data. Right, TOL data. CR dataset consists of 42 samples in total and 6864 genes, while TOL dataset consists 28 samples in total and 6864 genes. Reliability is given by the percentage of proven expression data by GENEPIX in such genes for 70 microarray samples. (Reprinted with the permission from Sivozhelezov et al., Immunosuppressive drug-free operational immune tolerance in human kidney transplants recipients: II Nonstatistical gene microarray analysis, Journal of Cellular Biochemistry in press, © 2008, Wiley-Liss, Inc., a subsidiary of John Wiley & Sons, Inc.).

Figure 2.23 shows the presence of the genes in the four categories. This limited validity does not appear however to introduce any bias in the "pool fullchip" data for either pro-TOL or pro-CR genes (Sivozhelezov et al., 2008), with the linear regression giving 93% correlation coefficient with the slope of the regression line close to unity. To further check if unreliability could be related to nomenclature problem, we calculated (Sivozhelezov et al., 2008) the fraction of genes not adhering to HGNC nomenclature in each of the four categories. If the nomenclature problems did not affect the reliability, we could expect the same fractions for the "bad" genes as for all genes. This is not the case. Even though the fractions of "bad genes" are close to those for all genes, occurrence of "bad" genes relative to all genes increases from category to category (Figure 2.23 bottom). The fact that the observed differences are small is readily explained by the fact that the disagreement with HUGO nomenclature does not necessarily mean that the deposited sample is unreliable. In fact, many of the genes obviously not adhering to the HGNC nomenclature in the "fullchip" microarray analyzed, can be assigned using parsing and database searches *i.e.,* siUNC13Celegans is resolved as the following gene, "Official Symbol: UNC13B and Name: unc-13 homolog B (*C. elegans*) [Homo sapiens]". Similarly, "siRAB11amemberR" is resolved as "Official Symbol: RAB11A and Name: RAB11A, member RAS oncogene family [Homo sapiens]". However, some gene specifications used in the "old fullchip" microarray contain sequences that have been revoked from GenBank presumably by their own authors, in which the GenBank record contains a note that it has been discontinued, one example being siVoltLOC121358. Such nucleotides do not necessarily contain gene sequences, and thus may well be the cause of the entire absence of expression data, as well as in poor reproducibility of the data when present. Our findings are in agreement with the reported generally poor (32–33% correlation coefficient) reproducibility of the microarray data across laboratories [Members of the Toxicogenomics Research Consortium, 2005], which, however, was increased to 56–59% after nomenclature and data handling was standardized.

Further increase (in some cases up to 97%) was indeed achieved by standardizing experimental procedures. This is indeed what appears in our individual "fullchip", which, in contrast to the pool "fullchip", has as much as 98% genes passing the 70% reliability criterium (Sivozhlezov *et al.*, 2008).

2.8.2 *Protein expression via Nucleic Acid Programmable Array*

Protein microarrays have found particular value in analyzing clustered protein expression, revealing co-regulated protein networks; protein expression analysis does readily predict protein abundance and does provide information about protein function (LaBaer and Ramachandran, 2005) and protein-protein interactions (Figure 2.24).

Is really the combination of mass spectrometry (see later) and protein microarrays that offer these features by allowing investigators to query thousands of targets simultaneously (Spera and Nicolini, 2008).

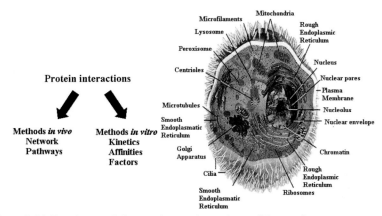

Figure 2.24. Protein-protein interactions and protein small interactions.

Given the central role that proteins play in biology and physiology, we need better methods to study protein abundance, structure and activity in HT.

Protein microarrays and mass spectrometry offer indeed such approaches that we intend to apply to deepen our study (Spera and Nicolini, 2007, 2008; Spera *et al.,* 2007) addressed to understand human cell cycle progression and reverse transformation, a prerequisite for the final goal of understanding and controlling human cancer at molecular level.

However, the development of a MALDI MS-compatible protein microarray is complex since existing methods for forming protein microarrays do not transfer readily onto to a MALDI target. Actually we are implementing a procedure to analyze by MALDI-TOF mass specrotrometry (see later) Nucleic Acid Programmable Protein Array (NAPPA) protein microarray (Ramachandran *et al.*, 2004, and Figure 2.25).

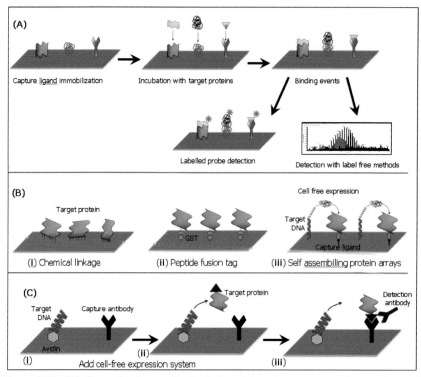

Figure 2.25. (A) General scheme of a typical protein microarray experiment. A set of capture ligands (proteins, antibodies, peptides) is arrayed onto an appropriate solid support. After blocking surface unreacted sites the array is probed by incubation with a sample containing the target molecules. If a molecular recognition event occurs, a signal is revealed either by direct detection (mass spectrometry, surface plasmon resonance, atomic force microscopy, quartz crystal microbalance) or by a labelled probe (CCD camera, DNASER). (B) Assembly methods used to produce function-based protein microarrays. (i) Expressed and purified proteins can be affixed directly to the surface of a chemically activated matrix. By this method, native protein can be used and the proteins will tend to position in random orientations, such that on average, each surface is likely to be exposed to the interacting sample. However, the close attachment to the surface may limit the overall solvent exposure of the protein and the chemical linkage may affect protein folding. Fusion peptide tags added at the N- or C-terminus affix the protein through an affinity capture reagent. Proteins are produced either by (ii) separate expression and purification or (iii) by simultaneous expression and capture of the protein on the array surface. The use of fusion tags allows the protein to be held at a distance from the matrix, exposing more overall surface area to solvent, but sterically blocking either the N- or C-terminus and requiring the addition of fusion tags to all target proteins. (C) NAPPA approach. (continue)

Biotinylation of DNA: Plasmid DNA is cross-linked to a psoralen-biotin conjugate with the use of ultraviolet light. (i) Printing the array. Avidin, polyclonal GST antibody, and Bis suberate are added to the biotinylated plasmid DNA. Samples are arrayed onto glass slides treated with 3-aminopropyltriethoxysilane and dimethyl suberimidate. (ii) *In situ* expression and immobilization. Microarrays were incubated with rabbit reticulocyte lysate with T7 polymerase. (iii) Detection. Target proteins are expressed with a C-terminal GST tag and immobilized by the polyclonal GST antibody. All target proteins are detected using a monoclonal antibody to GST against the C-terminal. (Reprinted with the permission from Spera and Nicolini, Nappa microarray and mass spectrometry: new trends and challenges Essential in Nanoscience Booklet Series, © 2008, Taylor & Francis Group/CRC Press, http://nanoscienceworks.org).

NAPPA protein microarray together matrix-assisted laser desorption-ionization time-of-flight (MALDI-TOF) mass spectrometry provides two powerful and independent tools for the study of protein function and structure in the human cell lines *in vitro* and *in vivo*. To study protein abundance and function and to obviate the need to express, purify and store the proteins we employ the self-assembling protein microarray technology called NAPPA (nucleic acid programmable protein array). As the proteins are freshly synthesized just in time for assaying, there is less concern about protein stability. This approach produces a sizable amount of protein per feature, averaging about 10 fmols. The microarrays are stable dry at room temperature until they are activated to make protein.

This approach has been optimized for the detection of protein-protein interactions and for the co-expression of both the target and query proteins, eliminating the need for any purified proteins. In a protein interaction mapping experiment recently reported by us among 30 human DNA replication proteins, 85% of the previously biochemically verified interactions were recapitulated. In NAPPA technology, full-length cDNA molecules are immobilized on a microarray surface and expressed *in situ* using a mammalian cell-free expression system (rabbit reticulocyte lysate).

A fusion tag present on the protein is recognized by a capture molecule arrayed (along with the cDNA) on the chip surface. This capture reaction then immobilizes the protein on the surface in a microarrayed format as shown in Figure 2.26 for the human kinase arrays introduced at the Harvard Institute of Proteomics (LaBaer and Ramachandran, 2005).

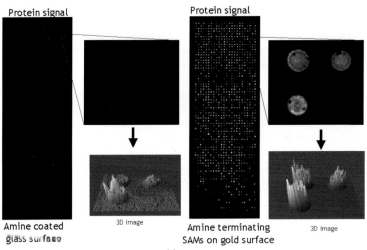

Figure 2.26. Human kinase array on gold surface (Reprinted with the permission from Prof. Joshua Labaer at Harvard Institute of Proteomics).

To identify genes and proteins and their interactions key to the specific cellular process, we integrate the above experimental observations with molecular modelling and bioinformatics (see next paragraph) utilizing existing database such as Gene, HomoloGene, MeSh, Nucleotide and Protein Sequence, along with various advanced software as String and MIM (Sivozhelezov *et al.*, 2006a).

This opinion article, without being exhaustive, will focus on the combined utilization of NAPPA arrays and mass spectrometry highlighting some of open key technical challenges and the new trends by means of a set of selected recent In house applications.

2.9 Biophysical Informatics

A new discipline has been emerging in the last few years to understand the complex processes becoming apparent with the introduction of nanobiotechnology in the study of proteins and genes, which we call here Biophysical Informatics being at the merging of Bioinformatics and Molecular Modelling based on Theoretical Biophysics.

2.9.1 *Bioinformatics*

Bioinformatics allows to identify the key genes or proteins involved in a given biological process by the iterative searches of gene-related or protein-related databases derived mainly from genes or proteins microarray experimentation, revealing and predicting interactions between those genes or those proteins, assigning scores to each of the genes according to numbers of interaction for each gene weighted by significance of each interaction, and finally applying several types of clustering algorithms to genes (or proteins) basing on the assigned scores (Sivozhelevov *et al.*, 2006a).

All clustering algorithms applied, both hierarchical and K-means, invariably selected the same six "leader" genes involved in controlling the cell cycle of human T lymphocytes. Six genes were identified in the cell cycle of human T lymphocytes, which appear to be uniquely capable to switching between stages of cell cycle of human T lymphocytes (Nicolini *et al.*, 2006a; Giacomelli and Nicolini, 2006). In the recent years mostly high-throughput, approaches has been used to identify key genes for particular cellular processes, usage of the already-existing knowledge bases on gene and protein interactions, combined from heterogeneous data sources, is rare. The "leader gene" search/statistics algorithm consists in:

(1) iteratively searching GenBank and PubMed databases to identify the genes with proven involvement in the given cellular process,
(2) query of the STRING (von Mering *et al.*, 2005) database to establish links between the genes,
(3) assigning STRING association-based scores to each gene, and
(4) clustering of gene list according to those scores to yield the final leader gene list.

The leader genes algorithm is being also applied to predicting genes involved in cell cycle progression of human T lymphocytes (Nicolini *et al.*, 2006a; Giacomelli and Nicolini 2006), in osteogenesis (Marconcini *et al.*, Leader genes in osteogenesis, in preparation), in inflammatory processes and in kidney graft rejection (Braud *et al.*, 2008; Sivozhelezov *et al.*, 2008).

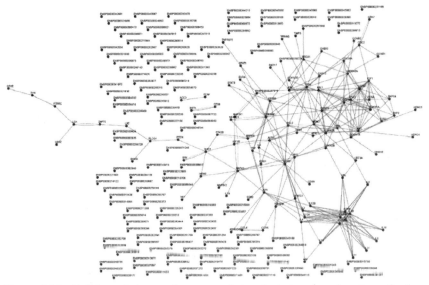

Figure 2.27. Relations between the genes induced during lymphocyte activation according to gene-gene interaction (induction or suppression) (Reprinted with the permission from Sivozhelezov et al., gene expression in the cell cycle of human T lymphocytes, Journal of Cellular Biochemistry 97, pp. 1137–1150, © 2006a, Wiley-Liss Inc., a subsidiary of John Wiley & Sons, Inc.).

Leader genes approach, with text-mining scoring option off, appear to provide a list of the few most important genes relevant for the given cellular processes according to the already-available experimental data, and should be useful in interpreting the microarray expression data and to guide clinical trials. A caution appear needed in the identification of leader genes obtained using either *text*mining or no text-mining scoring and clustering* which frequently do not have a single gene in common (Braud et al., 2008; Sivozhelevov et al., A new algorithm for Leader Gene identification", in preparation).

Indeed, *no text-mining* approach produces more valid results (Sivozhelevov et al., A new algorithm for Leader Gene identification, in preparation). To perform the above analysis, two softwares are required and need to be installed on a dedicated computer: MATLAB and GenePix software. The above bioinformatics algorithm, based on the scoring of importance of genes and a subsequent cluster analysis, allowed indeed us to determine the most important genes, that we call "leader genes" (Sivozhelezov *et al.*, 2006a) in human T lymphocytes cell cycle.

This particular cellular system is very well known and was quantitatively characterized time ago (Cantrell, 2002; Abraham *et al.*, 1980); therefore, it can be a good starting point to verify our algorithm. In particular, we identified 238 genes involved in the control of cell cycle. Most important, only 6 of them were previously identified to be the leader genes (Nicolini *et al.*, 2006a); interestingly, they actually are involved in the cell cycle control at important progression points, namely the most important four at the transition from G0 to G1 phase (MYC; (Oster *et al.*, 2002), at the progression in G1 phase (CDK4; Modiano *et al.*, 2000), and at the transitions from G1 to S (CDK2; Kawabe *et al.*, 2002), and from G2 to M phases (CDC2; Baluchamy *et al.*, 2003; Torgler *et al.*, 2004).

The two remaining "leader genes" (CDKN1A and CDKN1B) are inhibitors of cyclin-CDK2 or -CDK4 complexes and thereby contribute to the control of G1/S transition and of G1 progression (Jerry *et al.*, 2002; Chang *et al.*, 2004).

Bioinformatics can be used also in protein microarrays for the study of protein-protein and protein-gene interactions (Ramachandran *et al.*, 2004). Like for the DNA microarrays, the leader gene approach can simplify their analysis, by reducing the protein displayed to the most important ones to be subsequently tested by mass-spectrometry or by *ad hoc* experimentation.

2.9.2 Biophysical molecular modelling

Homology modelling (Sivozhelezov and Nicolini, 2005, 2006, 2007; Sivozhelezov *et al.*, 2006b) has been recently used to provide useful insights for the successful optimisation of numerous applications in Nanobiotechnology, namely the crystallization of octR rhodopsin (Sivozhelezov and Nicolini, 2006) and of cytochrome P450scc (Sivozhelezov and Nicolini, 2005; Sivozhelezov *et al.*, 2006b) and in the optical computation (Sivozhelezov and Nicolini, 2007).

Octopus (octR), bovine (bovR), and bacterio (bR) rhodopsins all belong to the structural super family of rhodopsin-like proteins sharing the overall 7 transmembrane helix topology of bR except for some details in distances and relative orientations of the helices (Faulon et al., 2003).

This allows to use bR as the primary template in the homology modeling for all structure/function studies with eventual applications, basing on the huge body of such structure/function data (Hirai and Subramaniam, 2003) with respect to bR, as well as biotechnology applications (Fischer et al., 2003; Fischer and Hampp, 2004; Hampp 2000; Hampp and Juchem, 2004; Hillebrecht et al., 2004; Wise et al., 2002). The structure of bR is typically compared with the homology model of octopus rhodopsin (octR), which is similar in topology to bR and as highly ordered in its native membranes as bR in purple membranes (Sivozhelezov and Nicolini, 2006).

2.9.2.1 Three-dimensional structure of octopus rhodpsin

While classical homology modeling is still successfully used (Miedlich et al., 2004), two other approaches were implemented for selecting a tool for modeling the 3D structure of octR, one based on intraprotein hydrogen-bond optimization (Pogozheva et al., 1997) and the other based on first principles of transmembrane protein assembly (Trabanino et al., 2004).

The first principles approach appeared more attractive considering that was already tested on bovine rhodopsin. We started by predicting positions of the helices, and then manually adjusting the resulting alignment of octR versus bR using sequence identity and similarity. The resulting approach (Sivozhelezov and Nicolini, 2006) is therefore a hybrid between homology modeling and first-principles modeling.

The sequence identity level in the eventual model resulted 24.4%, which is above the average 20% quoted for comparisons of vertebrate versus invertebrate visual pigments and therefore supports the reliability of our homology model (Figure 2.28), having as template the PDB entry 1U19 (Okada et al., 2002).

Model quality estimated by comparison of the model with its template appears quite good (Sivozhelezov and Nicolini, 2006). Indeed the model reproduces the positions of the residues surrounding the chromophore (retinal) correctly, particularly the Lys306 residue providing the Schiff base connection of the protein to the retinal.

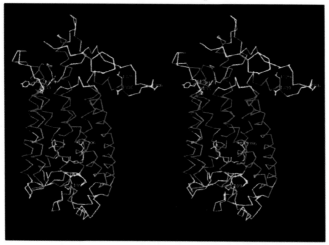

Figure 2.28. Stereo view of the homology model of octopus rhodopsin as a Cα trace, with the retinal shown as sticks. Invariant residues are numbered. Alpha helices are shown in green. (Reprinted with the permission from Sivozhelezov and Nicolini, Theoretical framework for octopus rhodopsin crystallization, Journal of Theoretical Biology 240, pp. 260–269, © 2006, Elsevier).

Notably, the model shows higher alpha helical content than its template (203 versus 190 alpha helical residues), a rare occasion in comparative modeling. This could be the basis for higher organization of octR into 2D arrays compared to the bovR, as observe *in vivo* experimentally (Davies *et al.*, 2001). The good quality of the model was also assessed by the model's ability to reproduce the distinctive features of all visual pigments described to date (Nathans, 1992), as the presence of the sequence (Glu/Asp)-Arg-Tyr, the lysine in the middle of the seventh putative transmembrane segment and the pair of cysteines forming a disulfide bond connecting the first and second extra-cellular loops. All are indeed present in our model.

Identity of the counterion is another important issue both with respect to both the structure/function relationships and biotechnology applications. The only other protein in addition to bR that we found to be structurally similar to our model of octR was the transmembrane cytochrome oxidase. This similarity is possibly related to similarity of function between bR and the oxidase in that both of them involve transfer of charged species, whether electron or proton, across the membrane. The mismatch of the position of Helix4 provides the source of most differences among the octR and bR structures, interestingly since most of the conformational mobility of octR is mediated by Helix 4.

2.9.2.2 Three-dimensional structure of cytochrome P450scc

Similarly for the crystallization (Sivozhelezov *et al.*, 2006b) and sensor (Sivozhelezov and Nicolini, 2005) optimisation a new homology model of bovine cytochrome P450scc has been obtained starting from the recently determined crystal structure of mammalian cytochrome P4502B4 (Sivozhelezov and Nicolini, 2005).

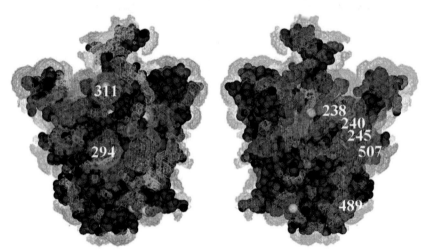

Figure 2.29. Hydrophobic patches around the molecule of cytochrome P450scc, according to the homology model based on cytochrome P4502B4. (Reprinted with the permission from Sivozhelezov *et al.*, Mapping of electrostatic potential of a protein on its hydrophobic surface: implications for crystallization of cytochrome P450scc, Journal of Theoretical Biology 241, pp. 73–80, © 2006b, Elsevier).

The new emerging structure (Figure 2.29) appears compatible with recent diffraction patterns of bovine P450scc microcrystals as obtained at the Microfocus Beamline of the European Synchrotron Radiation Facility in Grenoble (Nicolini and Pechkova, 2006). The same atomic structure is utilized thereby to correctly predict the mutations needed for obtaining both the modifying redox potential experimentally observed (Paternolli *et al.*, 2004) and the optimal LB assembly of the cytochrome obtained experimentally (Ghisellini *et al.*, 2004). Comfortingly these mutations being predicted for the given functional modification are quite different from what erroneously predicted by previous homology models present in the RCSB Protein Data Bank (Berman *et al.*, 2000).

2.9.2.3 *Protein crystallization*

Bacteriorhodopsin (bR) is presently a classical example of membrane protein crystallization.

Comparing its structure with the homology model of octopus rhodopsin (octR), the latter appears similar in topology and highly ordered in its native membranes as bR in purple membranes. Such comparison provides insights for optimization of octR crystallization (Sivozhelezov and Nicolini, 2006). Our results suggest that for optimal crystallization three new tryptophan residues should be introduced in the resulting mutant octR and/or a new protein (RALBP) added to otherwise non-crystallizable octR preparations. Experimentation is still under way. The molecular manipulation techniques already used in the process of 2D and 3D crystallization of both bovine rhodopsin (Caffrey, 2003; Okada *et al.*, 2002) and cytochromes (Hunte and Michel, 2003) are likely to eventually provide a general paradigm for membrane protein crystallization utilizing homology modeling. Indeed, they are promising with octR because, in addition to the transmembrane domain, it has an extensive soluble cytoplasmic domain that can potentially guide crystallization in aqueous medium once hydrophobic surfaces are shielded by detergent. Besides, techniques of molecular manipulation using Langmuir-Blodgett technology are already described for OctR and other photosensitive proteins (Pepe and Nicolini, 1996; Maxia *et al.*, 1995) allowing at least semi-qualitative testing of theoretical predictions. As the first step, we build a 3D model of octR using bovR as a template (Figure 2.28), and compare it with the structure of bR also in term of biotechnology applications that have been reported for bR (Nicolini *et al.*, 1998, 1999) but can also be implemented in octR.

Similarly calculation and combined visualization of electrostatic and hydrophobic properties of cytochrome P450scc based on two very different homology models allowed to identify extensive hydrophobic patches with neutral electrostatic potential and mutations removing such patches and thus expecting to facilitate crystallization of cytochrome P450scc (Sivozhelezov *et al.*, 2006b), especially for the nanotemplate crystallization method. These calculations allow to optimize crystallization and other aspects of protein surface properties and protein recognition. The most promising way for optimizing protein crystallization turned to be nanotemplate crystallization method (as shown later in chapter 3), as confirmed by both fluorescence labeling studies (Pechkova *et al.*, 2005) and by microGISAX spectra analysis (Nicolini and Pechkova, 2006; Pechkova and Nicolini, 2006). The resulting P450scc microcrystals diffraction rings were obtained by synchroton microfocus diffraction (Nicolini and Pechkova, 2006). However in order to further optimize P450scc crystallization for obtaining the necessary entire diffraction data set, is necessary to take into account that the plethora of the crystallized cytochrome P450 and the corresponding number of reported 3D atomic structures suggest the use of surface mutagenesis, specifically recommended for the case when the obtained crystals do not have the sufficient quality, but crystallizable homologs are available (Dale *et al.*, 2003). The general recommendation is to mutate residues predicted by homology models to be solvent-exposed to those favorable for crystal contact formation. Statistically, those desirable residues are arginine and glutamine (Baud and Karlin, 1999; Dasgupta *et al.*, 1997). By homology modeling we have recently proposed mutations that are expected to facilitate crystallization of cytochrome P450scc according to the above indications, but with several essential amendments (Sivozhelezov *et al.*, 2006b) allowing to obtain the more compact, stable and high ordered protein film, which can significantly optimize the protein nanotemplate method, guaranteeing the higher quality of nucleation for crystal formation. Here, we have combined the above concept with the notion that it is the electrostatic forces that primarily steer the molecules to each other in aqueous solution, even though the eventual energetics of protein-protein contact may be determined by the hydrophobic effect. The structure of cytochrome P450scc predicted by homology with cytochrome P4502B4 (Figure 2.29) shows striking difference from the pattern exhibited by cytochrome P4502B4 itself, with almost the entire surface of the protein covered by the green "clouds" designating hydrophobic surfaces of the proteins not screened by electrostatic potentials. Therefore, according to the logics of Patro *et al.* (1996), no ordering should be observed in the aggregates formed by the P450scc molecules, and thus very little chance for P450scc molecules in their native form to yield well-diffracting crystals. Our approach therefore suggests three options for altering the structure of P450scc to assist crystallization. The first and foremost is site directed mutagenesis. This approach has already a considerable record of success reviewed in Dale *et al.* (2003). These successful examples include, for instance, exhaustive mutagenesis of all 29 possibly exposed hydrophobic residues of catalytic domain of HIV integrase, resulting in just one mutant giving well-diffracting crystals (Dyda *et al.*, 1994). More rational approach (termed "crystal engineering") involved theoretical predictions and resulted in crystallizability introduced or improved by inducing smaller number of mutations. For example 9 mutations resulted in improved crystallizability on the case of 24 kDa fragment of DNA gyrase (D'Arcy *et al.*, 1999) while only one mutation was required for adding the propensity to crystallize in the case of human leptin (Zhang *et al.*, 1997a). This is the approach actually applied in this study. The size and occurrence of electrostatically unscreened hydrophobic patches (Figure 2.29) for the cytochrome P4502B4-based cytochrome P450scc model is larger than for P4502B4 (not shown). Some regions of the P450scc surface are completely covered by patches and the corresponding residues must be considered for mutations facilitating crystallization. They are marked on Figure 2.29: Tyr238, Met240, Val245, Phe294, Leu311, Ile489, Phe507, and Phe513. Following the guidelines of Baud and Karlin (1999) and Dasgupta *et al.* (1997), each of the proposed residues should be replaced either by arginine or glutamine, the latter two being most prone to crystal contact formation. The mutations suggested herein have showed a considerable effect on the hydrophobic patches decreased in size but, even more importantly, with quite fewer contiguous patches in the mutant protein. Therefore, the proposed mutation should further improve crystallization and the work is in progress.

2.9.2.4 Nanobiodevice implementation

Recent modeling has provided insights also for optimization of present octR experimentation for application in nanobiotechnology in a manner similar to bR, and possibly even superior in optical computation (Sivozhelezov and Nicolini, 2007).

Visual membranes of octopus, whose main component is the light-sensitive signal transducer octopus rhodopsin (octR), are extremely highly ordered, easily capture single photons, and are sensitive to light polarization, which shows their high potential for use as a Quantum Computing (QC) detector (Sivozhelezov and Nicolini, 2007). However, artificial membranes made of octR are neither highly enough ordered nor stable, while the bacterial homolog of octR, bacteriorhodopsin (bR), having the same topology as octR, forms both stable and ordered artificial membranes but lacks the optical properties important for optical QC. In a recent study (Sivozhelezov and Nicolini, 2007), we investigate the structural basis for ordering of the two proteins in membranes in terms of crystallization behavior. We compare the atomic resolution 3D structures of octR and bR and show the possibility for structural bR/octR interconversion by mutagenesis. We also show that the use of nanobiotechnology can allow (1) high-precision manipulation of the light acceptor retinal, including converting its surrounding into that of bacterial rhodopsin, the protein already used in optical-computation devices and (2) development of multicomponent and highly regular 2D structures with a high potential for being efficient optical QC detectors (Sivozhelezov and Nicolini, 2007). The key property of bR allowing to utilize it in optical memory devices (Fischer and Hampp, 2004), appears to be its existence in two stable forms giving rise to novel types of optical computation devices.

The implication of these studies has been successfully proved also with P450scc for optimal sensor construction (Paternolli *et al.*, 2004) and for LB nanoassembly (Ghisellini *et al.*, 2004). A previously reported

(Ghisellini *et al.*, 2004) molecular modeling prediction was indeed realized studying a hypothetical complex based on two P450scc molecules. The aim of the mutation proved successful was to increase the probability to form a pattern by improving electrostatic interactions and simultaneously adding hydrophobic amino acid residues into the surface regions included in the pattern. In particular, the mutation (K201E) was identified by electrostatic calculations and caused the enhancement of the stability/ordering during the protein immobilization.

2.10 Mass Spectrometry

The identification of different protein patterns in CHO-K1 cells grown with and without cAMP (Spera *et al.*, 2007) is performed with a RP-HPLC-ESI MS apparatus, a Thermo Finnigan (San Jose, CA) Surveyor HPLC connected by a T splitter to a PDA diode-array detector and to Xcalibur LCQ Deca XP Plus Mass Spectrometer (Figure 2.30).

The mass spectrometer is equipped with an electrospray ion source (ESI). The chromatographic column is a Vydac (Hesperia, CA) C8 column, with a 5 μm particle diameter (column dimensions 150 × 2.1 mm).

For HPLC-ESI MS analysis the protein solutions, stored at -80°C, were heated at room temperature and then lyophilized. The lyophilized proteins were immediately dissolved in aqueous 0.2% TFA and centrifuged at 12000 rpm for 5 minutes. After centrifugation, the supernatant was analyzed by HPLC-mass spectrometry. The following solutions were utilized for reversed-phase chromatography: eluent A, 0.056% aqueous TFA, and eluent B, 0.05% TFA in acetonitrile/water 80:20 (v/v).

The gradient applied was linear from 0 to 55% in 40 minutes, at a flow rate of 0.30 ml/min. The T splitter gave a flow rate of about 0.20 ml/min toward the diode array detector and a flow rate of 0.10 ml/min toward the ESI source. The diode array detector was set in the wavelength range 214–276 nm.

Mass spectra were collected every 3 msec in the positive ion mode. MS spray voltage was 4.50 kV, and the capillary temperature was 22 °C.

The deconvolution of the averaged ESI mass spectra was performed by Mag-Tran1.0 software. Preliminary proteins identification from the mass values of the intact protein was obtained from a search in Swiss-Prot Data Bank (http://www.expasy.org).

The identification of CHO-K1 proteins via mass fingerprint is performed coupling HPLC (Varian Inc) to separate the proteins, and MALDI TOF MS, a Bruker Autoflex (Bruker Daltonics, Leipzig, Germany) to analyze the tryptic digest of these samples (Spera *et al.*, 2007).

The HPLC measurements were carried out on a Varian Star HPLC system which includes: 9012 Gradient Solvent Delivery System, 9050 UV-VIS Detector, 9300 Refrigerated AutoSampler (fitted with a 20 µl loop), and a Star Chromatography Workstation (Varian Inc., USA). Proteins will be separated on a C8 (250 × 4.6 mm; 5 µm particle size) reverse phase column (Macherey-Nagel, Germany). For reversed-phase chromatography we utilized the following solutions: eluent A, 0.056% aqueous TFA, and eluent B, 0.05% TFA in acetonitrile/water 80:20 (v/v).

The gradient applied was linear from 0 to 50% in 30 minutes, at a flow rate of 0.30 ml/min. For HPLC analysis the protein samples stored at –80°C, were heated at room temperature and then 100 µl of solutions were injected. We used three samples of lysozyme from chicken egg white (Sigma Chemical Co., St. Louis, MO) at different concentrations to standardize the HPLC results and the protein amounts loaded to MS.

For protein fingerprint, the fractions collected from HPLC were digested with trypsin overnight, according to the enzyme supplier (Sigma-Aldrich).

For MALDI-TOF MS the tryptic digest samples were diluted to 4 nM protein concentration in a 0.1% TFA solution. The matrix used for the mass spectrometric analysis was a saturated solution of acid (α-Cyano-4-hydroxycinnamic acid for peptides and light proteins and sinapinic acid for heavy proteins, Bruker Daltonics) dissolved in 2/3 of 0.1% TFA and 1/3 of acetonitrile.

For the analysis 1.5 µl of matrix solution was mixed with 1.5 µl of sample, then 1 µl of this mixture was spotted onto a suitable aluminum plate and air-dried. MALDI-TOF MS was externally calibrated using protein and peptide calibration standard solutions (Bruker Daltonics), resulting in a mass accuracy < 100 ppm for intact proteins and < 10 ppm for peptides.

The mass lists obtained were submitted to a data bank search. We use a specific software for protein data interpretation, Biotools (Bruker Daltonics), that allows an automated protein identification via library search and that has a MASCOT Intranet search software (Matrix Sciences, Ltd. www.matrixscience.com) fully integrated.

MS analysis started from the HPLC-ESI MS analysis of nuclear proteins fraction from CHO-K1 cells and from cAMP treated cells (Figure 2.30).

The total ion current (TIC) and the UV absorbance at 214 and 276 nm (not reported) profiles obtained during chromatographic analysis of the two protein solutions are reported in Figure 2.30. TIC and UV profiles did not reveal significant signals between 0 and 10 min.

The highest variations are identifiable in the TIC plots (see enlargement in the box) and show an inhibition of the nuclear protein expression after treatment with cAMP.

In particular as shown in Spera *et al.* (2007):
- in the untreated CHO-K1 plot there is a peak at 42.45 min not observable in the cAMP treated CHO-K1 plot;
- the intensity of the 44.90 min peak in the untreated CHO-K1 TIC plot is double regarding the cAMP treated cells plot peak.

The total ion current of proteins of interest was used for their relative quantification.

Since the total ion current depends not only from concentration but also from the charge of the analyte, sample treatment was standardized both in dilution and HPLC-ESI-MS procedures, to ensure similar bias, namely same ion suppression, pH and organic solvent effects on the analyte charge.

Under these conditions, the total ion current can be roughly considered proportional to protein concentration, and it can be used to evidence correlations existing among different proteins in different samples (Messana *et al.*, 2004).

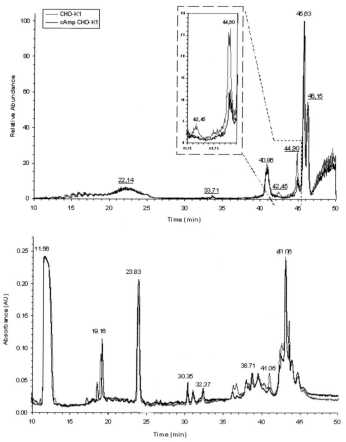

Figure 2.30. HPLC-ESI-MS TIC and UV profile of nuclear protein fraction of CHO-K1 cells (grey line) and of CHO-K1 treated with cAMP (black line). Upper plot: HPLC-ESI-MS TIC profile collected by ion-trap mass spectrometer (elution time 10-50 min). Enlargement of the 41.7–45.5 min elution range showing the TIC profile is represented in the box where the greater differences between the two plots are shown: the 42.45 min peak (absent in the treated CHO-K1 plot) and the 44,90 min peak (its relative abundance is double in the untreated CHO-K1 plot). Bottom plot: UV (214 nm) profile (elution time 10–50 min). (Reprinted with the permission from Spera and Nicolini, camp induced alterations of Chinese hamster ovary cells monitored by mass spectrometry, Journal of Cellular Biochemistry 102, pp. 473–482, © 2007, Wiley-Liss, Inc., a subsidiary of John Wiley & Sons, Inc.).

2.10.1 *Mass spectrometry of label-free NAPPA*

In order to analyze NAPPA array it was necessary to modify the standard MALDI target to host the NAPPA array, classically spotted on a standard microscopy slide. For this reason an adequate lodging with the dimension exactly the same of the array guarantees the immobility of the glass once putted in the lodging.

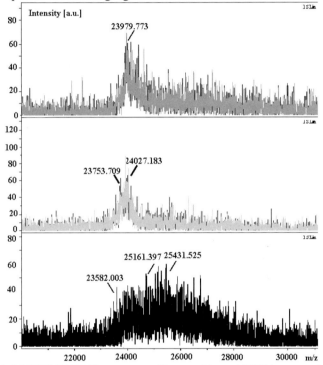

Figure 2.31. MALDI-TOF-MS spectrum of human kinase NAPPA array after protein synthesis (spot gene NA 7-A 12): high weights (20-35 kDa) for three different spots/genes (from the top: gene NA7-A 12, gene NA 7-E 12, and gene NA 8-D 12) (Reprinted with the permission from Spera and Nicolini, Nappa microarray and mass spectrometry: new trends and challenges Essential in Nanoscience Booklet Series, © 2008, Taylor & Francis Group/CRC Press, http://nanoscienceworks.org).

For an automated, reproducible and reliable MALDI identification of each NAPPA spots we have created a new geometry file in FlexControl (the Bruker software for MALDI-TOF control), since NAPPA geometry is different from all the standard MALDI target geometry. Utilizing the

"Manual Fine Control" window of the software we have manually identified each spots and then we have reordered its position in a new file. We analyzed by MALDI TOF MS the kinase gold NAPPA array (Figure 2.26). We followed expression of the NAPPA slides protocol by Harvard Institute of Proteomics (joint manuscript NWI-HIP in preparation) and once the proteins were synthesized we wash twice the NAPPA with bi-distillated water and we proceeded according to the previously described protocol (LaBaer and Ramachandran, 2005; Spera and Nicolini, 2008).

Some of the identified peaks of our resulting MS in the low-medium molecular weights are assimilable to that obtained by HIP MS analysis of NAPPA before protein synthesis (private communication), due to the components presents on the NAPPA (streptavidin, capture antibody, plasmid DNA). To be sure of this assumption we have analyzed two identical NAPPA, one before and the other after proteins synthesis (in progress). The peaks shown in the high molecular weights (Figures 2.31) are instead probably due to proteins synthesized on the spots. At the moment we have some problems to perform protein fingerprint due to the minimum drop dimension obtainable with the routine Pasteur available and for this reason we are printing newly designed NAPPA microarrays to overcome this problem.

The preliminary obtained results however encourage us to perform new experiments on NAPPA of opportune geometry. Future activities will be carried out on properly constructed NAPPA taking into considerations all reported fluorescence, MS and bioinformatics for cell-cycle study.

2.11 Synchrotron Radiation

The techniques of elastic scattering of X-rays are widely used as they provide valuable tools to probe the order properties of protein crystals.

The diffraction range is reached when the incident radiation wavelength is closed to the interatomic distances in a crystal and when the scattering angles are wide. By a careful analysis of the integrated diffracted intensities, one can access to the intimate atomic structure *i.e.*

the positions of nuclei and the spread of the electronic cloud around atoms. Usually, the range of "diffuse scattering" involves techniques, which allow to get statistical information at scales that are greater than the interatomic distances. This domain is restricted in between the first Bragg peak, which overlaps with the direct beam and the diffraction peaks. With X-rays with a wavelength of a few angstroms, this domain of small momentum transfer is reached in the small angle range in the range 1–100 nm (Figure 2.32 below).

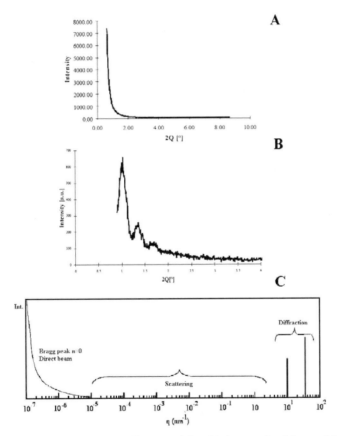

Figure 2.32. X-ray pattern of LB film containing 20 layers of wild-type (above) and recombinant (middle) cytochrome P450scc. Below is the scattering and diffraction ranges versus the wave vector transfer. (Part A,B: Reprinted with the permission from Nicolini et al., Supramolecular layer engineering for industrial nanotechnology, in Nano-surface chemistry, pp. 141–212 © 2001a, Marcel Dekker / Taylor & Francis Group LTD).

The scattering comes from strong variations of the mean electronic density for X-rays and up to fifteen years ago, was limited to three dimensional samples as the strong penetration depth of the radiations and the low signal to noise ratio hampered the surface sensitivity. Quite recently, thanks to synchrotron radiation, these techniques were extended to surface geometry using the phenomenon of total reflection of X-rays in the grazing incidence range. Several methods based on the utilization of X-rays are used to study the structure of LB film structure. Diffractometry and reflectometry are the most commonly used while diffractometry is mainly used when well-ordered periodic structure is under investigation and several Bragg reflections are present in the X-ray pattern. The position of these reflections is determined by the Bragg equation:

$$2D\sin\theta = n\lambda$$

where D is the thickness of the periodic unit (period or spacing), λ is the wavelength of the X-ray beam, θ is the incident angle, and n is the number of the reflection. Therefore, the thickness of the periodic unit (usually a bilayer) can be obtained directly from the angular position of Bragg reflection. The other information, which can be directly obtained from the X-ray pattern, is the correlation length (L). This parameter can be considered as a thickness until which the film can be still considered as an ordered one and it is determined by the following formula:

$$L = \lambda/2\sin(\Delta\theta)$$

where $\Delta\theta$ is the half-width of the Bragg reflections. More information can be obtained if several reflections were registered as in the case of LB of only recombinant cytochrome P450scc but not of the *wild-type* (Figure 2.32). Each Bragg reflection can be considered as a component in the Fourier row representing the electron density on the repeating unit of the film. Kiessig fringes correspond to the interference of the X-ray beam reflected from the air/film and film/substrate interfaces, and their angular position gives the information about the total thickness of the LB film. Similarly the Microfocus beamline of the European Synchrotron Radiation in Grenoble is utilized to probe for the structure of several protein systems in diffracting crystals.

2.11.1 *Diffraction*

Third generation synchrotron sources are now emitting synchrotron X-ray beams that are trillion times more brilliant than those produced by X-ray tubes, requiring quite smaller crystals for the 3D structure determination. One key experimental component of nanocrystallography is thereby an appropriately microfocussed Synchrotron Radiation (Cusack *et al.*, 1998), having microfocusing X-ray optics and a microgoniometer (Figure 2.33.a) in addition to intense beams capable to obtain X-rays diffraction patterns from microcrystals ranging in size between 5 and 20 microns (Figure 2.33b).

 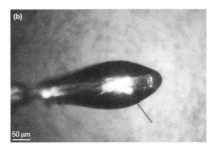

Figure 2.33. The needle human protein CK2α crystal in the mounted loop (b). In the left panel (a) is shown a photography of the crowded sample position on the microfocus beamline with collimator (beam enters horizontally from the right), microscope (vertical), cryo-cooling system and beam stop (not present) all very close to the crystal, which is moved by means of a microgoniometer (Part A courtesy of Prof. Christian Riekel at ESRF; Part B reprinted with the permission from Pechkova and Nicolini, Protein nanocrystallography: a new approach to structural proteomics, Trends in Biotechnology 22, pp. 117–122, © 2004a, Elsevier).

Protein microcrystallography requires indeed precision in alignment and high mechanical stability. This calls for high power microscopes to visualize both crystal and beam position and accurate motorized adjustments to bring everything into alignment (Pechkova and Nicolini, 2004, 2004a, 2003a). However, one cannot indefinitely compensate for small crystal size with increased beam intensity since at some stage so much X-ray energy is being deposited in a small volume that the protein structure and crystalline order will be destroyed by primary radiation damage very quickly. Radiation damage to crystalline proteins using X-rays is a problem which limits the structural information that can be extracted from the sample (Garman and Nave, 2002) and only the significant radiation stability induced in the crystal formed by our nanofilm template method open new avenues in structural proteomics (Pechkova et al., 2004). Recently, diffraction data could be collected on a single extrasmall human CK2α (of about 20μm in diameter) and its diffraction patterns were utilized to solve the structure of protein kinase CK2α catalytic subunit 2.4 Å (Figure 2.22c). Interestingly, the collection of data diffraction utilized for the subsequent 3D atomic structure determination (Pechkova et al., 2003) lasted for long time. The needle CK2α microcrystals appear quite stable to the incoming considerable radiation preserving their shape and size without any apparent damage. Radiation stability and high diffraction quality of nanotechnology-based crystals were apparent on a recent systematic study at the Microfocus Synchrotron Beamline (Pechkova et al., 2004), suggesting that protein crystals grown by our method are less sensitive to radiation than those obtained by classical method due to an intrinsic property of such nanotechnology-induced microcrystals.

Earlier experiments with Microfocus were carried out with channel receptors (Pebay-Peyroula et al., 1997) and with single bacteriorhodopsin (bR) crystal of about $30 \times 30 \times 5$ μm^3 (one good out of ten unutilizables), which produced a entire 2.4 Å data set and led to the first high resolution model of bR obtained by X-ray crystallography (Zhou et al., 2000). After that report, Ekström et al. (2003) were able to perform an analysis of microcrystals on the actin-binding domain of human actinin. Recently, microfocus has utilized with great success in some other experiments leading to new atomic protein structures determination (Luecke et al., 2001; Scheffzek et al., 2000; Berthet-Colominas et al., 1999; Brige et al., 2002; Hanzal-Bayer et al., 2002; Thom et al., 2002; Toeroe et al., 2001; Royant et al., 2000, 2001; Zouni et al., 2001).

As mentioned above, the utilization of Microfocus is restricted by the significant radiation damage to the crystal, which is one of the major sources of error in data collection. This inherent problem in X-ray microcrystallography has not so far been widely investigated. Radiation damage can be partially avoided by cryo-techniques implementation, but effect of ionizing radiation remains substantial even using moderate-intense synchrotron facility, and for brighter sources it becomes indeed a limiting factor (Burmeister, 2000). Beam defocusing can be used to decrease the available flux density (Walsh *et al.*, 1999), but the main problem remains unsolved, especially for protein microcrystals, often making the data collection impossible. Finally, although synchrotron radiation damage is usually presumed to be nonspecific and manifested as a gradual decay in the overall quality of data obtained, it was recently reported that synchrotron radiation can cause highly specific damage which can provide the information of structural and functional significance (Weik *et al.*, 2000; Ravelli *et al.*, 2003; Matsui *et al.*, 2002).

The curve obtained from the LB film of wild-type protein presents neither Bragg reflections nor Kiessig fringes. Such a result means that the film is not ordered and that there is not uniformity of the thickness along the sample area. In the case of recombinant protein, we see Kiessig fringes, whose angular position depends upon the number of deposited layers. The average monolayer thickness calculated from these data is about 6 nm, corresponding well to both the ellipsometric data and the protein sizes from the RCSB Protein Data Bank (Berman *et al.*, 2000).

2.11.2 *Grazing Incidence Small Angle X-ray Scattering*

µGISAXS is a technique of elastic scattering of X-rays widely used as they provide valuable tools to probe the order properties of protein crystals.

The diffraction range is reached when the incident radiation wavelength is closed to the interatomic distances in a crystal and when the scattering angles are wide. By a careful analysis of the integrated diffracted intensities, one can access to the intimate atomic structure, *i.e.*, the positions of nuclei and the spread of the electronic cloud around atoms. Usually, the range of "diffuse scattering" involves techniques, which allow getting statistical information at scales that are greater than the interatomic distances. This domain is restricted in between the first Bragg peak, which overlaps with the direct beam and the diffraction peaks (Nicolini and Pechkova, 2004). With X-rays with a wavelength of a few angstroms, this domain of small momentum transfer is reached in the small angle range in the range 1–100 nm (Nicolini, 1996a). The scattering comes from strong variations of the mean electronic density for X-rays and up to fifteen years ago, was limited to three dimensional samples as the strong penetration depth of the radiations and the low signal to noise ratio hampered the surface sensitivity. Quite recently, thanks to synchrotron radiation, these techniques were extended to surface geometry using the phenomenon of total reflection of X-rays in the grazing incidence range. The field of thin films growth brought a need of knowledge about layer morphology and sizes of quantum dots, supported islands or buried particles which, as shown in Nicolini and Pechkova (2004) as pushed the development of Grazing Incidence Small Angle X-Ray Scattering using the 5 micron beam size at ESRF in Grenoble (μGISAXS), providing information about the dependence of the electronic density perpendicular to the surface, like those due to the roughness of a surface, the lateral correlations, sizes and shapes of growing protein nanocrystals, of metallic islands or of the self organized dots superlattices (Roth *et al.*, 2003).

The protein templates were freshly plated in the hanging drop container during the three-day experiment.

The glass substrate with the thin film template and the droplet was removed from the container and the vacuum grease area was cleaned to ensure no contamination of the signal. The droplet existed for > 60min in air until complete evaporation. Hence an optimum timing had to be found concerning the preparation, adjustment in the beam and data acquisition time. The glass substrate was subsequently brought into the beam and the incident angle α_i adjusted to about 1°. This procedure was restricted to less than 20 min in order to avoid precipitation of salt or nanocrystals from the solution to the substrate, which would lead to a contamination of the μGISAXS signal.

The droplet diameter was about 2 mm. Hence the footprint of the X-ray beam will be fully within the droplet diameter. The small beam size allows avoiding excessive liquid scattering and provides therefore a reasonable signal-to-background ratio. The position of the drop relative to the beam was determined by an absorption scan with a photodiode. Experiments were performed with the beam at the center of the droplet at the contact area droplet, protein template in order to optimize the signal of the weakly scattering biopolymer samples. Data collection times for individual droplets varied from 7 min to 20 min. This is well below minimum droplet evaporation time of about 60 min. This "Stop" procedure described above interrupts therefore crystal growth at specific times and allows studying freshly grown thin films.

Figure 2.34 shows an enlargement of the Yoneda region of the corresponding μGISAXS pattern. This region is most sensitive to structural and morphological changes of the surfaces due to the interference effect involved in the occurrence of the Yoneda peak. The μGISAXS pattern is scaled to the same intensity. After 46 h clearly a new Yoneda peak (N) emerges next to the Yoneda peak existing at shorter times (G) with a critical angle below that of the substrate.

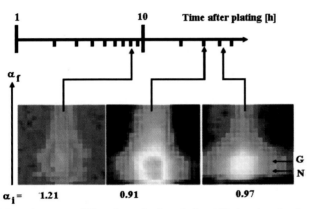

Figure 2.34. Development of Yoneda peak after plating (G) showing the development of a new peak (N) due to scattering from a protein layer (see text). (Reprinted with the permission from Pechkova *et al.*, μGISAXS and protein nanotemplate crystallization methods and instrumentation, Journal of Synchrotron Radiation 12, pp. 713–716, © 2005a, Blackwell Publishing).

One could suppose that salt (NaAc) is precipitating on the substrate thus leading to a new rough layer of small crystals. This is unlikely for two reasons: (i) in view of the rapid data collection after removal of the sample from the container. With the hanging drop method itself, salt precipitation on the substrate should be reduced. In addition this second peak does not appear at shorter times, (ii) one can calculate the nominal critical angle of NaAc using the known molecular mass and composition as α_c(NaAc)=0.23°, which is larger than the critical angles of the glass and the protein. Hence as a working hypothesis one can attribute the peak at higher α_f-values (G) to that of the glass substrate. The newly developing peak (N) thus seems to be related to the protein itself.

The occurrence of a second Yoneda peak indicates the development of a rough layer. Two models can be imagined both competing to increase layer roughness (Figure 2.35). In model A the roughness of the template layer is increased by holes.

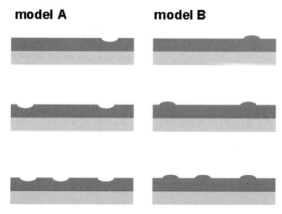

Figure 2.35. Two possible models for the increasing layer roughness with time. Model A favours increasing roughening via ablation of the nanostructured layer, while model B is nucleation-based (Reprinted with the permission from Pechkova *et al.*, µGISAXS and protein nanotemplate crystallization methods and instrumentation, Journal of Synchrotron Radiation 12, pp. 713–716, © 2005a, Blackwell Publishing).

That is to say this model assumes a decreasing density of the protein template layer as the layer itself is removed by the protein solution thus triggering or assisting the formation of new crystals in the solution. Model B is motivated by a nucleation and growth process of possible protein nanocrystals on the nanostructured template, whereby protein molecules out of the solution might be adsorbed on the nanostructured template. At the present state of experiments we cannot favour model A or model B, *i.e.*, a build-up of holes or a layer of small islands, acting as nucleation centers for crystals. This question will be addressed in future *in situ* µGISAXS experiments based on scaled intensity data. A feasibility study on the template-assisted protein crystal growth using µGISAXS has been reported in this paper. The small droplet size used in hanging drop experiments necessitates the use of small X-ray beams and hence the µGISAXS technique. The recording of diffuse scattering implies measuring times up to several minutes due to the intensity difference of several orders of magnitude between the specularly reflected and the diffuse scattering signal. As compared to bulk protein crystallography, the enlarged beam footprint results in less radiation damage. We have collected a time series of µGISAXS patterns, observing indeed a stable µGISAXS pattern at least up to the first few images being acquired for the experiments here reported. It appears then that we can exclude that the surface morphology changes are due to beam damage since the comparison among the different crystallization time intervals after plating is made among the corresponding pattern being immediately acquired. In future experiments one could also envisage to translate the beam laterally across the substrate surface in order to further limit radiation damage.

As compared to topography-sensitive methods (*i.e.*, AFM), µGISAXS exploits the penetration depth of X-rays. This allows investigating the interface solution-template where the significant signals were observed in the present case. Working with a droplet permits to avoid time-consuming cleaning of the sample, *i.e.*, salt removing. Thus the small remaining droplet is not hindering the measurement of the interface solution-template, where nanocrystal growth is expected to take place. The perspective to combine grazing incidence with wide-angle scattering in order to really see crystallisation (Bragg peaks) appear quite promising and we plan to do it in future *in situ* experiments.

Future experiments will be indeed extended to *in situ* investigations of template assisted protein growth, which provides several advantages. Thus crystal growth can be examined continuously. Furthermore, background subtraction will be facilitated and any contamination of the µGISAXS signal by cleaning or precipitating salt will be avoided. A possible set-up, namely a hanging drop container with thin entry and exit windows, is shown schematically in Pechkova and Nicolini (2003, 2004a). This set-up is inversed to the traditional µGISAXS set-up (Müller-Buschbaum *et al.*, 2003; Roth *et al.*, 2003) in order to obtain an upward-scattering geometry. This would allow for the first time following protein crystal growth and nucleation from its very early stages after plating. As compared to AFM, the hanging drop could be investigated by µGISAXS in its natural growth position without disturbing the growth process.

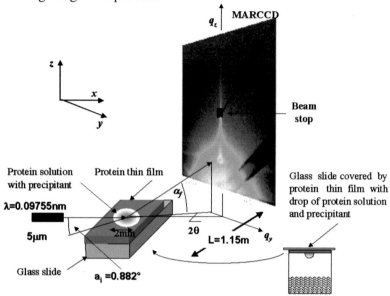

Figure 2.36. Schematic view of the microbeam grazing incidence small-angle X-ray scattering setup (µGISAXS) at ID13/ESRF. The sample is mounted on a xyz-gantry and a two-axis goniometer (ϕ_x, ϕ_y). The scan direction is y. α_i denotes the angle between the incident beam and the sample surface, α_f the corresponding exit angle, and 2ϕ the out-of-plane angle. The flight path (L=1.15 m) between sample and the 2-D detector (C) is evacuated (10^{-2} mbar). The typical 2-D µGISAXS signal of the cytochrome P450scc drop sitting on a LB layer is also shown. In the bottom right corner it is shown the experimental layout of the protein solution typical droplet sitting on the protein layer deposited over the glass. µGISAXS measured with 5 micron beam size points to the present of P450scc nanoscrystal in the early stage of the crystallization with LB ((Reprinted with the permission from Nicolini and Pechkova, Structure and growth of ultrasmall protein microcrystals by synchrotron radiation. I µGISAXS and microdiffraction of P450scc, Journal of Cellular Biochemistry 97, pp. 544–552, © 2006, Wiley-Liss, Inc., a subsidiary of John Wiley & Sons, Inc.).

The field of thin films growth brought a need of knowledge about layer morphology and sizes of quantum dots, supported islands or buried particles which, as shown in Figure 2.25 has pushed the development of Grazing Incidence Small Angle X-Ray Scattering using the 5 micron beam size at ESRF in Grenoble (μGISAXS), providing information about the dependence of the electronic density perpendicular to the surface, like those due to the roughness of a surface, the lateral correlations, sizes and shapes of growing protein nanocrystals, of metallic islands or of the self organized dots superlattices (Roth *et al.,* 2003). Such informations are of prime interest in understanding the link between the mechanisms of growing crystal morphology and its physical or chemical properties for both lysozymes (not shown) and cytochromes.

Chapter 3

Nanoscale Applications in Science and Health

This chapter overviews the present status of nanoscale applications of organic and biological nanotechnology to science and health, namely of those capable so far to yield a potential scientific and technological progress both to protein crystallography, medicine, genomics, proteomics and cell science, and to mechanics, optics and magnetism. Even if particular emphasis is placed on what has been accomplished in our laboratory in the last eight years, significant reference to the recent activity of numerous other groups is also given.

3.1 Nanobiocrystallography

One of the newest approaches to structural proteomics is the emerging field of protein nanocrystallography at the intersection between nanotechnology and proteomics (Pechkova and Nicolini, 2003, 2004a).

This new field results from the combination of advanced nanotechnologies – particularly atomic force microscopy (AFM), thin-film nanotemplate technology, nanogravimetry – and of advances in synchrotron radiation, namely micro-nanofocused diffraction and micro-nanoGISAXS. It should be noted that nanocrystallography as described here does not refer to the study of self-assembled nanocrystals (*i.e.*, crystals of nanometer size) of silver, cobalt, gold and/or nanoparticles created using reverse micelles or similar technologies. In addition, it does not refer to the nanodrop or microdrop crystallization technology resulting from the exciting advances made in microfluidic chips (Pechkova and Nicolini, 2003), which are now commercially available and show some accomplishement.

This progress is made now possible by the third generation synchrotron sources emitting synchrotron X-ray beams that are trillion times more brilliant than those produced by X-ray tubes, requiring quite smaller crystals for the 3D structure determination. One key experimental component of nanocrystallography is indeed an appropriately microfocused synchrotron radiation (Cusack *et al.*, 1998), having micro-focusing X-ray optics and a microgoniometer (Figure 3.1) in addition to intense beams capable to obtain X-rays diffraction patterns from microcrystals ranging in size between 5 and 20 microns (Figure 3.2).

Figure 3.1. Synchrotron radiation with micro-focusing X-ray optics and microgoniometer. Photograph of the crowded sample position on the microfocus beamline with a collimator (beam enters horizontally from the right), microscope (vertical), cryo-cooling system and beam stop (not present) all very close to the crystal, which is moved using a microgoniometer. (Courtesy from Christian Riekel at ESRF).

Protein microcrystallography requires indeed precision in alignment and high mechanical stability. This calls for high power microscopes to visualize both crystal and beam position and accurate motorized adjustments to bring microcrystals into alignment in order to obtain proper diffraction pattern (Figure 3.2).

However, one cannot indefinitely compensate for small crystal size with increased beam intensity since at some stage so much X-ray energy is being deposited in a small volume that the protein structure and crystalline order will be destroyed by primary radiation damage very quickly.

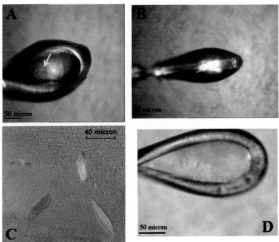

Figure 3.2. Human kinase and cytochrome P450scc microcrystals. Human CK2α microcrystal mounted on the nylon loop before (a) and after (b) exposure to the microfocus Synchrotron Beamline for the collection of complete diffraction data set. (c) Human CK2 microcrystals obtained by the protein nanofilm template. (d) Diffraction pattern of bovine P450scc ultramicrocrystal powder (A,C: Reprinted with permission from Pechkova and Nicolini, Atomic structure of a CK2α human kinase by microfocus diffraction of extra-small microcrystals grown with nanobiofilms template, Journal of Cellular Biochemistry 91, pp. 1010–1020, © 2004, Wiley-Liss, Inc., a subsidiary of John Wiley & Sons, Inc.; B: Reprinted with permission from Pechkova and Nicolini, Protein nanocrystallography: a new approach to structural proteomics, Trends in Biotechnology 22, pp. 117–122, © 2004a, Elsevier; D: Reprinted with permission from Nicolini and Pechkova, Structure and growth of ultrasmall protein microcrystals by synchrotron radiation: I μGISAXS and microdiffraction of P450scc, Journal of Cellular Biochemistry 97, pp. 544–522, © 2006, Wiley-Liss, Inc., a subsidiary of John Wiley & Sons, Inc.).

Radiation damage to crystalline proteins using X-rays is a problem, which limits the structural information that can be extracted from the sample (Garman and Nave, 2002) and only the significant radiation stability induced in the crystal formed by our nanofilm template method open new avenues in structural proteomics (Pechkova *et al.*, 2004).

A protein will stay in solution only up to a certain concentration. Once this limiting concentration is reached, the solution will no longer remain homogeneous, but a new state or phase will appear. This phenomenon forms the basis of all protein crystallization experiments. By changing the solution conditions, the crystallographer tries to exceed the solubility limit of the protein so as to produce crystals (McPherson, 1999). This plan rarely runs smoothly. After changing the solution conditions, one of several difficulties is usually encountered: (i) nothing happens, *i.e.*, the protein solution remains homogeneous; (ii) a new phase appears, but it is not a crystal. Instead, it is an aggregate or a liquid; or (iii) crystals do form, but they are unsuitable for structure determination because they give a poor X-ray diffraction pattern. It is often possible to overcome these difficulties by trial and error-repeated crystallization attempts with many different conditions – but this strategy does not always work. Even when it is successful, the lessons learned cannot be easily generalized; the conditions, which work with one protein, are not necessarily optimal for a different protein.

The problems associated with producing protein crystals have stimulated fundamental research on protein crystallization. An important tool in this work is the phase diagram. A complete phase diagram shows the state of a material as a function of all of the relevant variables of the system (Zemansky and Dittman, 1997).

For a protein solution, these variables are the concentration of the protein, the temperature and the characteristics of the solvent (*e.g.*, pH, ionic strength and the concentration and identity of the buffer and any additives). The most common form of the phase diagram for proteins is two-dimensional and usually displays the concentration of protein as a function of one parameter, with all other parameters held constant (Saridakis *et al.*, 1994).

Three-dimensional diagrams (two dependent parameters) have also been reported (Sauter *et al.*, 1999) and a few more complex ones have been determined as well (Ewing *et al.*, 1994).

When a protein crystal (Figure 3.3) is placed in a solvent, which is free of protein, the crystals will begin to dissolve. If the volume of solvent is small enough, the crystal will not dissolve completely; it will stop dissolving when the concentration of protein in solution reaches a specific value.

At this concentration, the crystal loses protein molecules at the same rate at which protein molecules rejoin the crystal - the system is said to be at equilibrium.

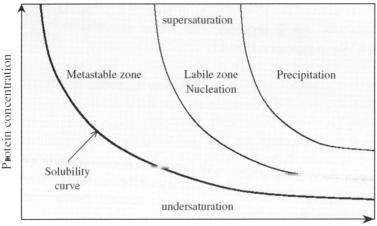

Figure 3.3. A schematic phase diagram of protein showing the solubility of a protein in solution as a function of the concentration of the precipitant present (Reprinted with the permission from Nicolini and Pechkova, Nanostructured biofilms and biocrystals, Journal of Nanoscience and Nanotechnology 6, pp. 2209–2236, © 2006a, American Scientific Publishers, http://www.aspbs.com).

The concentration of proteins in the solution at equilibrium is the solubility. The solubility of a protein varies with the solution conditions. A schematic diagram of a solubility curve, illustrating how the solubility varies with the concentration of a precipitant (*i.e.*, polyethylene glycol (PEG) or a salt), is shown in Figure 3.4.

Crystals dissolve in the undersaturated region-where the concentration is below the protein solubility – and grow in the supersaturated region. The three subdivisions of the supersaturated region are the metastable, labile, and precipitation zones) (Asherie, 2004).

In principle, crystals will form in any protein solution that is supersaturated, *i.e.*, when the protein concentration exceeds the solubility. In practice, crystals hardly ever form unless the concentration exceeds the solubility by a factor of at least three (Chernov, 1997). The large supersaturation is required to overcome the activation energy barrier, which exists when forming the crystal.

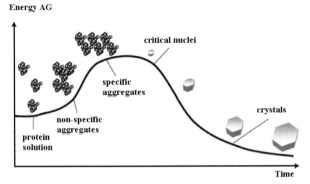

Figure 3.4 Crystallization process: free energy barrier overcoming (Reprinted with the permission from Nicolini and Pechkova, Nanostructured biofilms and biocrystals, Journal of Nanoscience and Nanotechnology 6, pp. 2209–2236, © 2006a, American Scientific Publishers, http://www.aspbs.com).

This barrier represents the free energy required to create the small microscopic cluster of proteins – known as a nucleus – from which the crystal will eventually grow (Figure 3.4). Since there is an energy barrier, nucleation (the process of forming a nucleus) takes time. If the supersaturation is too small, the nucleation rate will be so slow that no crystals form in any reasonable amount of time (Kashchiev, 2000). The corresponding area of the phase diagram is known as the "metastable zone". In the "labile" or "crystallization" zone, the supersaturation is large enough that spontaneous nucleation is observable. If the supersaturation is too large, then disordered structures, such as aggregates or precipitates, may form. The "precipitation zone" is unfavorable for crystal formation, because the aggregates and precipitates form faster than the crystals.

The three zones are illustrated schematically in Figure 3.3. Since these zones are related to kinetic phenomena, the boundaries between the zones are not well defined (this in contrast to the solubility line which is unambiguous description of the equilibrium between solution and crystal).

Even though the division in zones is qualitative, the different behaviors serve as guide when searching for the appropriate conditions to produce crystals (Saridakis and Chayen, 2003).

When a precipitant is added to a protein solution to help produce crystals, liquid drops sometimes form. These drops, also referred to as "oils" or "coacervates", can occasionally be observed by simply changing the temperature, pH, or other solution condition. Such drops generally contain a high concentration of protein. Under the influence of gravity, the drops may separate from the rest of the solution. Eventually, two liquid phases will form, and the concentrations of the various components of the original solution will be different in the two phases. This phenomenon is known as liquid-liquid phase separation (LLPS). Although LLPS is analogous to water condensing from steam, there is a significant difference. While liquid water is stable once it has formed, the protein-rich and protein-poor liquid phases are not. The phases may exist for days, even weeks, but LLPS is inherently metastable with respect to crystallization (Asherie *et al.*, 2001). In other words, the protein-rich liquid phase can convert into crystals. LLPS can promote crystal formation even without the formation of a macroscopic protein-rich phase (Kuznestov *et al.*, 2001). The precise mechanism by which LLPS promotes crystal nucleation is still not known. One factor is the high protein concentration, which exists in the droplets. This high concentration corresponds to a large supersaturation and so increases the crystal nucleation rate. Another factor may be the wetting of the surface of the crystal by one of the liquid phases (Wolde and Frenkel, 1997). Systematic studies have shown that in the case of lysozyme both factors are important (Galkin and Vekilov, 2000). Crystal formation can be initiated and its growth accelerated by a new method shown in Figure 3.5.

The protein thin-film nanotemplate is created using Langmuir-Blodgett (LB) technology or modifications of it (Pechkova and Nicolini, 2004a), and is subsequently deposited on a solid glass support, to be placed in the appropriate vapor-diffusion apparatus. This LB protein thin film assumes the role of the template for protein nucleation and crystal growth. During the screening procedure, the following parameters can be varied: protein monolayer surface pressure, precipitant nature and concentration, and number of protein thin-film monolayers. The protein surface pressure chosen for the deposition should correspond to that of the highly packed, ordered monolayer. In our experience, the surface pressure corresponding to the closely packed system is 25 mN/m^{-1} for chicken egg-white lysozyme, 20 mN/m^{-1} for human protein kinase CK2α and 15 mN/m^{-1} for bovine cytochrome P450scc.

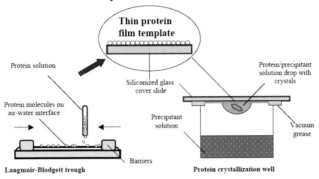

Figure 3.5. Vapour diffusion hanging drop nanotemplate method (Reprinted with the permission from Pechkova *et al.*, Three-dimensional atomic structure of a catalytic subunit mutant of human protein kinase CK2, Acta Crystallographica D59, pp. 2133–2139, © 2003, International Union of Crystallography).

It should be borne in mind that the surface pressure can also be influenced by subphase composition. It should be noted that a possible physical explanation for the increased crystal growth observed with our method (Pechkova and Nicolini, 2002a,b) comes from the variation in template surface pressure.

The increased anisotropy of the thin-film template associated with the increased surface pressure and protein molecule orientation (Nicolini, 1997) causes a dipole moment in the LB monolayer. Indeed, in previous studies (Facci *et al.*, 1993), the surface potential has varied from 20.2 mV in the self-assembly (randomly oriented proteins) to 280 mV in the LB film (ordered and oriented proteins).

Originally the nanotemplate induction of this dipole moment was offered as the possible cause of the controllable and predictable nucleation and growth of the protein crystal (Pechkova and Nicolini, 2002, 2002a).

Recently it was shown that the effect could also be due to the migration of thin protein film fragments formed by highly packed and ordered proteins into solution, in which they constitute the nucleation centers of the protein microcrystals (Pechkova *et al.*, 2005a).

Typically protein crystals are brittle and crush when touched with the tip of a needle, while salt crystals that can sometimes develop in macromolecule crystallization experiments will resist this treatment. This fragility is consequence both of the weak interaction between macromolecules within crystal lattices and of the high solvent content (from 20% to more than 80%) in these crystals (Ducruix and Giege, 1992). For that reason macromolecular crystals have to be kept in a solvent saturated environment, otherwise dehydration will lead to crystal cracking and destruction.

The high solvent content, however, has useful consequences because solvent channels permit diffusion of small molecules, as property used for the preparation of isomorphous heavy atom derivatives needed to solve the structure.

Further crystal structure can be considered as native structure as is indeed directly verified in some cases by the occurrence of enzymatic actions within crystals lattices upon diffusion of the appropriate ligands (Mozzarelli and Rossi, 1996).

Other characteristic properties of macromolecular crystals are their rather weak optical birefringence under polarized light: colors may be intense for large crystals (isotropic cubic crystals or amorphous material will not be birefringent).

Also, because the building blocks composing macromolecular are enantiomers (L-amino acids in proteins-except in the case of some natural peptides-and D-sugar in nucleic acids) macromolecules will not crystallize in space group with inversion symmetries.

Accordingly, out of the 230 possible space groups, macromolecules do only crystallize in the 65 space groups without such inversion. While small organic molecules prefer to crystallize in space groups in which it is easiest to fill space, proteins crystallize primarily in space groups, which it is easiest to achieve connectivity.

Macromolecular crystals are also characterized by large unit cells with dimension that can reach up to 1000 Å for virus crystals (Usha *et al.*, 1984).

In a crystalline form proteins are in a highly ordered three-dimensional array where the protein molecules are bound to each other with specific intermolecular interactions. Protein crystals contain not only protein molecules but also uniform solvent-filled pores that constitute 30–78% of the crystal volume depending on the protein and the crystallization conditions.

3.1.1 *Radiation resistance*

The emerging technology presented in Figure 3.5 produces nanostructured radiation-stable (Figure 3.6) biocrystals of any dimension by nanobiofilm template and characterizes them by the nanotechnologies earlier described in chapter 2, namely AFM, nanogravimetry and microfocused synchrotron radiation in terms of scattering and diffraction.

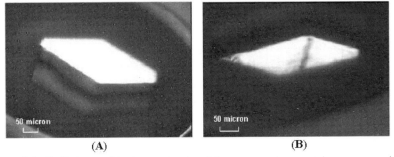

Figure 3.6. Radiation resistant nanostructured lysozyme microcrystal as compared to classical counterparts (Reprinted with the permission from Pechkova *et al.*, Radiation stability of protein crystals grown by nanostructured templates: synchrotron microfocus analysis, Spectrochimica Acta 59, pp. 1687–1693, © 2004, Elsevier).

This new technology is generalizable to all classes of proteins which can be immobilized in thin LB film at the air-water interface and is thereby providing a new route to structural proteomics with far reaching implications for the discovery of new drugs and for other interesting applications in material science and electronics (see later).

Primary radiation damage is linearly dependent on the X-ray dose even when the crystal is at cryogenic temperatures. Above a certain level an excessive damage of the crystal develops which is interpreted as the onset of secondary and/or tertiary radiation damage.

This upper limit of X-ray dose is compared with Henderson's limit (Henderson, 1990), and has profound implications for the amount of useful X-ray diffraction data that can be obtained for crystals of a given scattering power.

While the primary dose-dependent radiation damage is unavoidable, secondary radiation damage can be partially avoided by cryo-techniques implementation, but the effect of ionizing radiation remains substantial even using moderately intense synchrotron facilities, and for brighter sources it becomes indeed a limiting factor.

Radiation damage can cause specific changes in protein crystal structure. Disulphide bonds break up and acidic side chains become decarboxylated.

The unit cell volume increases, and the molecule might undergo small rotations and translations. Non-isomorphism is introduced, which can easily cause large differences in the structure factors.

The radiation stability induced in the biocrystal by the recently introduced nanofilm template method is suggested by a recent systematic study shown in Figure 3.6 (Pechkova *et al.*, 2004).

For the first time, highly diffracting and uniquely radiation stable protein microcrystals are consistently obtained by the nanotechnology-based method.

3.1.2 *New protein structures*

Until now in the last century the atomic three-dimensional structures of many proteins of different size up to the very large ribosomes have been solved (Figure 3.7), but they represent only a small fraction of the existing proteins in nature.

The significant radiation stability induced in the crystal formed by the nanofilm template method has opened new avenues in structural proteomics (Nicolini and Pechkova, 2004; Pechkova *et al.*, 2003).

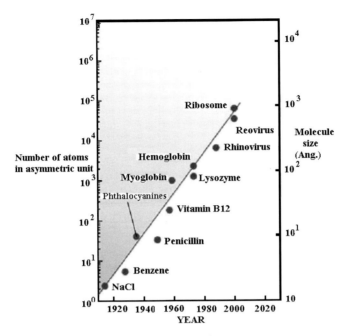

Figure 3.7. Protein crystal structures solved in the last century.

Indeed, diffraction data used for the subsequent determination of 3D atomic structures were collected on a single, miniscule human CK2α microcrystal (20 mm diameter) by the European Synchrotron Radiation Facility (ESRF) microfocus beamline (beam size 20 × 20 μm^2) and used to solve the structure of the protein kinase CK2α catalytic subunit 2.4 Å (Figure 3.8) (Pechkova *et al.*, 2003; Pechkova and Nicolini, 2004).

For what concerns the application of protein biocrystal it must be remembered that in early days the crystallization was an efficient protein purification method. Now is very rare and is replaced by chromatographic methods. However, protein purification by crystallization has many advantages: high yield, high purity in one step, unlimited scale up possibilities, and the product is highly concentrated protein crystal slurry ready for further formulation. Some industrial proteins have been purified by crystallization in large-scale. Furthermore a majority of small molecular weight drugs are produced in crystalline form because of the high storage stability, purity, and reproducibility of the drug properties (Hancock and Zografi, 1997).

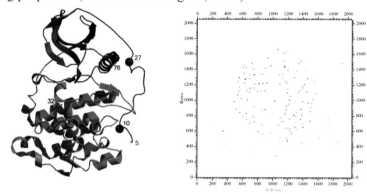

Figure 3.8. Atomic structure and diffraction patterns of human kinase microcrystal. Left: cartoon representation of the three-dimensional atomic structure of human CK2α. The N-terminal domain is in dark blue and the C-terminal domain is in light blue. Spheres mark the positions of the three point mutations. Right: synchrotron diffraction pattern of human CK2α microcrystals (Left: reprinted with the permission from Pechkova *et al.*, Three-dimensional atomic structure of a catalytic subunit mutant of human protein kinase CK2, Acta Crystallographica D59, pp. 2133–2139, © 2003, International Union of Crystallography; right: reprinted with the permission from Pechkova and Nicolini, Atomic structure of a CK2α human kinase by microfocus diffraction of extra-small microcrystals grown with nanobiofilm template, Journal of Cellular Biochemistry 91, pp. 1010–1020, © 2004, Wiley-Liss, Inc., a subsidiary of John Wiley & Sons, Inc.).

There are hundreds of macromolecular therapeutic agents used in clinical trials or approved as drugs (Thayer, 2003). However, only insulin is produced and administered in a crystalline form (Jen and Merkle, 2001). According to Margolin and Navia (2001) the crystallization of macromolecular pharmaceuticals can offer significant advantages, such as:
a) protein purification by crystallization as presented above;
b) high-stability of the protein product compared with soluble or amorphous forms;
c) crystals are the most concentrated form of proteins, which is a benefit for storage, formulation, and for drugs that are needed in high doses (*e.g.*, antibiotics);

d) the rate of crystal dissolution depends on the morphology, size, and additives; thus, crystalline proteins may be used as a carrier-free dosage form.

The most important applications of biocrystal nanofilm-based are presently the determination of protein structure at atomic resolution and the construction of new innovative 3D materials and devices. The whole process of structural determination by X-ray crystallography and the state-of-the-art methods in the field have been reviewed recently in the context of high-throughput crystallization, stressing the merits and caveats of nanodrop setups for both crystal screening and growth.

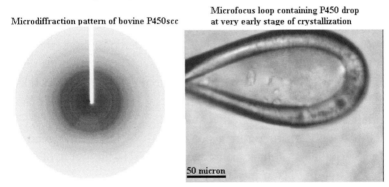

Figure 3.9. Diffraction pattern of bovine P450scc ultramicrocrystal powder (Left) apparently present in the microfocus loop (Right) containing the P450scc drop at 30 hours after the beginning of the crystallization process (Reprinted with the permission from Nicolini and Pechkova, Structure and growth of ultrasmall protein microcrystals by synchrotron radiation: I µGISAXS and µdiffraction of P450scc, Journal of Cellular Biochemistry 97, pp. 544–552, © 2006, Wiley-Liss, Inc., a subsidiary of John Wiley & Sons, Inc.).

The term "protein nanocrystallography" is indeed more appropriate (Pechkova and Nicolini, 2003, 2004a; Pechkova et al., 2004) and gave already the proof of principles with the three-dimensional atomic structure of a catalytic subunit mutant of human protein kinase CK2 (Figure 3.8) (Pechkova et al., 2003) and of few others (Pechkova and Nicolini, 2004a).

Microdiffraction was recently carried out at the Microfocus beamline of the European Synchrotron Radiation in Grenoble also to probe for the structure of protein system, such as P450scc cytochromes that has remained unsolved until now. This characterization has been recently done (Nicolini and Pechkova, 2004) in the form of "powder" P450scc microscrystal obtained by the homologous nanotemplate method. The periodic structure of the P450scc cytochrome appears evident from the diffracting rings present in Figure 3.9; similarly Tripathi *et al.*, (submitted) have determined the periodic structure of three different ribosomal proteins so far unsolved utilizing powder diffraction techniques.

3.1.3 Three-dimensional engineering

The construction of useful nanostructures via self-assembly of synthetic organic molecules has been successfully demonstrated for various applications, including molecular motors. The analogous use of biological molecules such as DNA or proteins, as "building blocks" for the *in vitro* construction of nanostructures is less developed and its potential still far from being fully exploited.

Self-assembly of three dimensional nanostructured functional arrays is a major challenge in nanotechnology, mainly in the fields of photonic crystals, quantum dots arrays and metallic nanoparticles. Exploitation of the inherent potential in the use of biological macromolecules, DNA and proteins, as "guiding template" for the construction of one and two dimensional nanoarrays of nanoparticles and quantum dots is just beginning to emerge. The array of inner spacing of protein crystals routinely prepared for X-ray crystallography attracted only little attention until now: it was noted that different crystallization conditions applied to same protein may result in different packing and attempts were made to analyze protein-protein interactions for better understanding of packing patterns. The use of protein crystal spacing as functional part for practical purposes was limited to the use of chemically cross-linked enzyme crystals as immobilized biocatalysts for a variety of synthetic applications with effective diffusion taking place via small channels. These studies have clearly demonstrated that protein crystals, routinely obtained either via common crystallization procedures or by the nanobiofilm templates here described, can be readily stabilized by chemical cross-linking, *e.g.*, with glutaraldehyde (Dotan *et al.*, 2001) (Figure 3.10).

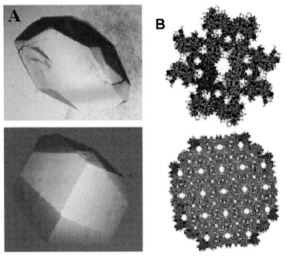

Figure 3.10. Inner water distribution in 3D lysozyme crystal and their replacement after glutaraldehyde staining with gold nanoparticles (Reprinted with the permission from Dotan *et al.*, Supramolecular assemblies made of biological macromolecules, in Nanosurface chemistry, pp. 461–471, © 2001, CRC Press LLC, Taylor and Francis Group LLC).

In this context it was also suggested that protein crystal spacing "channels" could be used as "natural zeolites" for chromatographic purposes. Particularly promising appears the use of directed self-assembly of a binding protein via specific cross-linking into pre-designed diamond-like protein crystal and the computer simulation of the array of the enzyme crystal cavities. In this case, the construction of electronic nanostructures has been successfully demonstrated (Dotan *et al.*, 2001; Pechkova and Nicolini, 2003), integrating in a novel "3D crystal engineering" approach protein "building block" selection (a tetrahedral lectin) and molecular modeling (for bi-ligand cross-linker design), thereby potentially providing a versatile tool for the preparation of novel composite nanostructured materials with their internal cavities as "guiding template" with potential applications in the area of photonic crystals and of semiconductors.

In summary, in many applications crystallized proteins are not suitable for use as such as a result of their fragility and solubility. In order to produce a crystalline protein matrix that is insoluble also in other conditions than those used in crystallization, the crystals have to be chemically cross-linked. In general, chemical cross-linking of protein crystals (Figure 3.10) creates an active and microporous protein matrix that can be used in catalytic and separation applications as molecular electronics. Self-assembled crystals of nanometer size with potential electronic applications can indeed be made of silver cobalt, gold and/or nanoparticles and are fabricated by using reverse micelles or similar technologies. Other examples of potentially interesting materials are calcium carbonate crystals and nanohedra, which use symmetry to design self-assembling protein cages, layers, crystals and filaments.

All the above materials based on organic and inorganic technologies compete with the protein-based technologies earlier described in this chapter, repeating the race to new electronic devices and systems which is under way world-wide between inorganic, organic and biological approaches.

A recent review by Nicolini and Pechkova (2006a) provides an overview of the field of nanostructured biomaterials and biocrystals, with particular emphasis on those being developed over the years by our Institute. It intends mainly to exemplify the tremendous advancement of this field and of their associated nanobiotechnologies (Gourley, 2005), which in recent time have been accelerating at a very rapid pace and are constantly redefining themselves worldwide (Zhao and Zhang, 2004; Zhang *et al.*, 2002; Wu and Payne, 2004; Fu *et al.*, 2005). As a result several areas of research, despite the numerous pending problems and limitations previously outlined, appear to confirm their validity and the various applications of the above technologies being here presented are showing a significant increase in the potential market size to an unprecedented rate.

3.1.4 Basics of crystal formation

μGISAXS information (see chapter 2) are of prime interest in understanding the link between the basic mechanisms of growing crystal morphology and its physical or chemical properties for both lysozymes (not shown) and cytochromes (Nicolini and Pechkova, 2006) (Figure 3.11).

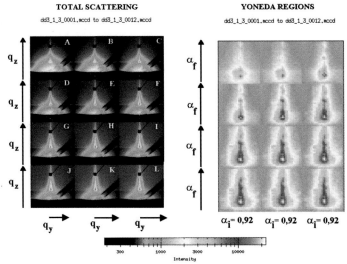

Figure 3.11. Full scattering and Yoneda regions of μGISAXS at the Microfocus beamline of the European Synchrotron Radiation in Grenoble of cytochrome P450scc LB. Raw μGISAXS patterns of a P450scc crystallizing drop sitting over one monolayer of homologous P450 at 44 hours after plating using nanotemplate-assisted hanging vapor diffusion method. Images were taken at 12 successive time intervals while salt and protein crystals were sedimenting on the glass and the solution was evaporating. (Left) total μGISAXS patterns of q_y versus q_z in A^{-1} unit, where, due to its high intensity, the specular peak is covered by a beamstop. (Right) Yoneda region of α_f dependency at $\alpha_i=0.92$ (Reprinted with the permission from Nicolini and Pechkova, Structure and growth of ultrasmall protein microcrystals by synchrotron radiation: I μGISAXS and μdiffraction of P450scc, Journal of Cellular Biochemistry 97, pp. 544–552, © 2006, Wiley-Liss, Inc., a subsidiary of John Wiley & Sons, Inc.).

The molecular mechanisms and the early steps of the growth of these protein crystals have been characterized by μGISAXS at the Microfocus beam line of the European Synchrotron Radiation Facility in Grenoble. This new technology was indeed being utilized to investigate protein crystal growth and nucleation mechanisms with and without nanotemplate, with quite interesting results pointing to a more clear understanding of the crystallization process (Nicolini and Pechkova, 2006; Pechkova and Nicolini, 2006). This new technology is indeed being utilized to investigate protein crystal growth and nucleation mechanisms with and without nanotemplate, with quite interesting results pointing to a more clear understanding of the crystallization process (manuscript in preparation).

It has been quite difficult so far to find the key to every existing protein for developing the general procedure for protein crystallization. Indeed, to be crystallized each protein requires its own specific conditions, which are often difficult to determine and that require empirical extensive searching (Thorsen et al., 2002; Rupp, 2003). That is why protein crystallography is often called an art instead of a science. For this reason the new approach based on nanotechnology had been introduced with potential applications in life sciences and drug industry. In addition to pharmaceutical industry, nanocrystallography can be also useful to fabricate nanoscale arrays, which up to now are based on lithographic techniques, utilizing protein crystals for the construction of next-generation electronic and photonic devices (McMillan et al., 2002).

3.1.4.1 *Cytochrome P450scc*

Ultrasmall P450scc cytochromes microcrystals are grown by classical hanging vapor diffusion and by its modification using homologous protein thin-film template displaying a long-range order. The nucleation and growth mechanisms of P450scc microcrystals are studied at the thin cytochrome film surface by a new μGISAXS technique developed at the microfocus beamline of the European Synchrotron Radiation Facility in Grenoble.

The scattering comes from strong variations of the mean electronic density for X-rays and up to fifteen years ago was limited to three dimensional samples as the strong penetration depth of the radiations and the low signal to noise ratio hampered the surface sensitivity. Quite recently, thanks to highly focused synchrotron radiation (Riekel, 2000), these techniques were extended to surface geometry using the phenomenon of total reflection of X-rays in the grazing incidence range (Müller-Buschbaum et al., 2003). The field of thin films growth brought a need of knowledge about layer morphology and sizes of quantum dots, supported islands or buried particles which has pushed down to the micrometer size for incoming beam the development of Grazing Incidence Small Angle X-ray Scattering (µGISAXS) (Müller-Buschbaum et al., 2000, 2003; Roth et al., 2003), providing information at the highest sensitivity down to the nanometer scale for the dependence of the electronic density perpendicular to the surface, like those due to the roughness of a surface, the lateral correlations, sizes and shapes of gold nanoparticles (Roth et al., 2003). Such information are of prime interest in understanding the link between growing ultrasmall crystal morphology and its physical, chemical, structural properties, the latter being also object of this work down to atomic resolution by X-ray microfocus diffraction (Riekel, 2000).

For understanding the basic physical aspects of the template induced P450scc cytochrome microcrystal nucleation and growth, the synchrotron microfocus has being here used both for diffraction and µGISAXS (Grazing Incidence Small-angle X-ray Scattering) experiments (Roth et al., 2003; Lazzari, 2002).

As scanning force microscopy (SFM), sensitive only to surface structures, grazing incidence small-angle x-ray scattering (GISAXS) appears an excellently suited method for structural and morphological studies of patterned thin films (Müller-Buschbaum et al., 1998, 2000) and of cytochromes P450scc crystal growth and nucleation at its very early stages, as induced by classical (Ducruix and Giege, 1999) and nanotemplate-based (Pechkova and Nicolini, 2002; 2002a) vapour-diffusion method. The innovative crystallization method is described in Pechkova and Nicolini (2002; 2002a).

The protein thin-film nanotemplate is created using Langmuir-Schaefer (LS) technology or modifications of it, and is subsequently deposited on a solid glass support, to be placed in the appropriate vapor-diffusion apparatus, namely the traditional hanging drop vapour-diffusion method.

The optical elements used for the microbeam preparation and the detector used is described in Roth *et al.* (2003) and in Pechkova *et al.* (2005a).

The layout of the scattering measurements using the reference cartesian frame which has its origin on the surface and is defined by its z-axis pointing upwards, its x-axis perpendicular to the detector plane and its y-axis along it. The light is scattered by any type of roughness on the surface. Because of energy conservation, the scattering wave vector q is the central quantity (A) to be monitored during the measurements. As shown by Roth *et al.* (2003) the q_y-dependence (out-of-plane scans) reflects the structure and morphology parallel to the sample surface plane (distances D, in-plane radius R) while the q_z-dependence (detector scans) reflects the height H of clusters, or the roughness parallel to sample surface with:

$$q_y = 2\pi/\lambda \, \sin(2\theta)\cos(\alpha_f)$$

$$q_z = 2\pi/\lambda \, \sin(\alpha_i + \alpha_f)$$

As shown earlier the scattering intensity is recorded on a plane ensuring that the angles are in the few degrees range and thus enabling the study of lateral sizes of a few nanometers. The direct beam is here suppressed by a beam stop to avoid the detector saturation as several orders of magnitude in intensity separate the diffuse scattering from the reflected beam. The protein solution droplets were placed in the hanging drop container during the three-day experiment. The glass substrate with the thin film template and the droplet was removed from the container and the vacuum grease area was cleaned to ensure no contamination of the signal. The 10 μliter droplet exists for about 60 min in air until complete evaporation since the glass substrate and the atmosphere were both cooled. Hence an optimum timing had to be found concerning the preparation, adjustment in the beam and data acquisition time. The glass substrate was subsequently brought into the beam and the incident angle α_i adjusted to about 1°. This procedure was restricted to less than 20 min in order to avoid precipitation of salt or nanocrystals from the solution to the substrate, which would lead to a contamination of the μGISAXS signal.

A characteristic feature of a GISAXS pattern is the Yoneda peak (Y) (Yoneda, 1963). This peak occurs at angles $\alpha_i, \alpha_f = \alpha_c$, where α_c is the critical angle of the sample. The critical angle depends on the material via the real part of the refractive index and hence on the density and roughness of the layer over the glass substrate. The relative intensities of the Yoneda peaks can hence be interpreted in terms of build-up over the glass substrate of protein crystal layers, islands of salts and protein crystals and/or of holes in the protein films.

The 10 µliter droplet diameter was about 5 mm. Hence the footprint of the X-ray beam will be fully within the droplet diameter. The small beam size allows avoiding excessive liquid scattering and provides therefore a reasonable signal-to-background ratio. The position of the drop relative to the beam was determined by an absorption scan with a photodiode. Experiments were performed with the beam at the center of the droplet at the contact area droplet, protein template in order to optimize the signal of the weakly scattering biopolymer samples. Data collection times for individual droplets varied from 7 min to 20 min. This is well below minimum droplet evaporation time of about 60 min. This "Stop" procedure described above interrupts therefore crystal growth at specific times and allows studying microcrystals freshly grown over thin films. The full description of the µGISAXS method is contained in the paper coauthored with the Grenoble beamline scientists (Pechkova *et al.*, 2005a). The quoted paper is one of a series of papers dealing with the same set-up, and it is the first one describing the experimental details sufficiently. Besides the technical details, the conclusions, results are not derived here and there on a level of the available theory of surface scattering. Several methods of interpreting the data are indeed only mentioned, but not implemented because of the extreme difficulty in utilizing *ab initio* considerations at the present stage of development. The interpretations was stopped there simply in showing images of the Yoneda peak, while here continue in more depth with in- and out-of-planes data being presented and discussed in both cytochrome (this paper) and lysozyme (Pechkova *et al.*, 2005a).

The 2D scattering patterns (Figure 3.12) of q_z, (detector scan) versus q_y (out-of-plane scan) allow to evaluate features like microcrystal cluster diameters, heights and distances for the P450scc cytochromes grown either by the nanotemplate-assisted method or the classical one.

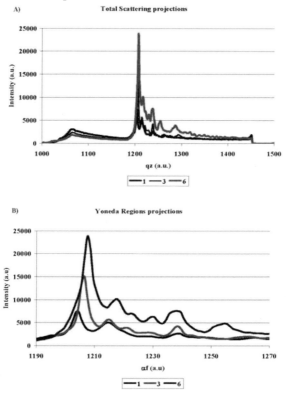

Figure 3.12. The projections along the center column are shown with respect to the total µGISAXS scattering (A) and the Yoneda regions (B) respectively as taken from Figure 3.11. (Reprinted with the permission from Nicolini and Pechkova, Structure and growth of ultrasmall protein microcrystals by synchrotron radiation: I µGISAXS and µdiffraction of P450scc, Journal of Cellular Biochemistry 97, pp. 544–552, © 2006, Wiley-Liss, Inc., a subsidiary of John Wiley & Sons, Inc.).

µGISAXS measurements of ten layers of P450scc cytochrome being here used as substrate for the subsequent deposition and measurements of crystallizing drop point to the existence of a long range order by plotting the logarithmic signal of intensity versus the q_y (not shown). Figure 3.12 shows the clear effect due to the P450scc nanocrystal being formed in the droplet at 44 hours after plating and sedimenting over the above P450scc multilayer. Interestingly while the lower peak in the Yoneda region appears to increase in magnitude with increasing acquisition time, the lower peak in the overall diffuse scattering region (likely corresponding the solution scattering) is drastically decreasing during the same time interval (Figure 3.12). The effect of sedimentation appears quite different in the droplet containing both buffer and proteins with respect to the one containing only buffer but without proteins: in the latter case the scattering signal is lacking in the lower region even before sedimentation (not shown).

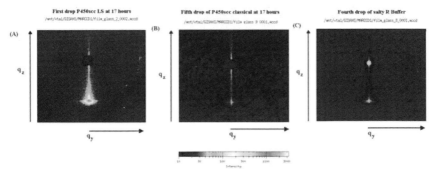

Figure 3.13. Raw µGISAXS patterns of q_y versus q_z of three different drops deposited over ten layers of P450scc. containing P450scc crystallizing solution at 17 hours after plating either nanotemplate-based (A) or classical (B). The drop containing only the buffer solution with precipitant as described in the text is shown in (C). (Reprinted with the permission from Nicolini and Pechkova, Structure and growth of ultrasmall protein microcrystals by synchrotron radiation: I µGISAXS and µdiffraction of P450scc, Journal of Cellular Biochemistry 97, pp. 544–552, © 2006, Wiley-Liss, Inc., a subsidiary of John Wiley & Sons, Inc.).

The raw µGISAX patterns of three different drops deposited over the same ten Langmuir-Schaefer layers of P450scc exemplify in Figure 3.13 the kinetics of protein crystallization in the nanotemplate method versus the classical one (not giving any signal), the role of buffer (not giving any signal) and of thin solid film (giving a signal) in the observed scattering profile. In summary, under the same conditions of buffer solution and temperature, the µGISAXS images of the ten µliters drop over 10 layers of P450scc acting as substrate display the following features: (1) lacks any peak and appears quite less intense with diffuse very low scattering signal at the critical angle of the substrate, when it contains only buffer solution with salt and precipitant; in the Yoneda region of the buffer drop three peaks do appear, with only the one with highest α_f being present before salt sedimentation but dominant and slightly off-axis after salt sedimentation; (2) appears to change significantly along the z-axis and the y-axis whenever successively acquired at equally spaced time intervals over 7 minutes, at any time after plating either with the "classical" method (see the lysozyme data reported in Pechkova and Nicolini, 2006) or (mostly) with the protein nanotemplate-based method, apparently due to sedimentation processes of protein crystals from the solution; (3) along the z-axis changes from the very low q_z solution scattering of the total µGISAXS to the very low α_f protein substrate peak in the Yoneda region.

For this reason in order to compare the crystallization process taking place with time in presence and absence of protein film as nanotemplate, we have kept constant the measuring time, carrying out the measurements quite rapidly immediately after removal from the container (Figure 3.14). Under these conditions it appears that:
- the overall µGISAXS pattern of the drop containing P450scc crystallizing solution based on "nanotemplate" at 44 hours after plating is quite more pronounced with respect to the corresponding "classical" one.
- in the Yoneda region of the corresponding µGISAXS pattern, being the most sensitive to structural and morphological changes of the surfaces due to the interference effect involved in the occurrence of the Yoneda peak, contrary to "classical" drops "nanotemplate-based" P450scc cytochrome drops taken at 17 and 44 hours after plating (Figure 3.14) clearly show at least two pronounced Yoneda peaks growing with time after plating.

A week Yoneda peak exists already at shorter times with a critical angle corresponding to that of the protein substrate, unlikely that of the salt precipitating on the substrate causing a new rough layer of small crystals which yields a second peak not present at shorter times. Hence all data are compatible with a working hypothesis, which attributes the peak at higher α_f-values to that of the glass substrate and/or to salt crystal with the newly developing peak at lower α_f-values being related to the protein itself.

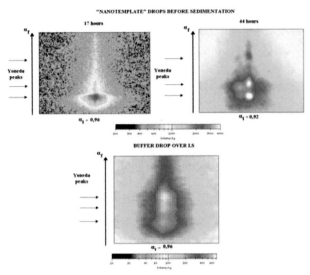

Figure 3.14. Yoneda regions of cytochrome P450scc "nanotemplate-based" drops taken at 17 (A) and 44 hours (B) after plating versus P450scc cytochrome, where the α_f dependency is given at α_i=0.96 and 0.92. Yoneda regions are indicated with arrows. The patterns are shown on a logarithmic scale to enhance the features in the Yoneda regions. With drops containing classical P450 crystallization solution or the buffer no significant signal is apparent under the same conditions with the acquisition being carried out immediately after drop deposition, After drop sedimentation with most liquid being evaporated a µGISAXS signal ten times less pronounced is apparent (see panel C for the corresponding buffer solution drop sitting on top of the same 10 layers of P450scc). (Reprinted with the permission from Nicolini and Pechkova, Structure and growth of ultrasmall protein microcrystals by synchrotron radiation: I µGISAXS and µdiffraction of P450scc, Journal of Cellular Biochemistry 97, pp. 544–552, © 2006, Wiley-Liss, Inc., a subsidiary of John Wiley & Sons, Inc.).

Finally, in the attempt to obtain structural information at the atomic resolution we have then performed X-ray microdiffraction with the largest microcrystals (about 5 micron in diameter) being obtained by the homologous nanotemplate method after 44 hours plating time in the form of "powder" P450scc microcrystals.

The kinetics and the structure of the ultrasmall P450scc cytochrome microcrystals here investigated proves the superiority of the nanobiofilm template method (Pechkova and Nicolini, 2003, 2004a) in inducing nucleation and growth up to a 5 micron crystal size of a protein system yet structurally unsolved. Our unprecedented approach with µGISAXS combines the powerful thin film characterization method with the micrometer-sized X-ray beam enhancing the spatial resolution used thus far by two orders of magnitude (Roth et al., 2003). The acquired scattering data along with the reproducible diffraction patterns of the ultrasmall P450scc microcrystals allows for a nondestructive and contact-free reconstruction of the crystallization process at the very early stage and of the three-dimensional structure and morphology of the nanocluster crystals. Such characterization and investigation on a submicrometric scale of extra-small crystals cluster on top of the protein film appears to allow a more clear understanding of the very early steps of crystallization in cytochrome P450scc. It is worth to notice that even if these data are quite illuminating, more conclusive data will be likely obtained with the in situ experimentation already being planned. The old paper of Yoneda (1963) has been frequently cited, and the underlying models concerning surface roughness and the analytical interpretation of the diffuse scattering has been utilized to reach a conclusion. The proper models and conclusions were then decided by performing a zero-control experiment using a template with a droplet containing the solvent alone (no protein). Furthermore the influence of the surface coating on the scattering has been measured separately, including a control diffraction experiment with the surface coating alone. The available diffuse scattering data have also been interpreted compatibly with state-of-the-art knowledge. Considering that the Yoneda peak is one of the major information extractable we have not made a simple generic comment, but pointed to the role of the density and of the roughness of the protein crystal and/or salt crystal layers. Detector scans are also shown in this context to interpreter the data where micro GISAXS pattern were provided and scaled to the same scattered intensity, taking into considerations the critical angles of the glass and of the proteins.

Interestingly, from the scattering data here obtained the occurrence and the time sequence of the various Yoneda peaks suggest the development of a layer with significant increase in roughness due to the significant increase in holes of the template layer as the P450scc crystal grows in the hanging drop. That is to say this model assumes a decreasing density of the protein template layer as the layer itself is removed by the protein solution thus triggering or assisting the formation of new crystals in the solution. In conclusion the data here presented favor a build-up of holes in the nanotemplate acting as nucleation centers for crystals formation in all protein systems including the ones so far impossible to crystallize as the human kinase (Pechkova *et al.*, 2004, 2003) and the P450scc here studied.

3.1.4.2 *Lysozyme*

To investigate the early steps of lysozyme crystallization and for this reason we concentrate on the µGISAXS method, which appear potentially able to utilize the out-of-plane cuts in the Yoneda regions of its 2D scattering profiles in order to detect ultrasmall lysozyme crystals quite before the light microscopy.

µGISAXS is indeed not restricted to the sample's surface, but is sensitive to structures within the penetration depth of X-rays impinging at small angles (grazing incidence) on a sample surface (Müller-Buschbaum *et al.*, 1998, 2000). In particular, its potential in nondestructively determining the structure and morphology of patterned thin protein films is exploited due to its order of magnitude increase in the achievable resolution compared to conventional SAXS experiments (Roth *et al.*, 2003).

As shown before (Nicolini and Pechkova, 2006) the protein thin-film nanotemplate is created using Langmuir-Blodgett (LB) technology or modifications of it (Nicolini, 1997a) and is deposited on a solid glass support, to be subsequently placed in the appropriate vapour-diffusion apparatus as a modification of the "classical" hanging drop vapour-diffusion method (Ducruix and Giege, 1999). This thin LB protein film acts as nanostructured template for protein crystal nucleation and growth up to 1000 micron size. This heterogeneous crystallization reduces the level of supersaturation, allowing acceleration of lysozyme crystal nucleation and growth (Pechkova and Nicolini, 2001). Moreover, this method seems to produce radiation stable crystals (Pechkova *et al.*, 2004), quite more resistant than those obtained by classical techniques.

This aspect concerns also the crystals of miniscule thickness dimensions (10–20 micron), such as those used for human kinase 3D structure determination by synchrotron microfocus diffraction (Pechkova et al., 2003). In the present work the data are obtained also with the standard vapor diffusion hanging method called "classical" (Ducruix and Giege, 1999). Every 10 µl drop contains equal proportion of lysozyme containing solution of CH_3COONa pH 4.5 and of CH_3COONa 0.9 M NaCl, pH 4.5. The experimental procedure and the schematic picture of the µGISAXS setup are shown in Figure 3.15 of the accompanying paper (Nicolini and Pechkova, 2006), where the sample is mounted on a xyz-gantry and a two-axis goniometer.

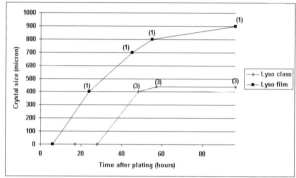

Figure 3.15. The best case for lysozyme crystal size measured by light microscopy (in microns) versus time (in hours) after plating utilizing the classical and the nanotemplate-based vapour-diffusion method. (Reprinted with the permission from Pechkova and Nicolini, Structure and growth of ultrasmall protein microcrystals by synchrotron radiation: II. µGISAXS and microscopy of lysozyme, Journal of Cellular Biochemistry 97, pp. 553–560, © 2006, Wiley-Liss, Inc., a subsidiary of John Wiley & Sons, Inc.).

The incoming monochromatic X-ray beam (X=0.9775 Å) is 5µ wide, and a two-dimensional (2D) high-resolution detector records the scattered intensity from the sample surface. In the 2D pattern, structural variations in z (depth of the sample), for example, a finite surface roughness, lead to a specular and off-specular scattered intensity in q_z-direction. Structures in the x-y plane lead to out-of-plane (with respect to the incoming beam and the sample surface normal) signals with finite q_y and $2\theta \neq 0$.

Thin films investigated by GISAXS thus far (Holy and Baumbach, 1994) were of limited Spatial resolution due to the large beam size available (several hundred micrometers) (Gehrke, 1992). As shown earlier (Roth et al., 2003) and in the accompanying paper (Nicolini and

Pechkova, 2006), to investigate our protein drops we utilized µGISAXS combining a micrometer-sized synchrotron beam with a specifically designed low-background GISAXS setup (Sennett and Scott, 1950; Riekel, 2000; Lazzari, 2002). Typically the data were acquired at 0.88° incident angle in a time ranging between 7 and 20 minutes to avoid amorphous protein precipitation and salt crystals. We systematically investigate the kinetics of the nanometer-sized crystals being formed with and without nanobiofilm template of homologous proteins. We used lysozyme as it can be easily crystallized and can thus be used as model system to monitor the time-dependent crystals growth in size and number as evaluated by light microscopy, in correlation with µGISAXS measurements.

During a series of parallel experiments the lysozyme crystals appear to grow in size with time, and significantly better for the nanotemplate-assisted method, but not in number, which appears to remain constant (Figure 3.16).

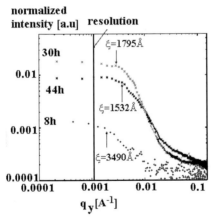

Figure 3.16. Out-of-plane scans of the Yoneda regions of the nanotemplate-assisted samples at 8, 30 and 43 hours after plating. Included are the template and the glass substrate. Clearly the enhancement of the diffuse scattering with increasing time is visible. The detector dark noise has been subtracted and the intensity normalized to the acquisition time. Curve 8h is shifted for clarity. The arrows indicate most-prominent in-plane lengths ξ. The resolution is marked by a vertical line. (Reprinted with the permission from Pechkova and Nicolini, Structure and growth of ultrasmall protein microcrystals by synchrotron radiation: II. µGISAXS and microscopy of lysozyme, Journal of Cellular Biochemistry 97, pp. 553–560, © 2006, Wiley-Liss, Inc., a subsidiary of John Wiley & Sons, Inc.).

3.2 Nanomedicine

Nanomedicine represents a growing field of wide interest but still with alternative results in clinical medicine and we take here only four representative stimulating applications of nanoscale technologies organic and biological to medicine.

3.2.1 *Carbon nanotubes biocompatibility and drug delivery*

Purified carbon nanotubes are new carbon allotropes, sharing similarities with graphite, that have recently been proposed for their potential use with biological systems, as probes for in vitro research and for diagnostic and clinical purposes.

However the biocompatibility of carbon nanotubes with cells represents an important problem that until recently remained largely uninvestigated. It was only in 2006 that Garibaldi *et al.* have shown how cardiovascular cells in vitro being exposed to purified carbon nanotube material appear to possess a high level of cytocompatibility. Coating or other functionalization procedures may render the single wall carbon nanotube (SWNT) fully compatible with living cells. However, in the perspective of an *in vivo* use of NT for clinical applications, it is important to take into account that there is the possibility of a not fully effective coating treatment, which could expose cells to a direct contact with carbon nanotubes material. Moreover, while superposed materials are often biodegradable, especially when using proteins or other biologic molecules, carbon nanotubes are not degradable and overtime, by losing functionalization, might come in direct contact with cells and tissues. In relation to that, our observations (Garibaldi *et al.*, 2006) point instead to a lack of short-term toxicity induced by carbon nanotubes when added to cardiomyocytes.

Figure 3.17 demonstrates that SWNT are biocompatible with cardiomyocytes in culture, suggesting that long-term negative effects are probably due to physical rather than chemical interactions.

Cardiac muscle cells (from rat heart cell line H9c2) viability in the first three days of culture was not different between NT-treated cells and untreated cells. However after 3 days of culture, cell death was slightly higher in NT-treated cells.

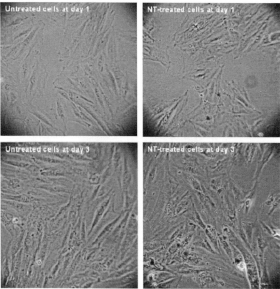

Figure 3.17. Photomicrographs of H9c2 cells untreated (on the left) and after treatment with carbon nanotubes (on the right): effect at day 1 (upper photomicrographs) and at day 3 (lower photomicrographs); magnification 40×; NT, nanotubes (Reprinted with the permission from Garibaldi *et al.*, Carbon nanotube biocompatibility with cardiac muscle cells, Nanotechnology 17, pp. 391–397, © 2006, IOP Publishing Limited).

Trypan blue exclusion confirmed these observations (Figure 3.17): compared to untreated after the corresponding time, trypan blue positive cells increased by 5 and 9% after 1 and 3 days of treatment respectively (p=ns). After reseeding non-viable cells coming from SWNT-treated samples increased by 25%, when compared to reseeded cells not treated with SWNT (p<0.05). Figure 3.18 shows the effect of SWNT treatment on apoptotic death of cardiomyocytes as determined by annexin/PI staining (Garibaldi *et al.*, 2006). Exposure of cardiomyocytes to 0.2 mg/ml of SWNT induced little change in the apoptotic pattern at day 1 and 3 after SWNT treatment (mean differences did not reach significance). After reseeding an increase in cell death could be observed in SWNT-treated samples.

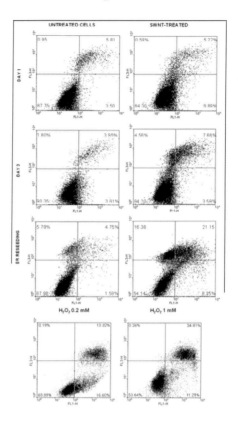

Figure 3.18. Flow cytometric analysis of annexin/PI staining; A) cells untreated or treated with SWNT for the indicated time; B) positive control of cells treated with hydrogen peroxide at the concentrations of 0.2 and 1 mM. FL1-H: relative Annexin V-FITC fluorescence intensity; FL3-H: relative PI fluorescence intensity. (Reprinted with the permission from Garibaldi *et al.*, Carbon nanotube biocompatibility with cardiac muscle cells, Nanotechnology 17, pp. 391–397, © 2006, IOP Publishing Limited).

This increase is accounted by annexin-positive/PI-negative cells (9.3% vs 2.9% in reseeded untreated samples, $p<0.05$), that are apoptotic cells (lower right quadrant), but at greater extent by annexin positive/PI positive cells (18.7% vs 5.2% in the reseeded untreated samples, $p<0.05$), that are late apoptotic or necrotic cells (upper right quadrant). Figure 3.18b shows a positive control, with hydrogen peroxide at the concentration of 0.2 with cells dying for apoptosis, while, at the concentration of 1 mM hydrogen peroxide, most cells are necrotic.

To determine whether SWNT bound to cell membrane affect repeated seeding of cardiomyocytes, subconfluent SWNT-treated cells were detached and reseeded (1:4) in petri dishes (Figure 3.19) to eliminate most of the unbound nanotubes material. At 24 and 72 hours after reseeding, untreated cells showed the average rate of exponential growth, while cells coming from trypsinized SWNT-treated sample showed a limited ability to proliferate, with a definite difference in shape, with a high degree of cell death. However, overtime cells from trypsinized SWNT-treated sample continued to grow, and partially recovered the original shape (Figure 3.19).

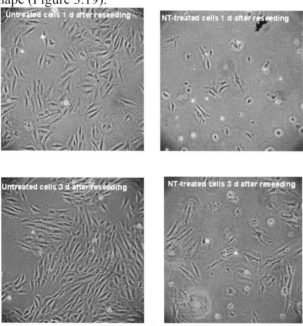

Figure 3.19. Photomicrographs of H9c2 cells untreated (on the left) and after treatment with carbon nanotubes (on the right): long term effect at day 1 (upper photomicrographs) and at day 3 (lower photomicrographs) after reseeding; magnification 40×; NT, nanotubes (Reprinted with the permission from Garibaldi *et al.*, Carbon nanotube biocompatibility with cardiac muscle cells, Nanotechnology 17, pp. 391–397, © 2006, IOP Publishing Limited).

In conclusion our results demonstrates that highly purified carbon nanotubes possess no evident short term toxicity and can be considered biocompatible with cardiomyocytes in culture, while the long term negative effects, that are evidenced after reseeding, are probably due to physical rather than chemical interactions. While more work is needed to establish biological consequences associated to long term interactions of SWNT with cells, our results encourages further research aimed to establish the use of purified SWNT for *in vivo* applications in cell systems, likewise drug delivery *in vitro* and *in vivo* systems (work in progress).

3.2.2 *Photosensitization of titanium dental implants*

Dental implants are now day successfully used in dentistry for oral rehabilitation supporting mobile or fixed prosthesis. Beside undisturbed osseointegration and an adequate prosthetic design, clinical success of dental implants can be jeopardized by bacterial infection inducing mucositis or periimplantitis (Mombelli *et al.*, 1987; Leonhardt *et al.*, 1992; Lindhe *et al.*, 1992). Various methods have been proposed for the treatment of periimplantitis including access flap procedures, the use of locally or systemically administered antimicrobial agents as well as decontamination of the exposed implant surfaces (Persson *et al.*, 2001; Kreisler *et al.*, 2002).

The eradication of pathogenic microorganisms from implant surfaces is a key step for successful treatment of failing implant. Several methods for cleaning of implant surfaces have been described in order to treat failing implants. Among others the laser treatment seemed to be effective in terms of bacteria elimination (Kreisler *et al.*, 2002, 2003; Dortbudak *et al.*, 2001; Shibli *et al.*, 2003; Karacs *et al.*, 2003; Swift *et al.*, 1995; Bereznai *et al.*, 2003). Implants are generally supplied in sterile vessels but infection may occur during the surgery. If bacterial infection of the interface between the implant and the bone or tissue occurs remedial surgery may be required. It has been demonstrated, in-vitro, that illumination of an implant using high intensity visible radiation can clean the surface. However, laser induced damage to the surface and undesired heating of the implant have been reported. On the other side, photosensitive dyes, for example toluidine blue, have also been employed in-vitro to sterilise implants (Shibli *et al.*, 2003). The use of such dyes reduces the photon flux required to sterilise the implant. When illuminated at the correct wavelength photosensitive dyes produce, with high yield, singlet oxygen. It is this singlet oxygen, which reacts with biological molecules and cleans the surface of the dental implants (Matysik *et al.*, 2002). Whilst the dye photosensitization strategy lowers the photon flux required to sterilise an implant, and eliminates laser induced structural damage, it necessitates the unwanted addition of an organic dye to the specially engineered surface of the implant.

It is well known that commercially pure titanium is used to produce implants owing to its excellent biocompatability. To promote bio-efficacy the surface of metal implants are roughened. Grit balsting followed by acid etching or titanium plasma spraying are employed to produce the roughened nanoscale surfaces. Titanium is a very reactive metal that is rendered corrosion resistant, and hence suitable for in-vivo applications, by a passive TiO_2 layer. Hence dental implants have nanoscale TiO_2 at their surface. TiO_2 is a wide band gap semiconductor that produces (hydr)oxyradicals when illuminated with UV light.

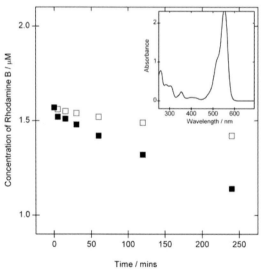

Figure 3.20. The rhodamine B concentration as a function of UV illumination time; ■ in the presence of a TiO_2 coated dental implant and □ with no implant present. The inset shows the UV-absorption spectrum of rhodamine B (Reprinted with the permission from Riley *et al.*, An in-vitro study of the sterilization of titanium dental implants using low intensity UV-radiation, Dental Materials 21, pp. 756–760, © 2005, Elsevier).

These free radicals, which are more reactive than singlet oxygen, may be successfully employed in redox chemistry, killing bacteria (Riley *et al.*, 2005). Our group indeed recently reported the photosensitization of titanium dental implants by TiO_2. Namely the photosensitization of titanium dental implants by TiO_2 was successfully reported (Riley *et al.*, 2005), whereby the investigations of the decomposition of rhodamine B indicate that the commercial implants are photoactive (Figure 3.20). Experiments were performed on dental implants supplied by Premium Implant System (Sweden & Martina, Padua, Italy). The implants were 3.75 mm in diameter and had an insertion depth of 8.5 mm. The implants were of commercially pure titanium and had been sand blasted and acid etched to promote osseointegration. All samples were unpacked from their sterile containers immediately prior to use. Indeed our *in vitro* studies of illuminated solutions containing *Escherichia Coli* and dental implants indicate that photosterilisation may be achieved at low UV light intensities (Figure 3.21) thereby suggesting a new possible utilization of titanium in oral surgery.

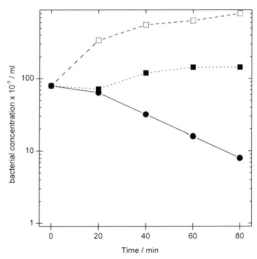

Figure 3.21. The concentration of *E. Coli* in solution as a function of time; ● under UV illumination in the presence of a TiO_2 coated dental implant, □ in the presence of a dental implant but no illumination and ■ under illumination but in the absence of a dental implant. The lines are plotted as guides to the eye. (Reprinted with the permission from Riley *et al.*, An in-vitro study of the sterilization of titanium dental implants using low intensity UV-radiation, Dental Materials 21, pp. 756–760, © 2005, Elsevier).

The photoreaction between TiO_2 and rhodamine B (RH) under UV-visible radiation is complex (Wu *et al.*, 1998). Essentially two mechanisms may operate in parallel. Absorption of UV radiation by the TiO_2 yields an electron and a hole at the particle surface. The electron reacts with molecular oxygen to form an $O_2^{\cdot-}$ which, on protonation yields $HOO^{\cdot}$. The $HOO^{\cdot}$ may be further reduced, via hydrogen peroxide, to a hydoxyl radical. Hydroxyl radicals are also formed by the reaction between water or OH^- and photogenerated holes. The photogenerated oxyradicals may then react with RH to give mineralised products; either,

$$RH + \{O_2^{\cdot-}, HOO^{\cdot} \text{ or } OH^{\cdot}\} \rightarrow \text{intermediate} \rightarrow\rightarrow \text{products}$$

or

$$R + h^+(TiO_2) \rightarrow R^{\cdot} + \{O_2^{\cdot-}, HOO^{\cdot} \text{ or } OH^{\cdot}\} \rightarrow \text{intermediate} \rightarrow \rightarrow \text{products}$$

Alternatively the RH may adsorb visible light and then transfer an electron to TiO_2. As above, the electron on the TiO_2 will react with molecular oxygen to form the (hydr)oxyradicals that breakdown the dye. This second process is termed sensitization.

3.2.3 *Biopolymer sequencing and drug screening chip*

Conventional DNA sequencing methodology is labour-intensive, requiring significant sample preparation and electrophoretic size-separation of radio-labelled DNA fragments. As an alternative, a chip comprising an array of short oligonucleotides can be used as a hybridization probe to analyse a fluorescently labelled DNA sample of unknown sequence.

The resulting hybridization pattern could then be lysed to determine the sequence of the target DNA. To test the feasibility of this approach, an array containing all the 8-mer combinations of the nucleosides guanine and cytosine (256 oligomers) was synthesized (Fodor *et al.*, 1993). After synthesis, the chip was incubated with 1 pmol of 5'-fluorescein-CCCAAACCCAA-3' and scanned at 15 °C. Hybridization of this target to the four complementary octanucleotides present on the array was observed as brightly fluorescent spots, as were many of their single-base mismatches. Methods and algorithms to reconstruct the target sequence from such hybridization data were developed time ago (Fodor *et al.*, 1993).

More than 100 million bases in the DNA from different sources have been already sequenced with various techniques. However, even this was absolutely insufficient, and the Human Genome Project introduced procedures able to sequence up to megabases. Four groups introduced originally the "sequencing by hybridization" (SbH) technique almost simultaneously: one in Russia (Lysov *et al.*, 1988; Khrapko *et al.*, 1989), two in the UK (Southern, 1996; Bains and Smith, 1988), and one in Yugoslavia (Drmanac *et al.*, 1996). This technique was based on sequence-specific DNA hybridization to a large set of oligonucleotides of specified length. By identifying overlapping sets of oligonucleotides that form perfect duplexes with the target DNA sequence, unknown DNA sequences were determined with a model SbH experiment introduced by Mirzabekov in 1994.

Incorporation of the SbH procedures into DNA sequencing microchips were manufactured time ago for detecting hybridization of the target DNA to immobilised oligonucleotides (Mirzabekov, 1994). The application of technologies with resolution threshold of 10 µm (such as printing or micromanipulation) permitted to reduce even further the dimensions of the chip. To overcome another drawback of the approach, the costly procedure of oligomer synthesis, methods of parallel synthesis on various surfaces in 2D arrays were developed (Cantor *et al.*, 1992). A matrix incorporating 256 purine 8-mers has been manufactured by solid-phase oligonucleotide synthesis (Cantor *et al.*, 1992). Spectacular opportunities became available through the use of addressable photo-activated chemistry in the parallel synthesis of oligonucleotides directly on a glass surface. This technology provided the means for effective industrial manufacture of highly complex sequencing microchips. When large numbers of small-dimension oligonucleotide matrices are to be produced industrially, and quality control for every immobilized oligomer was essential, then using robots to apply pre-synthesized oligomers to the matrix (Khrapko *et al.*, 1989) proved advantageous. This method is also applicable to the immobilisation of a much wider variety of oligonucleotide analogs, as well as other compounds, such as proteins, antibodies, antigens and low-molecular weight ligands, on matrices. Such microchips finded applications in many other areas of nanoproteomics and nanogenomics besides large-scale sequencing; for example, in diagnostics or molecular screening.

The cost of a microchip will incorporate the cost of manufacturing. By analogy with electronic microchip production, the per-unit cost should not be excessive for large-scale production. The challenge has been to establish manufacturing procedures able to produce, reproducibly, matrices incorporating the oligonucleotides synthesized to the degree of accuracy required. Traditional methods for generating and identifying biologically active compounds for pharmaceutical development are typically tedious and time-consuming. Screening natural products from animal and plant tissues, or the products of fermentation broths, or the random screening of archived synthetic molecules have been the most productive avenues for the identification of new, biologically active lead compounds. "Rational" drug-design is a more sophisticated approach, in which the tools of synthetic organic chemistry, X-ray crystallography and molecular modelling are employed to produce pharmaceutically active molecules (as shown in previous party of this chapter. This is an extremely time-consuming, labour intensive and expensive process, during which compounds are designed, synthesized and tested in an iterative fashion. Although successful, this approach has been slow to yield products with desired properties. Ideally, one would like to have access to an infinite source of compounds that could be screened rapidly for individual molecules with desired biological properties. The new field of research early referred to as combinatorial chemistry represented a first step towards this goal. For example, peptides are built up from the 20 naturally occurring, gene-encoded amino acids. The combinatorial assembly of these amino acids can generate up to 8000 (203) unique tripeptides, 160000 tetra peptides, 64×106 hexapeptides, etc. Practical technologies for achieving such random assembly of molecular subunits were first developed for the synthesis of peptides and oligonucleotides (Ellington and Szostak, 1990; Tuerk and Gold, 1990).

3.3 Nanogenomics

A new approach for medical diagnostics and therapy named "Nanogenomics" is emerging from the interplay of bioinformatics and biomolecular microarray to a previously unforeseeable level. This editorial summarizes here its major features with few key examples of molecular genomics application to medicine.

DNA microarrays have emerged as one of the most promising methods for the analysis of gene expression (Butte, 2002; Nicolini *et al.*, 2006a). This technique allows the study of an immense amount of genes (over 10000) with only one experiment and therefore can draw a picture of a whole genome. Anyway, the huge number of data coming out from microarray experiments may often raise experimental complications and difficulties in the analysis. Moreover, the greatest part of genes displayed on an array is often not directly involved in the cellular process being studied.

3.3.1 *Human T lymphocytes cell cycle*

Human lymphocytes gene expression is monitored before and after PHA stimulation over 72 hours, using DNA microarray technology. Results are then compared with our previous bioinformatics predictions, which identified 6 leader genes of highest importance in human T lymphocytes cell cycle.

Experimental data are strikingly compatible with bioinformatic predictions of the specific role and interaction of PCNA, CDC2 and CCNA2 leader genes at all phases of the cell cycle and of CHEK1 leader gene in regulating DNA repair and preservation. It does not escape our notice that the conception and use of *ad hoc* arrays, based on a bioinformatics prediction which identifies the most important genes involved in a particular biological process, can really be an added value in cell biology and cancer research alternative to massive frequently misleading molecular genomics.

Recently, we proposed a bioinformatics algorithm, based on the scoring of importance of genes and a subsequent cluster analysis, which allowed us to determine the most important genes, that we call "leader genes" (Sivozhelezov et al., 2006a) in human T lymphocytes cell cycle. The basis of the scoring system relies upon the calculation of interactions among genes, performed with software available in the web, such as STRING (von Mering et al., 2005). The number of links for each gene is then weighted and the final weighted numbers of links are clustered, in order to make a hierarchical classification of genes (Sivozhelezov et al., 2006a). In this way, it becomes possible to draw and to update maps of the major biological control systems, and to integrate them in a concise manner to discern common patterns of interactions between gene expression and their correlated coding of proteins. We chose, as a model system, human T cell lymphocytes stimulated to entry cell cycle with PHA (Nicolini et al., 2006a). This particular cellular system is very well known and was quantitatively characterized time ago (Cantrell, 2002; Isakov and Altman, 2002; Oosterwegel et al., 1999a,b; Abraham et al., 1980); therefore, it can be a good starting point to verify our algorithm. In particular, we identified 238 genes involved in the control of cell cycle. Most important, only 6 of them were previously identified to be the leader genes (see table 1 from Nicolini et al., 2006a); interestingly, they actually are involved in the cell cycle control at important progression points, namely the most important four at the transition from G0 to G1 phase (MYC; Oster et al., 2002), at the progression in G1 phase (CDK4; Modiano et al., 2000), and at the transitions from G1 to S (CDK2; Kawabe et al., 2002), and from G2 to M phases (CDC2; Baluchamy et al., 2003; Torgler et al., 2004). The two remaining "leader genes" (CDKN1A and CDKN1B) are inhibitors of cyclin-CDK2 or -CDK4 complexes and thereby contribute to the control of G1/S transition and of G1 progression (Jerry et al., 2002; Chang et al., 2004).

We also confirmed our results by analyzing changes in gene expression after 48 hours, using a newly developed and simple technology called DNASER, which is a novel bioinstrumentation for real-time acquisition and elaboration of images from fluorescent DNA microarrays developed in our laboratories (Nicolini et al., 2002; Troitsky et al., 2002). The validity of the DNASER measurements was confirmed by standard fluorescence microscopy equipped with CCD. This experimental analysis proved that DNASER is appropriate for monitoring gene expression during the human lymphocytes cell cycle (Nicolini *et al.*, 2006a).

The leader gene approach, validated by experimental analysis on a model system, can suggest a more rationale approach to experimental techniques and methods, as DNA microarray. The application of bioinformatics studies and the identification of leader genes can predict the most important genes in a particular cellular process. In this way, it becomes possible to design smaller microarrays, which display only the most interesting genes for a specific cellular process and thus are much easier to interpret.

We experimentally analyze the gene expression of human T lymphocytes treated with the mitogen compound PHA 24, 48 and 72 hours ("time series analysis", for a complete reference see Straume, 2004; Willbrand *et al.*, 2005) after the stimulation and compare the results with independent bioinformatics predictions, in order to give a further validation to leader gene approach and to identify co-expressions among genes involved in the cell cycle of human T lymphocytes.

Table 3.1. RNA extraction yield for the different lymphocytes samples (Reprinted with the permission from Giacomelli and Nicolini, Gene expression of human T lymphocytes cell cycle: Experimental and bioinformatic analysis, Journal of Cellular Biochemistry 99, pp. 1326–1333, © 2006, Wiley-Liss, Inc., a subsidiary of John Wiley & Sons, Inc.).

Time, in hours, after PHA stimulation	Number of cells	% of quiescent cells (in G_0+Q)	% of proliferant cells (in G_1,S,G_2)	Total RNA pg/cell	Fluorochromes utilized
0	10^7	90	8, 1, 1	30	Cy3 green
24	5×10^6	53	51, 8, 3	30.4 (35,7%)	Cy5 red
48	5×10^6	36	–	39.4 (+60%)	Cy5 red
72	5×10^6	17	45, 36, 2	38.4 (+72%)	Cy5 red

The employed array is the Human Starter Array by MWG Biotech, chosen on the base of the gene we are interested about (Nicolini *et al.*, 2006a).

It contains 161 oligonucleotides (designed to be specific for the respective human gene sequence), 32 replicas and 7 gene specific Arabidopsis control oligonucleotides, for 200 total spots, disposed in 10 columns and 20 rows, more an exact copy, in total 400 spots for array. The diameter of one spot is 100 µm and the distance between two near spots is 250µm.

As previously shown (Nicolini *et al.*, 2006a), in order to obtain total RNA extraction the cells pellet (minimal 1 x 10^7 cells) has been dealt with an extraction kit from Amersham Biosciences containing LiCl, CsTFA and an extraction buffer. The samples thus obtained have been conserved at -80 °C. For the estimation of the extracted RNA they have been used 200 µl RNAsi free cuvettes. In order to avoid contaminations of genomic DNA, the RNA samples have been subordinates to digestion with the enzyme Dnase I. For every experiment we obtained a good RNA total extraction yield, as shown in Table 3.1. Moreover the RNA spectrophotometrical analysis has evidenced a high purity degree, being the ratio 260/280 nm always more then 1.9. In total, 32 genes were identified to be expressed during the 72 hours of analysis. 8 of them (25%) were included in the list of 238 genes involved in the control of human T lymphocytes cell cycle (Sivozhelezov *et al.*, 2006a).

Among the 238 genes, we previously identified the 6 "leader genes" *i.e.*, those showing the highest number of interactions with other genes (Sivozhelezov *et al.*, 2006a) (Table 3.2). 3 of them (MYC, CDC2, CDK4) were present on the Starter Array and, as a confirmation of our prediction; they were all expressed during different experiments.

For instance, MYC is the gene with the highest number of interaction in our bioinformatic predictions. It is known as a very early gene in the proliferative response, since it regulates the entrance in G1 phase of the cell cycle (Oster *et al.*, 2002). It is interesting to notice that MYC is expressed, in smaller quantities, after 24 hours and 48 hours. Moreover, the absolute intensity of the corresponding spot on the array decreased during time, as expected from the entrance in the cell cycle and the progression along it.

The two other "leader genes" are CDK4 and CDC2. The former is known to regulate the progression in G1 phase, while the latter is involved in the G1/S and G2/M transition (Modiano *et al.*, 2000; Kawabe *et al.*, 2002; Baluchamy *et al.*, 2003).

Table 3.2. Leader genes in human T lymphocytes cell cycle (Reprinted with the permission from Sivozhelezov *et al.*, Gene expression in the cell cycle of human T lymphocytes: I. Predicted gene and protein networks, Journal of Cellular Biochemistry 97, pp. 1137–1150, © 2006a, Wiley-Liss, Inc., a subsidiary of John Wiley & Sons, Inc.).

Gene name	Weighted number of links	Gene description	Protein description	Apparent function in cell cycle
MYC	27.81	v-myc myelocytomatosis viral oncogene homolog; determines c-myc mRNA stability; v-myc avian myelocytomatosis viral oncogene homolog; v-myc myelocytomatosis viral oncogene homolog	Myc proto-oncogene protein (c-myc)	Entrance in G1 phase
CDK2	26.65	cyclin-depen. kinase 2; cdc2-related protein kinase; cell devision kinase 2; p33	Cell division protein kinase 2 (EC 2.7.1.-) (p33 protein kinase)	G1/S phase transition
CDC2	26.47	Cell division cycle 2, G1 to S and G2 to M; cell cycle controller CDC2; cell division control protein 2 homolog; cyclin-depen. kinase 1; p34 protein kinase	Cell division control protein 2 homolog (EC 2.7.1.-) (p34 protein kinase) (Cyclin-dependent kinase 1) (CDK1)	G2/M phase transition
CDK4	25.255	cyclin-depen. kinase 4; cell division kinase 4; melanoma cutaneous malignant, 3	Cell division protein kinase 4 (EC 2.7.1.37) (Cyclin-depen. kinase 4) (PSK-J3)	Progression in G1 phase
CDKN1A	25.08	cyclin-depen. kinase inhibitor 1A (p21, Cip1); CDK-inter. protein 1; DNA synthesis inhibitor; cyclin-depen. kinase inhibitor 1A; melanoma differentiation associated protein 6; wild-type p53-activated fragment 1	Cyclin-depen. kinase inhibitor 1 (p21) (CDK-interacting protein 1) (Melanoma differentiation associated protein 6) (MDA-6)	Inhibitor of cyclin-CDK2 or – CDK4 complexes
CDKN1B	23.90	cyclin-depen. kinase inhib. 1B (p27, Kip1); cyclin-depen. kinase inhibitor 1B	Cyclin-depen. kinase inhibitor 1B (Cyclin-depen. kinase inh. p27) (p27Kip1)	Inhibitor cyclin-CDK2 or - CDK4

Comfortingly CDK4 reaches its maximum expressed after 24 hour and it is lower after 48 hours. After 72 hours, the corresponding spot cannot be identified on the array. This can points that, after 24 hours, human T lymphocytes have entered the cell cycle (MYC and CDK4 expression) and are progressing along the G1 phase. CDC2 encodes for a member of the Ser/Thr protein kinase family. This protein is a catalytic subunit of the highly conserved protein kinase complex known as M-phase promoting factor (MPF), which is essential for G1/S and G2/M phase transitions of eukaryotic cell cycle (Kawabe et al., 2002; Baluchamy et al., 2003). Mitotic cyclins stably associate with this protein and function as regulatory subunits (Olashaw and Pledger, 2002). This gene regulates the progression from G1 to S and from G2 to M phase, and therefore represents an important signal of cell cycle progression (Modiano et al., 2000). Its expression varies during cell cycle, as shown by our results: it is indeed not expressed in resting cells and it starts to be expressed only after 24 hours. The expression decreases, so that cannot be identified, after 48 hours and then it is already detectable at the end of the experiments. These results and those derived by MYC and CDK4 expression are compatible with the expectastions: 24 hours after a mitogen stimulation with PHA resting T lymphocytes have started cell cycle (MYC), progressed along G1 phase (CDK4) and are preparing to replicate DNA (CDC2). Also, after 72 hours, T lymphocytes are about to enter mitosis (CDC2). The bioinformatic-based identification of leader genes as most important genes for each of the progression points in cell cycle is thus perfectly confirmed.

Our prediction identified also other genes whose importance was not as high as "leader genes", but slightly lower. In fact, we previously identified "leader genes" using clustering techniques (Sivozhelezov et al., 2006a). The "leader genes" were selected as the gene belonging to the cluster with the highest numbers of links (class A), but there were also other important genes in other high-ranked classes emerging from the clustering process (classes B, C and D, with decreasing importance). Of the 8 genes identified on the array and included in the list of Sivozhelezov et al. (2006a), 3 were of class A (leader genes), 2 of class B, 2 of class C and only 1 of class D. For instance, the gene encoding for proliferating cell nuclear antigen, PCNA, belongs to class B (weighted number of links = 17.59). The protein encoded by this gene is an auxiliary protein of DNA polymerase delta and appears to be requested for both DNA synthesis and DNA repair (Ohta *et al.*, 2002). This gene is present in low amount in resting normal human T lymphocytes and, upon mitogen stimulation, begins to increase in mid-G1 phase, approximately 12 to 15 hours before entry into S phase (Ohta *et al.*, 2002). PCNA continues to increase in amount throughout the cell cycle and remains high in proliferating cultures. This agrees with our experimental data shown in Figure 3.21. In fact, at the baseline we did not identify PCNA to be expressed on the array. After 24 hours, PCNA expression has increased and it reaches its maximum level after 72 hours. Interestingly, we were not able to identify its expression after 48 hours: this seems in contrast with the considerations reported above, but probably the missed identification may be due to experimental problems. This addressed our attention to have a deeper view of the behavior of this gene, using also bioinformatics resources.

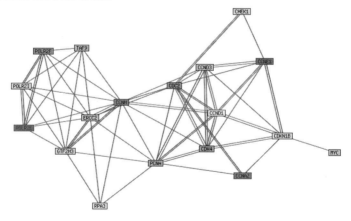

Figure 3.22. Final map of interactions among 8 high-ranking genes in cell cycle of human T lymphocytes and their neighbouring. (Reprinted with the permission from Giacomelli and Nicolini, Gene expression of human T lymphocytes cell cycle: Experimental and bioinformatic analysis, Journal of Cellular Biochemistry 99, pp. 1326–1333, © 2006, Wiley-Liss, Inc., a subsidiary of John Wiley & Sons, Inc.).

We used the online available software STRING (von Mering *et al.*, 2005) to formulate a detailed prediction of PCNA interactions with other identified genes, not considering text-based interactions. The results are shown in Figure 3.22, where PCNA is in the center of a complex map of interactions and is involved in many biochemical pathways. These interactions can also be identified from our experimental data. For instance, an interaction was predicted between PCNA and RFC2, a monomer of a heteropentameric protein complex consisting of the Rfc1, Rfc2, Rfc3, Rfc4, and Rfc5 subunits (Majka *et al.*, 2004; Majka and Burgers, 2004). This interaction is confirmed by our experimental data, which shows that RFC2 has a great level of expression after 72 hours. Also, other predicted interactions between PCNA and other identified genes confirm our experimental data, such as the co-expression with CCNA2 (Vendrell *et al.*, 2004). This protein was identified by us to be a class B gene for the control of cell cycle in human T lymphocytes (Sivozhelezov *et al.*, 2006a) (weighed number of links = 15.67). The protein encoded by this gene belongs to the highly conserved cycling family, whose members vary in protein abundance through the cell cycle and act as regulators of CDK kinases. This cycling binds and activates CDC2 and thus promotes G1/S and G2/M transitions along the cell cycle (Vendrell *et al.*, 2004). PCNA, CCNA2 and CDC2, which are genes of great importance in human T lymphocytes cell cycle, are thus bound by a very close link, which is fully confirmed in our experimental data.

Other high-ranked genes present on the Starter Array are CCNE1, CCNH (class C) and CHEK1 (class D). CCNE1, another cyclin, is the gene with the highest absolute intensity value in the 24 arrays (Sutherland and Musgrove, 2004). It is also expressed after 48 hours and 72 hours, but shows a much lower intensity. A STRING-based bioinformatics prediction suggests it is linked directly with CDC2 (Figure 3.22). Indeed, its expression reaches the maximum level in correspondence to one of the process regulated by CDC2, the G1/S transition after 24 hours. CCNH, also, is highly expressed 24 hours after the stimulation. In fact, the protein encoded by this gene is known to phosphorylate CDC2, thus contributing to the G1/S switching (Karan *et al.*, 2002). The last considered gene, CHEK1, is an essential kinase required to preserve genome stability. Very recent findings (Syljuasen *et al.*, 2005) proposed that CHEK1 is required during normal S phase to avoid aberrantly increased initiation of DNA replication, thereby protecting against DNA breakage. Our experiments show that CHEK1 expression can be identified 24 hours and 72 hours after the stimulation with PHA, in correspondence to DNA synthesis and mitosis. This can strongly confirm the importance of this gene on DNA repair and preservation mechanisms. Then, we calculated a final map of interactions among these 8 high-ranking genes in cell cycle of human T lymphocytes, which is shown in, representing also their neighboring genes. The other neighboring genes present in the map are not displayed on the Starter Array. Interestingly, one of them is CDKN1B, which we identified to be a "leader gene" in the control of human T lymphocytes cell cycle (Sivozhelezov *et al.*, 2006a; Chang *et al.*, 2004).

The Starter Array displays only 161 genes: therefore data can be easily analyzed. The use of more complex arrays, displaying a huge amount of genes (up to 10 thousands) often leads to a difficult and sometimes misleading analysis, due to the complexity of data. Human Starter allows a simpler analysis. Anyway, genes to be displayed must be chosen with a particular care. Presently we have in progress the construction of DNA chips based on genes identified by bioinformatics, namely the "leaders" of one particular process and "orphan" needing a more detailed attention (Sivozhelezov *et al.*, 2006a). In this way, it becomes possible to create *ad hoc* arrays, which can guarantee the best results in analyzing a particular cellular system. The here reported "time series" experimentation on human T lymphocytes stimulated with PHA confirm this hypothesis, but are limited by the fact that many interesting genes involved in the human T lymphocytes cell cycle, including 3 leader genes, are not displayed on the array.

In conclusion, the data on gene expression collected 24, 48, and 72 hours after the mitogen stimulus when compared with our bioinformatics prediction (Sivozhelezov *et al.*, 2006a) confirmed the theoretical prediction on leader genes. In particular, 24 hours after the stimulation, resting T lymphocytes have started cell cycle (MYC), progressed along G1 phase (CDK4) and are preparing to replicate DNA (CDC2); after 72 hours, T lymphocytes are about to enter mitosis (CDC2). Moreover, we got a deeper picture of gene expression considering 5 other genes, whose importance in human T lymphocytes cell cycle was theoretically proven, even if they were not included in the leader gene class. Our experimental data show a perfect agreement with theoretical ones, therefore further validating the leader gene approach, and also point out the importance of the interaction between PCNA, CDC2 and CCNA2 in controlling cell cycle and of CHEK1 in regulating DNA repair and preservation. At the same time the need of *ad hoc* array was once again confirmed.

The application of the leader gene approach, starting from the identification of involved genes and their subsequent ranking according to the number of interactions, should be extended to other cellular processes, which are known to a lesser extent if compared with human T lymphocytes cell cycle. Leader genes are defined as the genes with the highest number of interactions among those involved in a particular cellular process. By identifying leader genes of a given process, it becomes possible to design targeted microarray, whose analysis would allow to describe complex biomolecular pathways thorough the activity of a few, but highly important genes, which represent the real center of interactions maps. In this way, an easiest and more rationale approach to molecular genomics can shed new lights on complex cellular mechanisms.

3.3.2 *Organ transplants*

A French group (Jean-Paul Soulillou, Sophie Brouard *et al.*, at Inserm, Nantes) is systematically studying tolerance of kidney graft. In particular, they have examined patients tolerating a kidney graft without any treatment and patients with chronic rejection.

They also performed a microarray analysis and identified a list of genes able to classify the two different classes. While the French group data concern 35000 genes of a pangenomic array (Cantrell, 2002), the bioinformatic analysis being carried by a joint cooperation between INSERM and Genova University was conducted with two parallel approaches (Braud *et al.*, 2008, Sivozhelezov *et al.*, 2008) which run independently and, at the end, compared each other: *ab initio* analysis and experimental analysis.

The former concerns the identification of genes involved in kidney graft tolerance (Figure 3.23) and of their leader genes; the latter concerns microarray data and clinical data reduction. In the former our completely *ab initio* set of genes involved in kidney transplant tolerance will be indeed used (Cantrell, 2002; Abraham *et al.*, 1980).

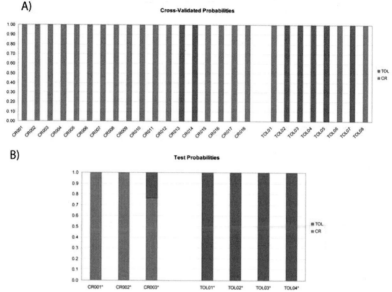

Figure 3.23. Classification probabilities of individual patients by Predictive Analysis of Microarray (PAM) based on the 343 differentially expressed genes between operationally tolerant kidney graft recipients (TOL) and patients with chronic rejection (CR). Each patient sample is shown by a bar, as labelled in the X-axis. The colour codes indicate the probability (0-1, as indicated in Y-axis) that the sample belongs to TOL (red) or CR (green). (A) Cross-validated probabilities on the 8 TOL and 18 CR that were used to set up the 2-class (TOL/CR) classification algorithm. Among the 26 patients, 5 samples (CR013, CR014, TOL01, TOL06 and TOL08) were misclassified. (B) Using the PAM algorithm defined with 8 TOL and 18 CR patients, 7 serially harvested samples (4 TOL and 3 CR) at a time interval of more than 1 year after the first sample were classified. The algorithm correctly classified all samples, with a probability of 100% for TOL and 92.0% for CR. (Reprinted with the permission from Braud *et al.*, Immunosuppressive drug-free operational immune tolerance in human kidney transplants recipients: I. Blood gene expression statistical analysis, Journal of Cellular Biochemistry, in press, © 2008, Wiley-Liss, Inc., a subsidiary of John Wiley & Sons, Inc.).

This effort is representing a big challenge to our search for automatic identification of "leader genes" previously described. In particular, it is important to notice that there is no obvious necessary direct correlation between leader genes we will identify by *ab initio* research and the quantitative changes in expression monitored by experimental analysis.

In a second paper on the nanogenomics of kidney transplants (Sivozhelezov *et al.*, 2008) we outline a microarray-based identification of key leader genes associated respectively to rejection and to operational tolerance of the kidney transplant in humans (Figure 3.24) by utilizing a non/statistical bioinformatic approach based on the identification of "key genes", either as ones mostly changing their expression, or having the strongest interconnections (Figure 3.25).

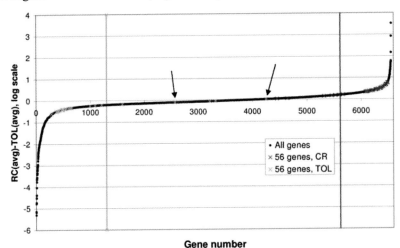

Figure 3.24. Genes from the "fullchip" plotted according to their tolerance or rejection propensity *i.e.* difference in expression in log scale between RC and TOL genes, with the 56 SAM-identified genes (Braud *et al.*, 2007) marked. Arrows indicate genes included in the 56 gene dataset but possibly unable to discriminate rejection/tolerance. Lines indicate thresholds used in selecting genes for "pro-tolerance" and "pro-rejection" leader gene calculations (Reprinted with the permission from Sivozhelezov *et al.*, Immunosuppressive drug-free operational immune tolerance in human kidney transplants recipients. II. Nonstatistical gene microarray analysis, Journal of Cellular Biochemistry, in press, © 2008, Wiley-Liss, Inc., a subsidiary of John Wiley & Sons, Inc.).

A uniquely informative picture emerges on the genes controlling the human transplant from the detailed comparison of these findings with the traditional statistical SAM analysis of the microarrays and with the clinical study carried out in the accompanying paper (Braud *et al.*, 2008).

The overall conclusion is that there are many genes in common in the highest interaction genes derived from individual fullchip and the 343 SAM-gene list. Notably, convergence between the SAM approach and our non-statistical approach becomes much better for the individual dataset, in which the CR/TOL fold changes are much lower, compared to the old dataset.

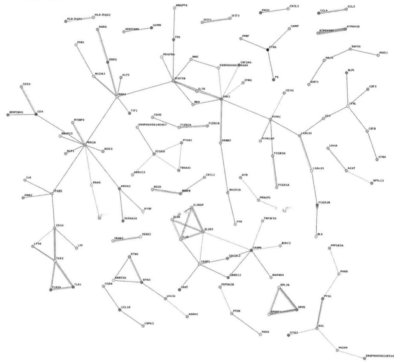

Figure 3.25. Complete interaction map "no textmining" interaction map calculated for the "Class 1" reliable genes and filtered by (CR-TOL) amplitude, as obtained from the new fullchip microarray datasets (see Table 2 for their names and ranking according to their number of interactions) (Reprinted with the permission from Sivozhelezov et al., Immunosuppressive drug-free operational immune tolerance in human kidney transplants recipients. II. Nonstatistical gene microarray analysis, Journal of Cellular Biochemistry, in press, © 2008, Wiley-Liss, Inc., a subsidiary of John Wiley & Sons, Inc.).

The primary reason are 1) we use expression levels in normal scale whereas SAM uses expression levels normalized by their errors (thus practically reducing the signal to noise level), and 2) SAM uses permutation (random shuffling) of the data, and then extracts significant genes by comparing the permuted and non-permuted set. The SAM-derived genes typically have shown small differences of expression levels explained by the fact that SAM package operates with "relative difference" d which is the actual difference D divided by a sum of its standard deviation s with an arbitrary constant s_0, $d=D/(s+s_0)$, which should make equally significant the small but highly reproducible change and the large but poorly reproducible change in gene expression. Since physical grounds of such an approach are unclear, we separated the two parameters, *i.e.*, the magnitude and the reliability, but using two independent filters, one based on percentages of valid samples, and the other on amplitude threshold. In this respect, our approach is less arbitrary because we calculate them using objective clustering and the actual experimental fluorescence distribution in the microarray. When proper microarray reliability and proper expression threshold are applied, we reached the conclusion that it was necessary to acquire and analyze a new, individual dataset, which proves quite adequate to the task. Poor compatibility between our approach and the SAM approach in the pool dataset is apparently caused by the essential difference between the two approaches in that our two filtering parameters are addressing the reliability and the amplitude of the expression levels independently. Indeed, two thresholds are present: one by amplitude, the other by reliability. Instead, both the SAM denominator parameter and the SAM significance threshold are related to reliability and amplitude in a complicated manner. Similarly to our approach, SAM has two adjustable parameters, namely the above-described arbitrary constant in the denominator $d=D/(s+s_0)$ for relative difference, and the significance threshold. In this respect, our approach is less arbitrary because we calculate them using objective clustering and the actual experimental fluorescence distribution in the microarray. When proper microarray reliability and proper expression thresholds are applied, the compatibility between the two approaches is very good (Sivozhelezov *et al.*, 2008). Furthermore the final leader genes map shed new light in the molecular mechanisms controlling human kidney transplant. Microarray experimentation becomes indeed much more targeted and significant, by comparing gene expression analysis with the analysis of gene networks and interactions. In this context, we successfully applied different variants of the leader gene identification algorithm, in order to identify the ones best representing real gene networks.

All the findings described here, regarding kidney transplant tolerance but also possibly being extended to other systems, confirm the existence of a small set of genes, having a higher number of interactions among all the genes involved in the cellular process and therefore playing a central role. The identification of most interacting genes can be of great importance in the systematization and analysis of data, since leader gene, considering also those largely changing expression in different patients, form a unique network: the mere changing in expression of a particular gene is not significant by itself, but only if it is put in a proper framework. This change can be often considered as a consequence of a more complex network of events, starting from leader genes, identified with bioinformatics predictions, which often do not vary their expression so much to be identified as significant using pangenomic arrays. However, microarray technology is a necessary confirmation of every prediction made by theoretical network analysis. On the other hand, statistically-processed microarray data can serve as the starting point for network analysis. We introduce a non-statistical approach to processing microarray data, in which we apply K-means clustering to microarray data only after independent filtering by both amplitude and reliability that we define as percentage of valid data for each gene. The need for non-statistical treatment, in addition to statistical treatment of microarray data, has been recognized time ago (Affymetrix Inc., 2004) because "microarrays are the unusual statistical case where the number of tests greatly exceeds the number of samples, so standard statistical methods for multiple comparisons are pushed to their limit". To our knowledge this is the first step in that direction. Results of the non-statistical approach of microarray data interpretation are widely different from the statistical (SAM) approach for the pool dataset, but are similar for the individual dataset. At the moment, none of the three approaches (Sivozhelezov *et al.*, 2008), namely the "*ab initio*" approach, the microarray-based statistical approach and the microarray-based non-statistical approach, has proved superior in identifying the key genes responsible for kidney graft rejection and/or tolerance, and showing that those approaches must be used in a complementary manner, considering also that reasons for divergence of those approaches have been identified. Moreover, Sivozhelezov *et al.* (2008) showing average number of pro/tolerance and pro/rejection genes respectively with TOL and CR patients for the three different microarray-based estimates provides a basis for combined sets of genes to be used in such forthcoming studies. Besides, identification of a pathway possibly important in controlling mechanisms of tolerance and rejection has demonstrated a high potential for combination of approaches used herein. Genomics do however suffer many pitfalls (Nicolini *et al.*, 2006a) and only functional proteomics (LaBaer, 2006) represents the long-range answer to the basic molecular understanding and to the clinical control of the human kidney transplants.

3.3.3 *Osteogenesis*

We tentatively identified 7 leader genes in the osteogenic process (Covani *et al.*, 2008). They were classified according to their involvement in osteogenesis subprocesses (cell adhesion and proliferation, ossification, skeletal development, Calcium ion binding). On this base we are presently constructing a simple array ($2 \times 4 \times 4$ genes, for a total of 16 genes repeated 2 times) using APA technology and displaying the leader genes and an equivalent number of controls, in order to begin to study the osteogenetic process at molecular level.

In summary, putative "leader genes" undergo some changing in gene expression, but the amount of this changing is not necessarily related to their leader gene status. In fact, leader genes are the most-interacting genes involved in a process, not necessarily the most varying ones as apparent also in human T lymphocytes cell cycle where the most changing gene is CCNH, which is not one of the leader genes. *Ab initio* "leader plus class B" genes were scanned against the list of experimental genes changing their expression levels (quite more numerous than the leader genes identified *ab initio*). These very preliminary results show that text-mining approach is dangerous to use for interpreting microarray expression data. Indeed "theoretical" and "experimental" leader gene sets are clearly different and for this reason we are attempting to check links between the two sets of leaders: links between theoretical and experimental leader genes using text-mining, and the same without text-mining. The final results are still in progress and very preliminarily leader genes identified with text-mining appear to have almost no interactions shown; the validity of this approach in relating bioinformatics predictions and experimental data is therefore much lower than the "no text-mining" approach. Leader genes identified with a completely independent bioinformatics prediction appear closely interacting with genes changing expression in experimental analysis. An important change in expression of these leader genes between the two conditions is not necessary for their function. This was also proved in our previous microarray study on T lymphocytes cell cycle: MYC was overall the most important gene in the whole process (it is necessary to enter the G1 phase, indeed), but it changes its expression very slightly from quiescent to replicating cells. Many questions have still to be answered before we reach a conclusion on this open question, but in any cell system the changing in expression of a particular gene could be considered as the ultimate consequence of a complex network of biochemical interactions, whose most important nodes could be the "leader genes" as identified with *ab initio* prediction. Several questions emerge: which should be considered the true leader gene set, the theoretical only or theoretical + experimental ? The latter answer proves correct (Nicolini *et al.*, 2006a; Sivozhelevov *et al.*, 2006a; 2008; Giacomelli *et al.*, 2006) suggesting that the mere changing in expression of a particular gene is not meaningful by itself, but only if it is put in a proper framework. This change can be often considered as a consequence of a more complex network of events, starting from leader genes, identified with bioinformatic predictions, which often do not vary their expression until now identified as significant using pangenomic arrays. The work in progress in osteogenesis suggest that the leader gene approach need to be validated by experimental analysis using DNA microarrays and by independent clinical data.

3.4 Nanoproteomics

Two major lines of nanoproteomics, namely NAPPA Microarray and Mass Spectrometry, gave independent new results aiding in the understanding of cell cycle progression.

3.4.1 Cell cycle

Regarding the first variation identified by Mass Spectrometry we analysed the averaged ESI mass spectrum (300–2000 m/z range) recorded in the elution range 42.25–42.68 min, shown in the 2nd panel of Figure 3.26). From the Gaussian deconvolution of this spectrum, performed by MagTran1.0, we identified three protein species. We performed first tentative protein identification on the basis of these weights through a search in Swiss-Prot data bank (Boeckmann et al., 2003).

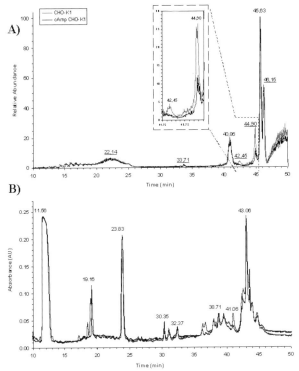

Figure 3.26. (A) 1st panel (from the top), HPLC-ESI-MS TIC profile (elution range 41.5-45.5 min). 2nd panel, averaged ESI mass spectrum (300–2000 m/z range) of CHO-K1 nuclear protein fraction recorded in the elution range 42.25–42.68 min. (B) deconvoluted spectrum of averaged ESI mass spectrum reported in the second panel. (Reprinted with the permission from Spera and Nicolini, cAMP induced alterations of chinese hamster ovary cells monitored by mass spectrometry, Journal of Cellular Biochemistry 102, pp. 473–482, © 2007, Wiley-Liss, Inc., a subsidiary of John Wiley & Sons, Inc.).

We restricted our search to rodent proteins. In Table 3.3, we report the experimental and theoretical weights and the function of the identify proteins. Among these proteins, Guanine nucleotide-binding protein (Gna11) and Myosin heavy chain 10, non-muscle (Myh 10), are particularly interesting because they are both involved in cellular regulation. Heterotrimeric guanine nucleotide-binding proteins (G proteins) are integral to the signal transduction pathways that mediate the response of the cell to many hormones, neuromodulators, and a variety of other ligands (Strathmann and Simon, 1990).

Table 3.3: Proteins detected in the CHO-K1 protein nuclear fraction correspondently to the 42.54 min TIC peak (Reprinted with the permission from Spera and Nicolini, cAMP induced alterations of chinese hamster ovary cells monitored by mass spectrometry, Journal of Cellular Biochemistry 102, pp. 473–482, © 2007, Wiley-Liss, Inc., a subsidiary of John Wiley & Sons, Inc.).

Exp. mass (Da)	Th. mass (Da)	Protein	Function
31133	31133	**Myh10** (**fragment B**) Myosin heavy chain -B (Fragment), non-muscle	Cellular myosin appears to play a role in cytokinesis, cell shape, and specialized functions such as secretion and capping.
42024	42024	**Gna11** Guanine nucleotide-binding protein, alpha-11 subunit	Guanine nucleotide-binding proteins (G proteins) are involved as modulators or transducers in various transmembrane signalling systems. Acts as an activator of phospholipase C.
52136	52134	**S61A1** Protein transport protein Sec61 alpha subunit isoform 1	Plays a crucial role in the insertion of secretory and membrane polypeptides into the ER. Required for assembly of membrane and secretory proteins. Tightly associated with membrane-bound ribosomes, either directly or through adaptor proteins.

Nonmuscle myosins play a role in diverse cellular functions, for instance cytokinesis, proliferation, secretion, and receptor capping. Two isoforms of the nonmuscle myosin heavy chain (nmMHC), chain A (nmMHC-A) and chain B (nmMHC-B) have been identified. The nmMHC-B in particular is involved in cell growth regulation and transformation (Strathmann and Simon, 1990).

Regarding the 44.90 min TIC peak variation -relative abundance passed from 26.6 for the untreated cells plot to 12.3 for the cAMP treated cells plot (Figure 3.27a, *1st panel)*- we analyzed the averaged ESI mass spectra recorded in the elution range 44.40–45.20 min (Figure 3.25a, *2nd panel)*. The two spectra are clearly identical: the difference of intensity in the TIC peaks can be explained by a different amount of the same proteins in the two samples. We performed a Gaussian deconvolution of this spectrum and identified three protein species (Figure 3.27b).

Through a search in Swiss-Prot data bank (Boeckmann *et al.*, 2003), we have been able to identify only the 29628 Da peak as CD82_MOUSE protein. This protein, associates with CD4 or CD8 glycoprotein delivers co-stimulatory signals for the TCR/CD3 pathway (Itoh and Adelstein, 1995) involved in apoptosis regulation. For 6632 Da peak and 20739 Da peak, it was not possible to identify any protein. Moreover we analyzed seventeen other proteins of the nuclear envelope present in the same content in the cells before and after cAMP exposure. The relative molecular weight are 2.4 kDa, 11.9 kDa, 17.3 kDa, 29.55 Da, 35.6 kDa, 41.2 kDa, 46.7 kDa, 52.3 kDa, 52.9 kDa, 58.9 kDa, 64.2 kDa, 70.7 kDa, 60.9 kDa, 82.5 kDa, 93 kDa, 96.2 kDa and 99.3 kDa. It has been possible to identify none of these proteins only from their molecular weight.

We subsequently performed a series of experiments, coupling HPLC (Table 2 in Spera and Nicolini, 2007) and MS (Nagira *et al.*, 1994), to identify all these proteins and to confirm the identity of the proteins via mass fingerprinting (that allows to identify the protein with a very high probability).

This second step was conducted on a MALDI-TOF Mass Spectrometer. For protein fingerprint, the HPLC fractions were digested and the tryptic digest samples were analyzed by MALDI-TOF MS. Until now we have obtained from this analysis, the confirm of identification of 31133 kDa protein as "Myosin heavy chain -B (Fragment), non-muscle", present only in the CHO-K1 sample (HPLC fraction 3, Table 2 in Spera and Nicolini 2007).

The study here reported on the effect of cAMP on the protein expression of the CHO-K1 cells is continuing structural and functional work started many years ago at the level of nuclei and genes (Nicolini and Beltrame, 1982; Vergani *et al.*, 1992, 2001). Our RP-HPLC-ESI MS results, confirmed by HPLC measures, show a different protein content in the nuclear protein fractions of the cells after exposure to cAMP, possibly linked to the above early observations.

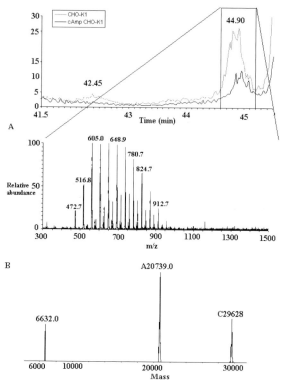

Figure 3.27. (A)1st panel (from the top), HPLC-ESI-MS TIC profile (elution range 41.5–45.5 min). 2nd panel, enlargement of averaged ESI mass spectra (300–2000 m/z range) of CHO-K1 (grey line) and cAMP CHO-K1 (black line) nuclear protein fraction recorded in the elution range 44.4–45.20 min. The spectra are identical. (B) deconvoluted spectrum of averaged ESI mass spectra reported in the second panel. (Reprinted with the permission from Spera and Nicolini, cAMP induced alterations of chinese hamster ovary cells monitored by mass spectrometry, Journal of Cellular Biochemistry 102, pp. 473–482, © 2007, Wiley-Liss, Inc., a subsidiary of John Wiley & Sons, Inc.).

In particular we focused our attention on the two main differences in the TIC plots of the nuclear protein fractions of untreated and treated cells that show a decrease in the nuclear protein amount after cAMP treatment Through ESI MS analysis we identified a group of three proteins, Myh10, Gna11 and S61A1, present only in the nuclear protein fraction of untreated cells and involved in cellular regulation processes. The Myh10 protein is involved in the catalysis of movement along a polymeric molecule such as a microfilament or microtubule, coupled to the hydrolysis of adenosine 5'-triposphate (ATP), while the Gna11 protein regulates the cascade of processes by which a signal interacts with a receptor, causing a change in the level or activity of a second messenger.

Mascot Search Results - Protein View

Match to: Q62707_RAT Score: 88 Expect: 0.0048
Nonmuscle myosin heavy chain-B (Fragment).

Nominal mass (M_r): **31113**; Calculated pI value: **7.22**
Cleavage by Trypsin: cuts C-term side of KR unless next residue is P
Number of mass values searched: **12**
Number of mass values matched: **9**
Sequence Coverage: **30%**

Start - End	Observed	Mr(expt)	Mr(calc)	Delta	Miss	Sequence
1 - 11	1307.6680	1306.6607	1306.6629	-0.0022	1	KDHNIPGELERQ
12 - 27	1744.0110	1743.0037	1742.9202	0.0835	0	RQLLQADPILESFGNAKT
31 - 39	1024.5100	1023.5027	1023.4733	0.0294	1	KNDNSSRFGKF
66 - 71	672.4190	671.4117	671.4078	0.0039	1	RAVRQAK.D
104 - 118	1676.8150	1675.8077	1675.8569	-0.0492	0	RFLSNGYIPIPGQQDKD
196 - 200	572.4010	571.3937	571.3805	0.0132	1	RIKVGRD
201 - 205	652.3280	651.3207	651.3228	0.0020.	0	RDYVQKA
210 - 221	1291.6600	1290.6527	1290.6455	0.0072	0	KEQADFAVEALAKA
222 - 226	639.3110	638.3037	638.3024	0.0014	0	KATYERL

No match to: 715.4450. 994.2610, 1605.3320

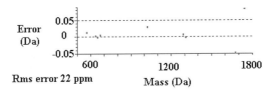

Figure 3.28. Mascot results relative to mass fingerprint of HPLC (Reprinted with the permission from Spera and Nicolini, cAMP induced alterations of chinese hamster ovary cells monitored by mass spectrometry, Journal of Cellular Biochemistry 102, pp. 473–482, © 2007, Wiley-Liss, Inc., a subsidiary of John Wiley & Sons, Inc.).

Moreover we identified another group of three proteins present in both samples but in a double concentration in the nuclear protein fraction of untreated cells. Through ESI MS analysis we were able to identify only one of these three proteins, Cd82Mouse, that results involved in apoptosis regulation processes (Figure 3.27).Other seventeen proteins present in equal amount before and after cAMP exposure have been identified.

From preliminary HPLC and MS experiments we had a confirmation of the inhibition of the nuclear protein expression after exposure to cAMP; moreover by protein fingerprinting we confirmed the identification of 31133 kDa protein as Myh10 (Figure 3.28).

The same analysis on the other protein fractions (cytosolic, membrane and membrane organelle and cytoskeleton fraction) is in progress to analyze the entire proteome in presence of, to obtain a comprehensive understanding of the reverse transformation (Spera and Nicolini, 2007; Spera et al., 2007).

3.4.2 Cell transformation and differentiation

The focus of function-based microarrays is to study the biochemical properties and activities of the target proteins printed on the array.

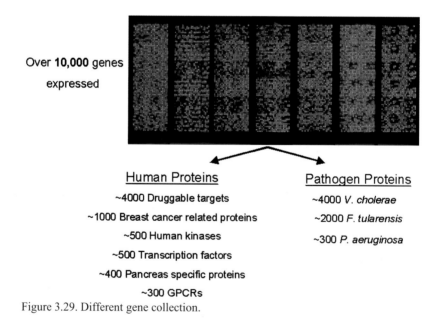

Figure 3.29. Different gene collection.

Function-based microarrays can be used to examine protein interactions with other proteins, nucleic acids, lipids, small molecules and other biomolecules (LaBaer and Ramachandran, 2005; Ramachandran *et al.*, 2004).

In addition, function-based microarrays can be used to examine enzyme activity and substrate specificity associated to cell transformation and to cell differentiation. These microarrays are produced by printing the proteins of interest on the array using methods designed to maintain the integrity and activity of the protein, allowing hundreds to thousands of target proteins to be simultaneously screened for function using a wide range of gene collections (Figure 3.29).

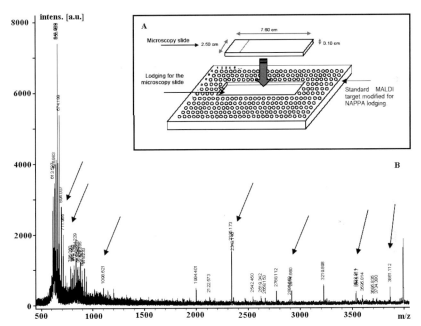

Figure 3.30 (A) MALDI-TOF target modification project. (B) MALDI TOF MS spectrum of Human Kinase NAPPA array acquired after protein synthesis (spot gene NA 7-A 12), low masses region. The arrows indicate the peaks identified also in the Dr. Fuentes and LaBaer spectrum (namely m/z =711 ± 12, m/z =865 ± 18, m/z =1090 ± 20, m/z =2900 ± 40, m/z =3540 ± 40, m/z =3860 ± 60) (Reprinted with the permission from Spera and Nicolini, Nappa microarray and mass spectrometry: new trends and challenges Essential in Nanoscience Booklet Series, © 2008, Taylor & Francis Group/CRC Press, http://nanoscienceworks.org).

The list of potential applications of such microarrays is large. A microarray of a particular class of enzymes such as kinases could be screened with a candidate inhibitor to examine binding selectivity. A candidate drug could be used to probe a broad range of enzymes to look for unintended binding targets that might suggest possible toxicities. Proteins expressed by pathogenic organisms can be screened with serum from convalescent patients to identify immunodominant antigens, leading to good vaccine candidates.

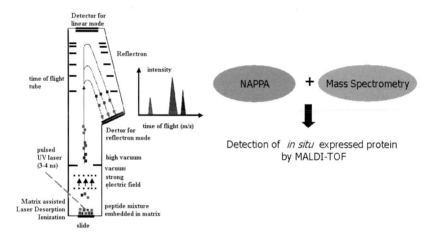

Figure 3.31 *In situ* protein detection (Reprinted with the permission from Prof. Joshua LaBaer at Harvard Institute of Proteomics).

Protein interaction networks, including the assembly of multiprotein complexes, can shed light on biochemical pathways and networks in the control of cell transformation and cell differentiation. Eventually, it may even be possible to use with proper inert surface (Figure 3.30) these high-density microarrays as a MALDI source for mass spectrometry, allowing users to probe complex samples for binding partners to many proteins simultaneously.

However, as with the abundance-based microarrays, there still remain challenges in building and using function based protein microarrays. First, the notorious lability of proteins raises concerns about their stability and integrity on the microarray surface. Second, it is time consuming and costly to produce proteins of good purity and yield, and many proteins cannot be purified at all. Finally, the methods used to attach proteins to the array surface may affect the behaviour of the proteins. Despite these challenges, there has been some success in building and using function-based protein microarrays in medicine (LaBaer and Ramachandran, 2005; Ramachandran *et al.*, 2004). In this Volume we have described different approaches, protein spotting microarrays and self-assembling microarrays, including their recent most promising advances utilizing Label Free technologies (Figure 3.31).

3.5 Nanomechanics and Nanooptics

Several new sectors have been emerging in the area of nanomechanics and nanooptics, which are summarized in this paragraph as an example of their very promising nature in terms of our basic understandings of matter and of future challenging new technological applications in health and science.

3.5.1 *Nanocontacts for addressing single-molecules*

Of particular interest is how sensitivity, selectivity, and switching may be improved by directly contacting single molecules (for a review see Carrara *et al.*, 2005). To achieve this goal the fabrication of electrodes separated by nanometric sizes gaps that may be bridged by single-molecules is necessary.

Two approaches to fabricate nanometric-sized contacts were recently proposed in literature. The first involves the movement of electrodes pairs. The second relates to the formation of electrodes pairs via etching. In the field of biosensors they first appeared in the late 1990s, when papers concerned with the organization of monomolecular layers of sensing molecules (Bykov, 1996) that either provide stable networks to benefit electrochemical detection of bound molecules at nanostructured electrodes (Tiefenauer *et al.*, 1997) or permit single molecules to be addressed using devices such as Scanning Probe Microscopes (Göpel, 1998) were published. Recently, system based on cantilever micromechanical technology were proposed (Raiteri et al., 2001) but more complex nanostructured systems have been employed in biosensors (Vo-Dinh et al., 2001): for example, carbon nanotubes and fullerenes have been used to stabilize electrochemical mediators and enzymes (Sotiropoulou et al., 2003) and gold nanotube structures have been shown to enhance sensitivity to glucose (Delvaux and Demoustier-Champagne, 2003). A future goal of nanotechnology in the field of biosensors must be the development of simple nanoscale devices for addressing the single molecule. Till now in the field of biosensors such devices have been built on the micro-scale (Gorschlüter et al., 2002). It has been demonstrated that it is possible to study DNA molecules that bridge interdigitated electrodes (Hoölzel et al., 2003). DNA strands, greater than 5000 base pairs in length, have been investigated by placing them across a two micrometer gap (see Figure 1(h) in Hoölzel et al., 2003).

Similarly, sensors based on the change in magnetoresistance when two micrometer magnetospheres coated with streptavidin are adsorbed in a six micrometer biotin coated gap have been demonstrated (Graham et al., 2003). The feasibility of addressing single short chain molecules has been established in the emerging area of the molecular scale electronics. Conductivity measurements on single benzene-1,4-ditiol molecules (Reed et al., 1997) have been performed. It has also been demonstrated that single molecule conductivity measurements require chemical binding of the molecule of interest to both contacts (Cui et al., 2001). To achieve such a configuration requires that facile methods of placing electrodes at nanometer scale separation be developed. Very recently, across the field of nanoscience different methodologies of obtaining such nanocontacts have been proposed.

Literature presents two completely different approaches to the fabrication of nanocontacts. The first is to mechanically align a pair of electrodes at nanometer separation. These methods require special drivers to accurately position the electrodes. The second uses nanometric level etching of pre-structured conductive materials. The success of this second approach is dependent on to what degree the etching processes may be controlled. In the following part of this paper we review these methods.

In this section we review methods related to the movements of electrode pairs in order to obtain nanocontacts. The general idea of all methods presented here is to bring together two electrodes and place them at nanometer scale separation. García and co-workers (2002) developed a method in which the tips of two wires are brought in to contact. In their work, which is primarily concerned with magnetoconductivity, they describe how nanoscale contacts may be formed between magnetic and non-magnetic materials. The metals studied include nickel, copper, aluminum, gold and iron. The nanocontacts were prepared by first placing two rounded wire tips in the apertures of a teflon tube (Figure 3.32).

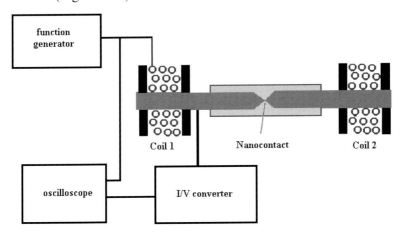

Figure 3.32. Nanocontacts were prepared taking a couple of wires tightly bound within a Teflon tube. A bias voltage was applied to the wires. The wires were approached till a current flow is established between them. The resulting contact shows quantum conductivity.

A bias voltage was then applied to the wires and the wires approached till a current flow was established. In such a manner, quantum magneto conductivity was observed between the nanocontacts at a field of tens of Oe. The as-produced nanocontacts showed quantum conductivity and the conductivity remained stable even when fields of more than one hundred Oe were applied (Figure 3b in García et al. 1999). However, the mechanism by which the force was applied to the wires in order to bring them into contact is not detailed. Further, the fact that the nanocontacts obtained are housed within a Teflon tube suggests that the fabrication technique cannot easily be employed to fabricate biosensors.

Two $(La_{0.7}Sr_{0.3})MnO_3$ single crystals have been placed at nanometer separation (Versluijs *et al.*, 2000). The quantum conductance across the resultant nanoscale gap has been recorded. Initially the crystals were placed in mechanical contact. They were then pulled apart until quantum conductivity was established. The authors have employed both a mechanical relay and a piezo-device to separate the crystals. Quantum conductance has been realized with both a relay and a piezo-mover. In the case of piezo-driven separation 20 steps in the conductance versus time curve were observed. However, even using piezo drivers the maximum time for which the nanocontact remained stable was of the order of seconds, precluding the use of this methodology of forming nanocontacts in biosensor application. It is noteworthy that ceramic crystals are better suited than metals to nanocontact formation by this technique as the former undergo brittle fracturing whilst the latter display plastic deformation. However, temporal nanocontacts have been formed between vibrating macroscopic wires (Costa-Krämer *et al.*, 1997). Using a piezo-driver two wires are brought into contact and then set in vibration. As the wires vibrate a nanocontact is momentarily formed. Again, the temporary nature of the nanocontact means this method of production is of limited use in biosensor applications.

A piezo-driver offers the possibility of controlling on the nanoscale the approach of a wire to a planar macroelectrode. It has been shown (Facci *et al.*, 1996) that using the tunneling current flowing through the contact as a feed-back signal it is possible to bring a wire tip to a few tens of nm from a flat electrode. This is achieved by stopping piezoelectric movement when the tunneling current reaches a pre-defined value. Using this technique the distance between electrode and wire has been controlled to within a nanometer (Carrara *et al.*, 2006). This method was used for addressing single cadmium sulfide (Facci et al, 1996) and lead sulfide (Erokhin *et al.*, 1997) nanoparticles. The resultant metal-insulator-nanoparticle-metal configurations displayed behavior characteristic of single-electron junctions. It is noted, however, that in the above configuration it is not possible to realize two chemically bonded contacts, *i.e.*, the criteria required in order to obtain reproducible measurements of single molecules conductivity are not met (Cui *et al.*, 2001).

Alternative approaches to the formation of stable gaps between conducting materials that are of small enough dimension that they may be bridged by biological molecules to form nanocontacts are based on etching. That is, starting from a single piece of conducting material a nanoscale gap is cut by controlled dissolution of material. The major constraint to the wide spread use of this approach is the problem of controlling etching on the nanoscale. In this section we review methods of fabricating nanocontats in preformed structures. Both physical and chemical etching is considered.

A chemical method to form nanocontacts is via a combination of controlled electrochemical etching and electrochemical deposition. The method has been termed "ELENA" (ELEctrochemical NAnodeposition) (Céspedes *et al.*, 2002). The technique has been used to fabricate metallic electrodes with a separation on the nanometer scale. Morpurgo and co-workers (1999) first developed this two-step process. Using conventional lithography two electrodes at micrometer separation were prepared, the electrode pair. At this point the separation of the electrodes was not a critical parameter. Then, by applying a DC potential relative to a counter electrode, metal is electrodeposited on to the electrode pair. During the deposition an AC potential is applied across the electrode pair allowing the resistance between the electrodes to be measured. As the electrodes grow towards each other the resistance changes as follows; at large separation the high resistance of the electrolyte is monitored, at nm separation a tunnelling current is observed and on contact a step decrease in resistance is observed. The tunnelling region is characterized by an exponential relationship between electrode separation and current. At slow deposition rates, steps may be observed in the resistance versus time plots at the point when the two electrodes first make contact. The step features are due to atom-by-atom growth of the contact. A very important feature of this method is the reversibility of the process. Hence it is possible to deposit metal until the two electrodes are in contact and then electro-dissolve the metal and reopen the gap. SEM images of contacts prepared using this methodology show in Figure 3.33 nanosized gaps in the range 20 nm to 100 nm.

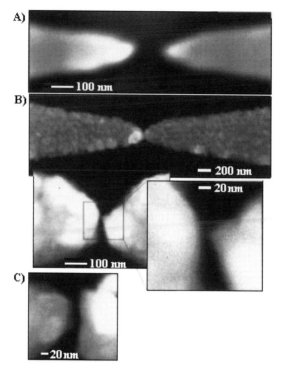

Figure 3.33. SEM images after the formation of the separation with conventional lithography (A) and after the electrodeposition (B), electrodes in which the gap was reopened by electrodissolution, by reversing V_{dc} following an intentional short circuiting (contacting) in a previous electrodeposition process (C). In the case of these measurements the S.E.M. resolution was only 5 nm and, therefore, the gap visualized could be even smaller (Reprinted with the permission from Morpurgo et al., Controlled fabrication of metallic electrodes with atomic separation, Applied Physics Letters 74, pp. 2084–2086, © 1999, American Institute of Physics).

Recently it has been shown that using the resistance across the electrode pair as a feedback signal it is possible to obtain nanocontacts of pre-defined dimension. Li and co-workers (2000) using e-beam lithography, prepared two Au electrodes with an initial separation of 60 nm on a silicon dioxide substrate. The structure was covered by polymeric resist and silicon dioxide. This was to minimize conduction through the electrolyte and hence permit accurate monitoring of the feedback signal.

The tunnelling current was observed to increase during Cu deposition. Comparison with STM experiments indicated that it is possible to monitor the gap width at a resolution of 0.5 Å resolution. Thus by switching off etching when a pre-defined tunnelling current is achieved it is possible to fabricate nanocontacts with sub-nanometer precision. He and co-workers (2002) have shown that expensive lithographic pre-etching steps are not required for the ELENA technique. They demonstrated that starting from a thin copper wire, insulated except for 1 micrometer region, it is possible to prepare electrode pairs separated by sub-nanometer gaps. Céspedes *et al.* (2002) compared the quality of nanocontacts prepared using the ELENA methodology with those fabricated using Focused Ion Beam. It was found that nanocontacts fabricated with the ELENA method were better defined than those prepared using the FIB method. In addition smaller gap sizes could be achieved using the ELENA technique (Figure 3.34).

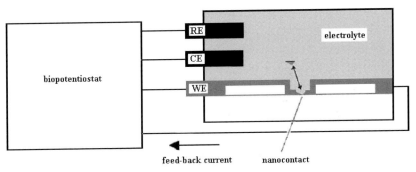

Figure 3.34. In the modified ELENA method, the electrochemical etching or deposition is controlled by the current flowing through the nanocontacts by monitoring the quantized conductance through the electrodes in contact with nanometric area or by monitoring the tunnelling current trough the electrodes gap.

The methods detailed above allow nanosized gaps to be fabricated between electrodes. In a vacuum a tunnelling current will flow between such electrodes. The change in current that accompanies bridging of the gap by an analyte biomolecule will form the basis of future generations of biosensors.

The physics of conductivity across such a nanocontact remains poorly understood and the development of biosensors based on this technology will require further studies in this area. Several models have been proposed to explain the non-linear I-V characteristics observed in nanocontacts: blocked conductance channels, parallel metallic and tunnelling channels and the Luttinger liquid (Versluijs et al., 2000). In addition to the physics, the chemistry of the molecule-metal interface must be considered (Cui et al., 2001). For example, it has been postulated (Hipps, 2001) that the nature of the contact determines whether a DNA "wire" acts as an insulator, a semiconductor, a conductor or a super-conductor. In addition, mechanical instabilities related to fractures (Sotton, 1996) and to rearrangements (Rodrigo, 2002) could dominate the behaviour of nanocontacts.

Above we have considered the fabrication of electrode pairs separated by nanoscale gaps that may be bridged by analyte molecules. However, we note that this is not the only nano-electrode configuration that may be employed in analysis. Most of the fabrication techniques described above may be used to prepare devices in which two electrodes are in direct contact, the area of the contact being on the nanoscale. As detailed elsewhere the conductance of such contacts is quantised. Adsorption of single analyte molecules on such contacts leads to large changes in the resistance of the contact, i.e., a large change in signal is observed for a single molecular event.

For example a conductivity change of 50% has been observed when mercaptopropionic acid adsorbed on a quantum wire (Bogozi et al., 2001).

Sensors and biosensors devices need to address single molecules in order to increase the sensitivity of the devices. Nanotechnology offers technical solutions to fabricate nanocontacts. The aim of this paper was to review all the methods for fabrication of nanocontacts that have been presented recently in literature. Two main classes of methods were identified. The first is based on the possibility of moving one electrode close to another at nanometric distances. The second is based on the etching of a couple of electrodes with gap at nanometric scale. The second class of methods seems to be the most reliable and return the most stable nanocontacts. The electrochemical etching and deposition appears to be the most economic method. This method also allows control of the electrode gap with a precision of 0.5 Å. The next question that nanotechnology must answer of the road to nanoscale biosensors is what are the physical and chemical parameters that must be managed.

3.5.2 *Nanofocussing*

The Nanofocus extension is now becoming operational at ID13 of ESRF (Riekel, 2000; Riekel *et al.*, 2000) as the third experimental hutch (EH3). EH3 will be a dedicated nanofocus facility, offering sub-µm monochromatic X-ray beams in the energy range 12–13 keV as a matter of routine.

A pink beam option will be added later. The target beam size once established will be 50 nm or less. The availability of nano-focused X-ray beams on the ID13 beam-line will open up many new avenues of research. Whilst existing scanning X-ray SAXS/WAXS studies will be enhanced by the higher spatial resolution available, one can expect that highly coherent nano-beams will find complimentary diffraction and imaging applications. In addition, many experiments will also benefit from the higher flux density available for microscopic sample volumes.

EH3 will offer an ultra-stable sample end-station with nm-resolution translation and tilt options. An integrated rotation axis will allow single-crystal and texture experiments. It will also provide solutions to the problems associated with visualizing (and aligning) microscopic samples. This includes the development of complementary, integrated nano-tools, calibrated to the X-ray beam's position. The challenges of preparing and isolating such small samples will be met by micromanipulation and laser-cutting equipment accessible to users on-site. The use of dedicated beam-line software will also help ensure the most efficient use of experimental beam time.

EH3 will offer an ultra-stable sample end-station with nm-resolution translation and tilt options. An integrated rotation axis will allow single-crystal and texture experiments. It will also provide solutions to the problems associated with visualizing (and aligning) microscopic samples. This includes the development of complementary, integrated nano-tools, calibrated to the X-ray beam's position. The challenges of preparing and isolating such small samples will be met by micromanipulation and laser-cutting equipment accessible to users on-site. The use of dedicated beamline software will also help ensure the most efficient use of experimental beam time.

The commissioning of EH3 in February 2007 has herald the start of a new era for ID13. As an innovative experimental tool, nano-beams appear to give a new insight into many materials with nm-sized heterogeneities (Nicolini *et al.*, in preparation; Pechkova *et al.*, in preparation). They will also ensure that ID13 remains at the cutting edge of science. The ID13 beam-line will indeed shortly offer *in situ* micro- and nano-focus Raman Spectroscopy facilities.

A custom-built MicroRaman system was recently designed in collaboration with Renishaw PLC. The system consists of a spectrometer, fibre-optically coupled to a remote probe positioned within the EH2 sample environment. This coaxially delivers a focused laser spot to the same position on the sample as the X-ray beam. The laser spot at the focal position is approximately 1 µm in diameter, comparable to the X-ray beam size from several different ID13 optics. Using a common trigger signal between the beam-line control system and spectrometer, it is possible to simultaneously collect WAXS/SAXS and Raman spectra from the same position on the specimen at the same time. Raman spectroscopy and nanofocused X-ray scattering are complementary techniques on many different levels. They can provide structural information on a range of different length scales, from molecular bonds up to tens of nanometers (for SAXS). They complement each other in their phase- or volume-selectivity, whilst their non-destructive nature (for many materials) makes them ideal for coupling with other methods. They therefore have a diverse range of potential applications, from studying deformation micromechanics and monitoring chemical reactions to characterizing materials over both macro- and microscopic length scales.

3.5.3. Optical tweezers

An optical tweezer is a device for the non-contact trapping and manipulation of micron-sized objects under a microscope using a laser beam.

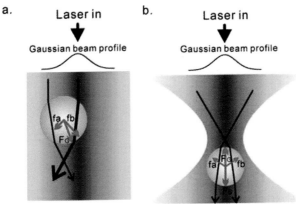

Figure 3.35. Optical tweezers. The microscopic sphere is pulled into the brightest part of the focused laser beam. Schematics showing the principle of optical tweezers based on ray optics. The ability to trap and manipulate small objects, such as polystyrene beads, results from light possessing momentum which is in the direction of propagation of the beam. Here, a bead is illuminated by a Gaussian profiled laser beam. The representative laser paths are shown as black lines with arrows indicating the direction of beam propagation. The thickness of the black lines indicates the intensity of laser beam. The forces are shown as green and blue lines with arrows indicating the direction of forces. The length of the lines indicates the intensity of forces. (A) Gradient forces which are generated upon refraction (F_G, f_a and f_b, green lines). The beam is refracted on the surface of the bead, resulting in the change of momentum of the beam. Gradient forces (F_G) result to compensate the momentum changes on the surface of the bead. The gradient forces from the inner region (f_b) are larger than that from outer region (f_a) of the beam, due to the profile of the laser. Consequently, the net gradient force in the lateral direction directs particles to center of the beam (F_G). (B) Stable 3D trapping. The laser beam was focused by a lens of high numerical aperture. The scattering forces (F_s, blue line) are in the direction of propagation of the laser beam (*i.e.*, downward in this figure), while the gradient forces are directed toward the focused spot. Consequently, the bead is trapped slightly beyond the focused spot where the gradient force and scattering force are in equilibrium (Reprinted with the permission from Kimura and Bianco, Single molecule studies of DNA binding proteins using optical tweezers, Analyst 131, pp. 868–874, © 2006, Royal Society of Chemistry).

The microscope objective focuses the laser light to a very small spot, about 1 μm in diameter. The focal spot works like a trap for microscopic objects, which are pulled into the brightest part of the beam. The microscopic sphere shown has a higher refractive index than the water that surrounds it, so it works like a tiny lens bending the rays of light away from the axis. As the light has momentum, and total momentum must be conserved, the sphere is then pushed by the light towards the axis, becoming trapped in the focus. The trapped sphere can then be moved by moving the laser beam, just as tweezers can be used to pick up and move small objects.

The magnitude of the force exerted by the laser light is typically a few pico-newtons (10^{-12} N), which is comparable to that produced by biologically interesting molecular motors, and so optical tweezers have found several applications in interdisciplinary science and biophysics.

Many experiments have been done on trapped biopolymers commonly used in ultrasound scans to improve the contrast of the image. However to fully understand their properties single microbubbles must be studied in order to compare the response to that predicted by theory. Optical tweezers are the ideal tool for isolating a single bubble from a sample and observing it while being irradiated with ultrasound, as recently shown by Sarah Skoff at UCL on the bubble's protein shell, either to improve the signal in an ultrasound scan, or to make the bubble break open when exposed to ultrasound waves. This would be useful if the bubbles were filled with a drug, which could then be targeted to the exact location needed.

An alternative method for trapping a microbubble is to use a "hollow" laser beam that has a dark spot on the center, surrounded by a bright ring. A class of laser beams that has this property is the Laguerre-Gaussian beams. Most lasers produce a beam with a Gaussian intensity profile - the brightest part is in the center. In order to make a beam with a dark center we use a computer-generated hologram which when illuminated with an ordinary laser beam works like a diffraction grating. The difference is that the diffracted orders have the dark center characteristic of a Laguerre-Gaussian beam (see the review by Neuman and Block, 2004). Optical tweezers are a useful and important tool with many applications in the physical and life sciences. The techniques being used will help us to understand some of the fascinating physics that controls biological systems. The technology could one day be used to make self-assembling electronic components or chemical factories-on-a-chip.

Bernard Yurke and his colleagues (2000) at Bell Laboratories in New Jersey made a tweezer-like structure out of three strands of DNA. When a fourth strand is added to the mixture, it joins the loose ends of the tweezers, pulling them shut. Adding yet another strand of DNA pops the tweezers open in order to get controllable motion on a nanometre scale. Single molecule studies of DNA binding proteins using optical tweezers. Optical tweezers have become a versatile tool in the biological sciences. Combined with various types of optical microscopy, they are being successfully used to discover the fundamental mechanism of biological processes. Recently, the study of proteins acting on DNA was aggressively undertaken at the single-molecule level, providing detailed mechanistic insight that could not be revealed, at least not easily, using bulk-phase or ensemble approaches (Kimura and Bianco, 2006).

3.5.4 *Magnetism*

Ever since the German physicist Max von Laue's 1913 insight that X-rays could be used to unravel crystal structure, they have been an essential tool for the study of matter. Some neighborhoods, however, have been off limits to X-rays, such as materials' fine-scale magnetic structure and the fleeting molecular alliances within disordered materials such as liquids and glasses.

Those have been the domains of neutron beams since the first research reactors were built in the 1950s. With the advent of third-generation synchrotron sources, however, X-ray scattering is making inroads into neutron territory. "The special points are very high brightness, good-quality polarization, and very high energy x-rays" says Hiroshi Kawata of the Photon Factory at the KEK high-energy physics lab in Tokyo. Because of these properties, "you can start thinking about experiments it would not have been possible to do a few years before" says physicist Michael Krisch of the European Synchrotron Radiation Facility (ESRF) in Grenoble, France, the first of the new machines. Indeed, the third-generation sources are teasing out information about magnetic properties and disordered materials that neutrons could not reveal and researchers are beginning to answer what is the real nature of the magnetization. And the X-rays' brightness and tightly controlled energy have opened the way to studies of disordered materials that capture, for example, a high-speed form of sound in water.

Those results are only the first in what is expected to be a torrent, says David Laundy, an ESRF user from Britain's University of Warwick, because "X-rays give different information from that of neutrons." Neutrons are sensitive to magnetism, for example, because they are scattered not only by collisions with atomic nuclei, but also by magnetic interactions with atoms as a whole. Neutrons, although they lack charge, nevertheless have their own magnetic field. But neutrons cannot distinguish the two contributions to an atom's magnetic field, which come from the inherent spin of its electrons and from the magnetic effect of those electrons orbiting the nucleus. Separating out the spin and orbital parts really lies at the heart of understanding magnetic properties, and X-rays offer a way to untangle these two effects. Scattered by an atom's electrons, x-rays respond mainly to the electrons' electric charge, but the magnetic part of the photon's electromagnetic wave also interacts feebly with the electrons' magnetic field if they are aligned advantageously. This is possible with the new X-ray sources; because they have beams whose polarization, the alignment of their electric and magnetic fields, can be controlled. And it turns out that X-rays are also sensitive to the two different components of magnetism (Suortti at ESRF).

X-rays offer other advantages over neutrons, tending to be more sensitive to the surface of the material than neutrons for studying exotic magnetic structure. X-rays are also flexing their muscles in another domain that was once the preserve of neutrons, so called inelastic scattering. Inelastic scattering studies with neutrons have unraveled the dynamics of a wide range of systems, from liquids to metallic glasses. But neutron studies often require samples to be made from rare isotopes, rather than the common ones. Water, for example, has to contain deuterium rather than hydrogen for neutron studies. Neutron beams are also dim, and their energy range is limited. ESRF's X-rays, however, have a wide range of energies and momenta, enabling Francesco Sette and his colleagues recently to study just this type of fast excitation process in determining the dynamical properties of disordered materials in a region that was not accessible before. These transitory get-togethers by molecules are called collective excitations, and they may affect everyday properties of a liquid, such as chemical reactions, thermal properties, and the way sound waves propagate.

However, Sette and colleagues (1996) from ESRF and from the University of L'Aquila in Italy confirmed the latter effect by setting off fast sound waves in water with inelastically scattered X-rays.

Solid but disordered materials such as glass are also coming to be studied by inelastic x-ray scattering. Despite their newfound abilities, x-rays are not about to replace neutrons as a tool for studying matter, but to be complementary. Even though x-rays can probe magnetism in ways neutrons cannot in many situations, neutrons are still the probes of choice to determine a magnetic structure, at least for the present.

3.6 Cell Nanobioscience

Few key examples are here quoted of the usefulness of Nanobiotechnology in the study of intact cell component, namely nucleosome core and proteins *in situ*.

3.6.1 *Nucleosome core*

Protein crystallography is currently turning out novel atomic structures of biological macromolecules at the rate of 400-500 per year (Hendrickson and Wüthrich, 1997). These structures range in size from proteins comprising under 100 amino acids to virus particles of several million molecular weight.

An area of intense interest currently concerns macromolecular complexes containing multiple components such the combination of interphase chromosomes named chromatin that must periodically assemble and disassemble in order to carry out their biological function in the cell (Baserga and Nicolini, 1976; Diaspro *et al.*, 1991;Kendall *et al.*, 1977; Nicolini, 1983; Nicolini and Kendall, 1977; Pepe *et al.*, 1990; Zietz *et al.*, 1983). A classic example in all higher cells is the nucleosome, which is the fundamental repeating unit of DNA organization in chromosomes and accounts for the two most fundamental levels of higher order chromatin structure (Nicolini, 1983; Nicolini et al., 1975, 1976, 1977, 1977a, 1982, 1983, 1983a, 1984, 1991). Not only do nucleosomes efficiently package DNA, they are also intimately involved in the gene expression mechanisms that allow only selected regions of the vast store of genomic information to be read out as a consequence of signaling processes. These two functions require about 25 million nucleosomes in each human cell nucleus. The nucleosome core particle structure explains in atomic detail how DNA is kept untangled in the cell nucleus and clarifies the unique role of the nucleosome in maintaining and controlling the expression of genetic information. Nucleosomes do not exist as isolated particles in the cell, but are packed into arrays with an internal repeat of 157–240 bp. The dynamic assembly and disassembly of the higher-order structures made from these arrays helps determine the functional and dynamic state of DNA, and this will be most likely solved by the combined utilization of optical tweezers and nanofocused beam-line at the Synchrotron Radiation (previously described) in the study of intact mammalian nuclei and cells. This endeavor will elucidate the structures of higher-order arrangements of nucleosomes in the chromatin fiber and to relate this information to the way these assemblies and their super-coil participate in gene regulation (Nicolini, 1986).

In 1984, the structure of the nucleosome core particle (NCP), the larger part of the nucleosome, was published at 7 Å resolution (Richmond et al., 1984). X-ray data was collected before synchrotron radiation was generally available for protein crystallography by using a single detector diffractometer and a rotating anode source. The spatial resolution of this first structure was limited by the material itself as it was prepared from whole nuclear chromatin and was therefore heterogeneous in composition. Eventually, it became technically feasible to assemble NCP in the test tube from homogeneous components made individually in bacterial cells (Luger et al., 1997a).

After a long period of experimentation for developing homogeneous preparations of NCP, crystals were obtained that diffracted to high resolution. Nevertheless, the Bragg intensities from these crystals are extremely weak. Fortunately, the ESRF had opened for business with ID13 and as a result, the structure of the nucleosome core particle was published in 1997 at 2.8 Å resolution (Figure 3.36) (Luger *et al.*, 1997b). At 206 kDa, the NCP is the largest and most universal protein/DNA complex solved in atomic detail.

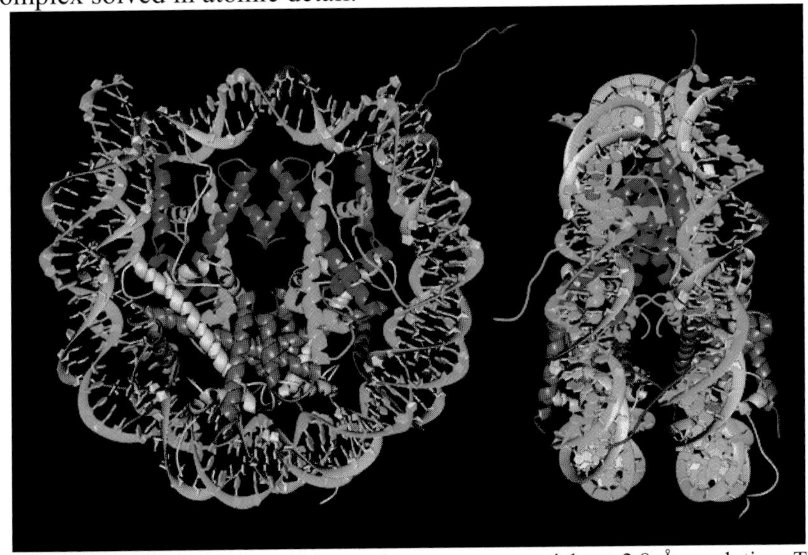

Figure 3.36. Crystal structure of the nucleosome core particle at 2.8 Å resolution. The DNA double helix (146 base pairs in two chains: turquoise and brown) is wound around the protein histone octamer (two copies each of H2A: yellow, H2B: red, H3: blue, and H4: green) in 1.65 left-handed superhelical turns. This is the form of DNA, which predominates in higher living cells. The left view is down the superhelix axis. The right view is orthogonal to the superhelix and overall pseudo-twofold axis (Reprinted with the permission from Luger *et al.*, Crystal structure of the nucleosome core particle at 2.8 Å resolution, Nature 389, pp. 251–260, © 1997, Macmillan Publishers Ltd).

Our best nucleosome core particle crystals show Bragg intensities to 1.9 Å and have measurable data to 2.0–2.1 Å spacings (Figure 3.36). The NCP contains pairs of the four core histone protein molecules named H2A, H2B, H3, and H4, and a roughly equal mass of DNA in 147 nucleotide pairs (we used 146 bp). Compared to the nucleosome, the NCP is missing only the "linker histone" H1 and the short stretches of DNA that connect the nucleosome cores to each other in chromatin (Nicolini and Kendall, 1976). The core histones are arranged in an octameric unit around which the DNA is wrapped in 1.65 turns of a left-handed superhelix (Luger *et al.*, 1997b).

This arrangement necessitates a substantial deformation of the DNA, bending the 22 Å diameter double helix to a mean radius of 42 Å in the nucleosomal superhelix.

The histone protein chains are divided into three types of structures: 1) rigid, folded alpha-helical domains named the histone-fold, 2) histone-fold extensions which interact with each other and the histone-folds, and 3) flexible "histone tails".

The histone-fold domains are structurally highly conserved between the four types of core histones and have also been discovered in an increasing number of other molecules involved in the regulation of gene read-out or transcription. They form crescent-shaped heterodimers, which have extensive interaction interfaces in the pairings H3 with H4 and H2A with H2B.

The histone-fold domains are responsible for organizing 121 base pairs (bp) of DNA in the superhelix, not the entire 147 bp. It is the responsibility of the extensions just prior to the H3 histone-folds to bind the first and last 13 bp of DNA.

The flexible tails of the histones reach out between and around the gyres of the DNA superhelix to contact neighboring particles. About one-third of these flexible histone tails can be observed in the electron density map, the remainder is too disordered to be interpreted. The implication from the structure is that these flexible regions are meant to make inter-nucleosomal interactions, perhaps facilitating the formation of nucleosome higher-order structures (HOS).

There are 14 regions of contact between the histone proteins and DNA: three by each of the four histone-fold dimers and two by histone-fold extensions. This construction allows the DNA molecule in a single nucleosome core to come loose over one-half of the superhelix while the histones maintain their grip on the other half, permitting the genetic information stored in the DNA to be read out without complete dissociation of the DNA from the histone octamer.

The nucleosome core was previously thought to be held together simply by electrostatic attraction: the negatively charged DNA molecule wound as yarn around a positively charged histone spool.

Although this type of interaction does occur, equally many interactions of other kinds, such as hydrogen bonds and hydrophobic interactions, are also important.

3.6.1.1 *DNA deformation*

The path of the DNA around the histone octamer deviates from that of an ideal superhelix, displaying strong bends in some regions, while being nearly straight in others. This path is determined predominantly by the histone/DNA contacts and is probably largely independent of the DNA sequence of nucleotides.

The close spatial proximity of the two turns of the DNA superhelix with a pitch of 24 Å, and the periodic variation of double helix parameters with a mean of 10.3 bp per turn, result in an alignment of major and minor grooves from one superhelical gyre to the next (Figure 3.36).

The resulting narrow channels formed by the aligned minor grooves serve as the exit points for four of the eight basic histone tails, whereas the large pores formed by the aligned major grooves are, in principle, free to make base-specific contacts with other proteins.

The Debye-Waller B-factors show that the mobility of the DNA backbone varies greatly, having low values when it is bound to the histone octamer and high values when it is facing away from it.

3.6.1.2 *Water and ions*

The 2.0 Å diffraction data have allowed us to locate a large number of ions and well-ordered water molecules (see next paragraph). We expect to gain valuable general information on the role of water molecules in mediating protein-DNA and protein-protein interactions.

The binding of divalent metal ions, such as manganese or magnesium, to the DNA appears to favor the distortion of the DNA seen in the NCP. The presence of ordered water molecules at the interface between the protein subunits may provide a means to favor their disassembly and thus could be important to the processes of DNA transcription and replication.

3.6.2 Protein stability to heat and radiation

Physico-chemical basis of thermal stability of proteins has been a subject of a large number of both fundamental and applied studies, the former exploring the many physical mechanisms of thermo-stability and even linking them to evolutionary relationships between the proteins thus arriving at physics of evolution, while the latter primarily aiming at design of thermo-stable enzymes and optimizing their catalytic properties, not only via rational design but also via directed evolution or combination of the two approaches.

The general belief is that protein stability is due to many small effects, and that different factors stabilize different proteins. An understanding of the molecular motifs of structural resistance in thermo-stable proteins is pursued by site-directed mutagenesis experiments. Generally, comparative studies are abundant in this field of research, which belong to either of the two types. In one, thermo-stable protein is compared to its mesophilic counterparts in great detail, allowing to formulate the mechanism(s) of thermal stability for the case of those particular proteins. In most studies, the specific factors of thermal stability generally include aminoacid composition, proline residue occurrence, exposed/buried hydrophobic/hydrophilic areas, insertions/deletions, disulfide bridges, ion pairs, and hydrogen bonds. (Pechkova et al., 2007c). However, such approach is usually based on analysis of individual factors contributions but fail to generalize those contributions on a unified scale.

The other class spawned by the rapidly growing number of 3D protein structures resolved using crystallography or NMR, comprises theoretical or statistical studies of large numbers, tens to hundreds, of thermostable proteins compared to their homologous mesophilic proteins. However, the statistical characteristics are in general difficult to interpret physically, so they mostly remain descriptive. Particularly, the analysis of *Thermotoga maritima*/mesophile protein pairs, led the conclusion that the discriminating parameter is the average number of atomic contacts within the protein (contact order), which characterizes protein compactness. Besides, a combination of statistical and simulation studies suggested that one of thermophilic adaptation mechanism was related to more compact structure of the globule. In both cases, the thermophilic/mesophilic discriminating parameter is related to compactness, and in both cases the authors explicitly specify that it is not caused by decreased loop contents, *i.e.*, increased secondary structure contents.

The two other remaining factors that might cause greater compactness are the following: 1) increased density of the hydrophobic core which should manifest itself in the change of the structure of the peptide backbones (however at variance with the classical viewpoint of Argos *et al.* (1979); and 2) increased density of the side-chains at the hydrophilic exterior of the protein that should lead to reduction of the number is the water molecules bound to the protein surface. From the other hand such a protein compactness has been observed in the LB film, which shows the thermo stability of the structure and function of immobilized proteins (Nicolini, 1997, 1998b), or in crystals growing in presence of Langmuir-Blodgett (LB) film template (Pechkova and Nicolini, 2004a; Pechkova *et al.*, 2004), both related to a decrease water contents and a slight increase the alpha-helix content, maintaining the native folding present in solution.

To clarify this issue, in this report, we compare the 3D structures in protein pairs from thermophilic/mesophilic species in terms of overall fold similarity (via percentage of structurally aligned amino acids) and average local similarity (via RMS deviations of backbone coordinates).

It is worth to notice that usually the explored properties are essentially intraprotein and do not explicitly include interactions of the protein with water, with the exception of exposed areas. The latter, however, can be misleading because of specific interactions of water molecules with the highly intricate patterns of protein surface. From our earlier studies, nanogravimetry revealed a minor content of bound water in thioredoxin (Trx) from *Bacillus acidocaldarius* with respect to *E. coli* Trx, with the difference of transition temperature of approx. 10 °C. Moreover, highly ordered protein monolayers with defined structures and function, generated using appropriate modifications of original Langmuir-Blodgett method, shows the thermostability of the structure and function of immobilized proteins (Nicolini, 1997, 1998b). It is well known, that the water content in the LB film inside the molecule is low.

Thus, the mechanisms by which LB film can induce protein stability are via altering patterns of aqueous environment of the given protein. This is in agreement with the fact that smaller amount of water should enhance thermo stability, taking in the consideration that the limited availability of water hinders thermal denaturation. Several additives decreasing the amount of water such as charged polymers (Foremant et al., 2001) were found to increase protein thermo stability. Thus, our previous data and the large body of literature data suggest a marked role of aqueous environment in protein thermal stability. This prompted us to compare the number of water molecules present in the thermophilic species/mesophilic species protein pairs in crystals, and to compare the 3D *backbone* structures in protein pairs from thermophilic/mesophilic species in terms of overall fold similarity (via percentage of structurally aligned amino acids) and average local similarity (via RMS deviations of backbone coordinates).

3.6.2.1 *Bioinformatic analysis*

The selection of the first set (representatively-based) constructed with only wild-type proteins, was based on statistical examination of sequence and structural parameters in families of homologous thermophilic and mesophilic proteins.

The conditions used were as follows. Each pair consisted of a protein from a thermophilic or hyperthermophilic species and its most similar, in sequence and structure, mesophilic homolog. The other condition was the maximum dissimilarity within each of the thermophilic proteins dataset. Therefore the proteins for thermophilic species were selected so as to be maximally dissimilar from each other, which was used a criterion for representatively of the data set. Among those, only high-resolution (R<2.5 Å) crystal structures for wild-type proteins available in the RCSB Protein Data Bank (PDB). Selection of the other dataset of protein pairs (filter-based) initially was formulated starting from the growth temperatures of microorganisms. Those with growth temperatures above 45 °C were defined as thermophilic (Pechkova et al., 2007c). The corresponding structures from PDB from these source organisms were filtered with respect to resolution and nativity, so that only the ones with highest resolution and non-mutant proteins were retained. For each of the proteins from thermophilic species, their mesophilic homologs were found using the FSSP database (Holm and Sander, 1996). FSSP entries containing the thermophilic proteins also contain all their structural homologues by definition. Pairs that did not contain proteins from mesophilic species were excluded. The next filter was identity of the sequence for which the cut off was set at 30%, which always selects proteins with the same fold. Also excluded were structure with missing atoms and chain breaks. The final filter was crystallographic data quality from the PDBREPORT database. Pairs with resolution of 2.5 Å or worse, as well as those qualified as 'bad' in the quality report, were removed.

Finally, the pairs with no direct experimental evidence for thermal stability of the thermophilic-species protein were excluded using the ProTherm database, resulting in 20 filter-based protein pairs. Some of the pairs found using the above-described procedures were manually excluded on a case-by-case basis. Such cases included firstly the falsely detected homologues as in ferredixins whose homology is limited to the immediate vicinity of the iron-sulfur cluster, which is the mail structure-forming element.

Secondly, multidomain proteins were excluded, but only those for which relative positions of domains very strongly varies in the course of function and consequently in crystallographic structures, as exemplified by the pair of elongation factors Tu (1EFT) from *T. aquaticus* versus Ef-Tu from *E. Coli* (1EFU). Other excluded cases contained strains of *B. subtilis* as the mesophilic source of protein. For *B. subtilis*, both mesophilic and thermophilic strains were reported. We failed to find experimental evidence that, even though the particular strain is mesophilic, the corresponding protein is not thermally stable. The relations between thermal stabilities from various strains of *Bacillus subtilis* therefore requires further study. However, an indirect evidence that the "mesophilic" proteins can possess thermo-stability comes from the fact that there are as much as 217 examples of various *B. subtilis* proteins showing marked thermophilicity according to the ProTherm database (see Pechkova *et al.*, 2007c for further details and references), while no such data is available for other mesophilic bacterial species.

The eventual protein pairs from the two datasets combined are presented in Figure 3.37, and are separately shown in Figure 3.38, top, where their water contents are shown.

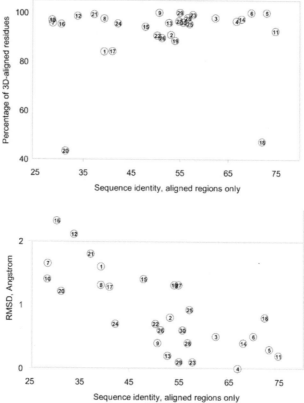

Figure 3.37. Similarity of the overall fold (top) and local geometries (bottom) depending on sequence homology (in terms of sequence identity %) between the thermostable and mesophilic proteins. Each point represents a thermostable/mesophilic pair and numbered, with numbers encoding the following PDB ids: 1-1PCZ/1VOK, 2-1BDM/4MDH, 3-1BMD/1B8P, 4-1CAA/8RXN, 5-1CIU/1CDG, 6-1CYG/1CDG, 7-1EBD/1AOG, 8-1GTM/1HRD, 9-1HDG/1GAD, 10-1LDN/1LDG, 11-1LNF/1NPC, 12-1OBR/1AYE, 13-1QEZ/1OBW, 14-1Q9H/1GPI, 15-1THM/1BH6, 16-1TMY/3CHY, 17-1VJW/1FXD, 18-1WB8/1JA8, 19-1WL7/1UV4, 20-1XGS/1BN5, 21-1XYZ/1CLX, 22-1YNA/1XNB, 23-1YNA/1ENX, 24-1ZIP/1AKY, 25-2BMM/1NGK, 26-2PRD/1SXV, 27-3MDS/1D5N, 28-3TGL/1LGY, 29-4PFK/1PFK, 30-1YNR/451C. (Reprinted with the permission from Pechkova *et al.*, Protein thermal stability: the role of protein structure and aqueous environment, Archives of Biochemistry and Biophysics 466, pp. 40–48, © 2007c, Elsevier).

3.6.2.2 Structural comparisons

The protein parameters were computed using VAST software, and described below. Namely, the *percentage identity* refers to residues in the aligned sequence region. This is a raw measure of sequence similarity in the parts of the proteins that have been superimposed.

The *RMSD* signifies the root mean square superposition residual in Angstroms. This number is calculated after optimal superposition of two structures, as the square root of the mean square distances between equivalent $C\alpha$ atoms. Note that the RMSD value scales with the extent of the structural alignments, so RMSD number should always be used in combination with Aligned Length, as done in this study.

The *Aligned Length* parameter is the number of structurally equivalent pairs of $C\alpha$ atoms superimposed between the two structures. In other words, this is the maximal number of residues possible to superimpose for two proteins.

In this study the percentage Aligned/Total is used as the characteristic of the overall fold similarity, since obviously Aligned Length scales quasi linearly with the total length. Incidentally, we calculated also the measure *Loop Hausdorff Metric (LHM)*, which is a loop similarity measure showing how well two structures conform to each other in the loop regions, once structurally superposed, loop regions being the parts of the structures between aligned secondary structure elements (helices and strands).

The LHM value is in Angstroms, with a smaller value showing greater similarity. Similar arrangements of loops typically have LHM smaller than 3 Angstrom, while considerable differences in loops have LHM over 5 Å. The loop similarity may be undefined if there are too many residues with missing coordinates in the loops, which never occurred in this study.

3.6.2.3 Structural comparisons of homologous thermophilic/mesophilic pairs

Structural similarity parameters of protein pairs from the representative thermophilic/mesophilic species are shown in Table 3.4.

Here we aimed to characterize it qualitatively using the fraction of structurally aligned residues in the entire sequence of each protein, in addition to RMSD comparison of the protein backbones.

Table 3.4. Numbers of water molecules per subunit in crystals of several thioredoxin. (Reprinted with the permission from Pechkova et al., Protein thermal stability: the role of protein structure and aqueous environment, Archives of Biochemistry and Biophysics 466, pp. 40–48, © 2007c, Elsevier).

PDB id	Organism	Waters molecules per subunit
1thx	Mesophilic E. coli	222
1t00	Mesophilic S. coelicolor	150
2cvk	Thermophilic T. thermophilus	43
1nw2, 1nsw	Thermophilic A. acidocaldarius mutant[a]	72
2trx	E. coli with artificial water removal	70

[a] Thermophilic but with mutations introduced canceling thermophilicity (Bartolucci et al., 2003). Average value is shown.
[b] Mesophilic but crystallized in presence of MPD, which is known to displace water from protein surface (Anand et al., 2002)

Besides, since structural alignment requires that amino acids are superimposed in 3D, the RMSD parameter does not include the amino acids that cannot be aligned, as for example in the case of large inserts.

As seen from Figure 3.37, most pairs have the "crude" similarity parameter above 80%. Besides, no correlation is observed between sequence identity/similarity and structural similarity, suggesting that the overall fold remains the same in the thermophilic/mesophilic pairs for both higher and lower sequence identity pairs. Loop lengths were compared using the above-quoted VAST software, and it was found that the differences were marginal (data not shown), with one exception of 26 extra amino acids in the single insertion loop in the mesophilic protein, which is unlikely to affect thermal stability since it is presence of multiple requires multiple extra loops rather than a single long loop that strongly affects (thermo)dynamics of protein folding. As seen from Figure 3.37, the RMSD ranges from 0.6 Å (practical identity, in case of rubredoxins) to 2.3 Å (hydantoinases) which indicates significant similarity, considering also that the thermo-stable hydantoinase is L-hydantoinase while the mesophilic is D-hydantoinase. The RMSDs are plotted against sequence identity, but only a weak correlation is observed (Figure 3.37). There are at least as many protein pairs not following the expected correlation pattern, which is high sequence identity for lower RMS, as the ones following it. Overall, relation between two parameters, the percentage of aligned residues and the RMSD values, allows to conclude that the conformations of loops were different in the pairs, loops were further compared using the new option LHM in the pairs using the newly available option of the VAST software, but no significant difference within the mesophilic species protein/thermophilic species protein was observed.

As seen from Figure 3.37, the RMSD ranges from 0.6 Å (practical identity, in case of rubredoxins) to 2.3 Å (hydantoinases) which indicates significant similarity, considering also that the thermo-stable hydantoinase is L-hydantoinase while the mesophilic is D-hydantoinase.

The RMSDs are plotted against sequence identity, but only a weak correlation is observed (Figure 3.37). There are at least as many protein pairs not following the expected correlation pattern, which is high sequence identity for lower RMS, as the ones following it. Overall, relation between two parameters, the percentage of aligned residues and the RMSD values, allows to conclude that the 3D structures of proteins from thermophilic and mesophilic species remain the same, regardless of their sequence identities.

3.6.2.4 *Water comparisons of homologous thermophilic/mesophilic pairs*

For the representatively-based dataset, overall numbers of water molecules per protein molecule scaled by the water-accessible surface area are shown in Figure 3.38, top left. It is evident that the number of water molecules in all cases is larger in the mesophilic species proteins than in their thermophilic homologs, and several cases is drastically larger.

In all cases, resolution is similar in the pairs, typically within 0.2 Å, with the only exception of the 2PRD/1SXV pair, where the difference is 0.6 Å, but the 2PRD/1SXV pair from the same dataset has the difference of 0.3 Å, but shows practically the same difference in water content.

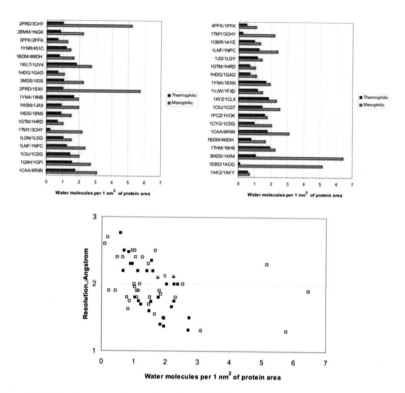

Figure 3.38. Numbers of water molecules scaled by water-accessible surfaces of protein subunits contained in a unit cell in crystals of thermophilic species protein versus mesophilic species proteins (light gray). Top left, the representative dataset taken from Kumar and Nussinov (2001). Top right, dataset verified by experimentally determined thermal stability of all proteins studied (Szilágyi and Závodszky, 2000). Bottom, Dependence of scaled number of water molecules on crystallographic resolution. (Reprinted with the permission from Pechkova et al., Protein thermal stability: the role of protein structure and aqueous environment, Archives of Biochemistry and Biophysics 466, pp. 40–48, © 2007c, Elsevier).

To clarify the issue, we performed the same analysis of the water content for the dataset in which all protein pairs were verified by experimentally determined thermal stability for both the "thermophilic" and the "mesophilic" members of the pair (Szilágyi and Závodszky, 2000). Results (Figure 3.38, top right) show the same pattern as our original dataset. Considering that the number of crystallographically detected water molecules essentially depends on resolution of the crystallographic structure, we also plotted the dependence resolution on water molecules scaled by protein surface, and found very weak correlation (Figure 3.38, bottom) for both thermo-stable and mesophilic proteins. Therefore our results, either from the representativity-based or filtering-based datasets, are not likely caused by a crystallographic artifact.

Comfortingly the number of internal (buried) water molecules, even if not found in several protein pairs present in PDB, is far larger in the mesophilic species proteins than in their thermophilic homologs (Figure 3.39) for the vast majority of the protein pairs selected by representatively (mostly) or by experimental verification. The plot in Figure 3.39 includes only data for those protein pairs, scaled to show the numbers per 10 kDa of molecular weight of proteins, which is 0.18% w/w.

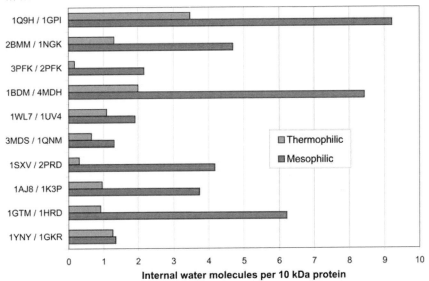

Figure 3.39. Numbers of internal water molecules per 10 kDa molecular weight of protein in crystals of thermophilic (red) versus mesophilic proteins (green). (Reprinted with the permission from Pechkova *et al.*, Protein thermal stability: the role of protein structure and aqueous environment, Archives of Biochemistry and Biophysics 466, pp. 40–48, © 2007c, Elsevier).

The data with two and less water molecules are probably insignificant and thereby not included. Conditions of crystallization and data acquisition were similar in each protein pair with respect to the following parameters potentially able to affect the quantity of crystallization water: concentration of amphiphiles (although not the exact formula of amphiphiles), ionic strength (not exact composition of the buffer), presence of organic solvents (never used), and temperature (room throughout the datasets). The data collection parameter, apart from the parameters that are accumulated into the resolution, also affecting the number of water molecules, is the electronic density threshold for water detection, but this parameter is very seldom reported or varied during data processing. Wherever it was reported, it was the same within the pairs.

3.6.2.5 Detailed comparison of mesophilic versus thermophilic thioredoxin

One special case, for which two mesophilic homologs correspond to two thermophilic homologs, is bacterial thioredoxin. Particularly, thioredoxin from *Alicyclobacillus acidocaldarius* (BacTrx) is homologous to thioredoxins of *T. termophilus* (both thermophilic), *E. coli* and *S. coelicolor* (both mesophilic).

Their water contents are shown in Table 3.4. The highest water contents are shown by two mesophilic proteins thioredoxins from *E. coli* and *S. coelicolor*. If thioredoxin from *E. coli* is crystallized in presence of a strong protein dehydration agent, MPD (2-methyl-2,4-pentanediol), then the number of water molecules drops to 70 (Line 5 of the Table 3.4). Similar water content is shown also by crystals of thermophilic thioredoxin from *A. acidocaldarius*. Those crystals, however, contain the protein that has been mutated in such a manner as to reduce thermostability (about 10 degrees drop in denaturation temperature). Finally, thioredoxin from *T. termophilus* shows the lowest water content. As follows from Table 3.4, water contents in bacterial thioredoxins can be arranged in the following row: mesophilic > mesophilic with artificially reduced hydration ≈ thermophilic with mutation-reduced thermal stability > thermophilic, which is in agreement of our analysis of water content in crystals.With respect to thioredoxin, we also measured the water contents experimentally.

The thioredoxin from *Alicyclobacillus acidocaldarius* BacTrx purified by anion-exchange chromatography (11577 Da) has in primary structure an identity ranging from 45 to 53% with all sequences of known bacterial Trxs. Nanogravimetry experiments (Figure 3.40.) showed a lower content of bound water in BacTrx than in *E. coli* Trx, and a transition temperature approx. 13 °C higher for BacTrx either in solution or in non-oriented self-assembled film, while in LB highly oriented film the same *E. coli* Trx, displays a transition temperature more than 25 °C higher with a dramatic decrease in desorbed water (Figure 3.40).

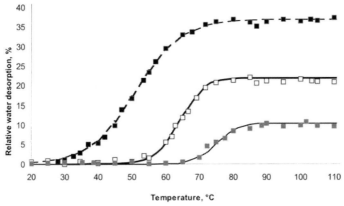

Figure 3.40. Desorbed water versus temperature for non-oriented self-assembled film 7 of *E. coli* TrxEc (black squares), BacTrx in non-oriented self-assembled (white squares) and LB (gray squares) film. The water amount has been normalized to the initial protein amount in the sample. Thioredoxin samples were deposited on the quartz crystals in 50 mM sodium phosphate ph 5.8 and dried under vacuum for 30 minutes (Bartolucci *et al.*, 1997; Facci *et al.*, 1994b). (Reprinted with the permission from Pechkova *et al.*, Protein thermal stability: the role of protein structure and aqueous environment, Archives of Biochemistry and Biophysics 466, pp. 40–48, © 2007c, Elsevier).

Comfortingly, the circular dichroism signal at 193 nm versus temperature, as shown earlier by differential scanning calorimetry, demonstrated that BacTrx in self-assembled non oriented film is endowed with a higher conformational heat stability than the self-assembled film of Trx from *E. coli*, which instead when immobilized in LB film displays an even more dramatic increase in heat stability up to 200 °C. As indeed shown earlier in solution, *E. coli* Trx displayed a melting temperature at 81 C against 94 C for BacTrx, which in turn is much lower than the 200 °C shown here for *E. coli* Trx in LB film (Figure 3.41). Incidentally, BacTrx in LB film displays the same dramatic increase in heat stability up to 200 °C (not shown) and a similar dramatic decrease in desorbed water (not shown).

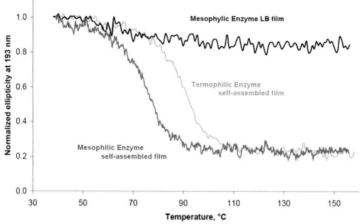

Figure 3.41. Temperature dependence of circular dichroism signal at 193 nm LB film (black line) and in non-oriented self-assembled film of BacTrx (light gray line). Measurements have been taken in 50 mM sodium phosphate ph 5.8 The CD signal for non-oriented self-assembled film of *E. coli* EcTrx displays a melting temperature about 13°C lower than the corresponding BacTrx one (dark gray line). Instead, the LB film of *E. coli* EcTrx even after heating up to 200°C does not display a melting transition (black line). The CD signal has been normalized to the initial molar ellipticity value of the BacTrx self-assembled film sample at 25°C (Reprinted with the permission from Pechkova *et al.*, Protein thermal stability: the role of protein structure and aqueous environment, Archives of Biochemistry and Biophysics 466, pp. 40–48, © 2007c, Elsevier).

Note that the water desorption rate which is calculated relative to the weight of protein is used herein as a parameter characterizing thermal phase transition of protein with increasing protein, the main parameter being the transition temperature. The empirical equation used for curve fitting and thus determining transition temperatures (end of Methods) was derived from the general phase transition theory, implicitly assuming that the transition in question in protein unfolding. It does indeed follow from comparing Figures 3.40 and 3.41 that water desorption and protein unfolding are closely correlated. However, the actual amount of water contained in the film per protein weight, which is roughly proportional to the height of each desorption curve, is progressively decreasing in the row "self assembled mesophilic " – "self assembled thermophilic" – "LB mesophilic", in perfect agreement with the discussed correlation of increased thermal stability and lower water contents.

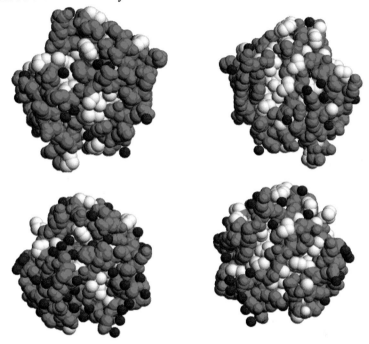

Figure 3.42. Positions of internal and first hydration shell water (red) around the *B. acidocaldarius* thioredoxin (top) and the *E. coli* thioredoxin (bottom). Left and right panels differ by 180° rotation around a horizontal y-axis. Hydrophilic surface is in green, hydrophobic surface is in white (Reprinted with the permission from Pechkova *et al.*, Protein thermal stability: the role of protein structure and aqueous environment, Archives of Biochemistry and Biophysics 466, pp. 40–48, © 2007c, Elsevier).

In an attempt to capture atomic-level mechanisms of the relationship between thermal stability and water content, we compared atomic-level structures of *E. coli* and *B. Acidocaldarius* thioredoxins (Figure 3.42) where we show positions of internal first hydration shell water over the surface of the protein, relative to positions of hydrophilic and hydrophobic surfaces of the protein. It is evident that not only the overall quantities but also the first hydration shell contains fewer water molecules in the case of a thermo stable protein. Note that, in the case of the thermo stable protein, water molecules do not assemble into clusters. Lower water content is consistently apparent in thermophilic proteins with respect to their mesophilic counterparts possibly just a consequence of the well-known mechanisms of stabilization as electrostatic interactions or compactness. We infer from combining the representative PDB data with the fact that the protein backbone remains unchanged in the mesophilic/thermophilic protein pairs that the compactness, when determining the protein thermal stability, is related as to the protein inner aqueous environment. Relation between protein stability and water content is confirmed by the dramatic thermal stability induced in thioredoxin by the LB immobilization, which strikingly correlates with the similarly dramatic decrease in the amount of inner water.

The other possibility could be related to decreased sizes of water-accessible cavities in the protein, even if no water is crystallographically observed therein. To clarify, we applied two different compactness criteria, one known as packing scores, and the other being the volume of water accessible cavities in the protein. Neither was found to correlate within the thermostable/mesophilic protein pairs. Interestingly the three-dimensional model of the two thioredoxins displays minor differences in the tertiary atomic structure. Indeed, a comparison with the EcTrx structure and analysis derived by X-ray crystallography of stabilizing factors in terms of the global folding indicates that the BacTrx overall fold is very similar to that of EcTrx and only small differences in the molecular architecture can be observed.

The minor NMR structural differences suggested that protein stability could be due to cumulative effects, the main factor being an increased number of ionic interactions cross-linking different secondary structural elements and clamping the C-terminal alpha-helix to the core of the protein. The superposition of all the backbone atoms (N, Cα, C) (residues 5–104) of the 20 final BacTrx structures onto the *E. coli* crystal structure yields an average RMSD of 0.14 nm, while the RMSD is reduced to 0.12 nm if the secondary structure elements are compared. This indicates that some external factor is responsible for the thermal stability of thioredoxin, which could well be related to smaller amount of water contained in the BacTrx compared to the *E. coli* Trx.

In an attempt to capture atomic-level mechanisms of the relationship between thermal stability and water content, we compared atomic-level structures of *E. coli* and *B. acidocaldarius* thioredoxins where we show positions of internal first hydration shell water over the surface of the protein, relative to positions of hydrophilic and hydrophobic surfaces of the protein. It is evident that not only the overall quantities but also the first hydration shell contains fewer water molecules in the case of a thermo stable protein. Note that, in the case of the thermo stable protein, water does not assemble into clusters. This allows to suggest that thermo stable proteins have more self-neutralized charge distributions over its surface, which would be related to lower contents of crystallization water.

Another pertinent observation being made recently, compatible with the above data and with *Pyrococcus furiosus* chromosome reassembly following gamma irradiation and with the heat shock enhancing radiation resistance, revealed that the synchrotron radiation resistance versus time of Lysozyme crystal prepared by LB nanotemplate was larger than the corresponding classical vapour diffusion, as previously discussed. Here we find that lower water contents also correlates with higher thermal stability, while placing a mesophilic protein in a LB film leads to both lower water content and higher thermal stability, thus possibly linking together water contents, thermal stability, and LB film effects. Since entropy-driven water release, as recently shown, is the main thermodynamic driving force of protein crystallization, our data indicate that thermophilic proteins possibly have better structured protein/water interface than their mesophilic homologs.

Considering that structuring of water caused by increased hydrophobic group exposure has been traditionally assumed to be the cause of heat capacity increase upon protein unfolding, our data indicate that, for thermophilic proteins, the variation in C_p should be lower than those for their mesophilic homologs, which is the tendency in fact observed. Alternatively since solvent water structuring can result either from the hydrophobic effect or from more regular occurrence of polar protein groups along the protein-water boundary leading to higher hydrogen bond propensity, our finding is in agreement of the two apparently conflicting concepts.

The very well known phenomenon of increased numbers of salt bridges in thermo-stable proteins (even if apparently not sufficient to justify the increased thermal stability) could also lead to higher compactness of protein structure not necessarily via a denser hydrophobic core. If these considerations are true, some charged amino acid residues may have exceptionally high contributions to the free energy of the thermo stable protein's electric field so their mutation may lead to drastic effects, which apparently follows from the above quoted work. We have shown, by data mining, the systematically lower water contents in crystals of thermophilic protein compared to their mesophilic counterparts, especially for the case of thioredoxin, for which we also experimentally show that the low water content in the thermophilic homolog and in the LB film of mesophilic homolog corresponds to their higher thermal stability. In combination with our earlier data, this indicates that the decrease in surface or inner bound water may be the factor connecting experimentally observed changes in protein stability made in solution (as shown in all papers here cited and here), thin Langmuir film and nanotemplate-grown crystals, namely of the thermal stability in thermophiles, the thermal stability and storage stability for LB film, and the radiation resistance in nanotemplate-grown crystals. As shown before, indeed the significant radiation resistance of protein crystal derived from the protein thermo-stable LB patches acting as nucleation centers. Water content in LB-based protein crystal indeed decreases with respect to "classical" along with increase in alpha helix content, thereby suggesting that decrease in bound water is one the sources of thermal stability in thermophiles, thermal and storage stability in LB films, and radiation resistance in crystals (Table 3.4).

Chapter 4

Nanoscale Applications in Industry and Energy Compatible with Environment

This chapter overviews the present status of nanoscale applications of organic and biological nanotechnology to industry and energy, namely of those capable so far to yield a potential technological progress to electronics, energy and catalysis.

Particular emphasis is placed on what has been accomplished in our laboratory in the last eight years, whereby the references to numerous other groups can be found in recent complete reviews (Nicolini *et al.*, 2001; 2005; Nicolini and Pechkova, 2006).

The data presented here point to the successful engineering of nanotechnology based on supramolecular layer engineering of potential industrial relevance. In fact, as emphasized, the filmation process was able to induce high thermal stability and associated high lifetime and recycling, which represent a prerequisite for several processes of industrial interest.

Therefore, although work is still in progress to further optimize the parameters and to evaluate in more detail, case by case, the temporal stability of thin layers within required cost effectiveness and the reproducibility within a highly competitive industrial context, this methodology clearly represents a promising general-purpose tool for the design of new industrial products and processes.

4.1 Nanobioelectronics

This subchapter of the volume on Nanobiotechnology comes ten years after a comprehensive and well received treaty on *Molecular Bioelectronics* (Nicolini, 1996a) and eleven years after a special issue of Bioelectronics and Biosensors (volume 10, 1995) as a result of a Workshop called in Bruxelles by the European Commission to launch a Program on Bioelectronics, comprehensive of neural VLSI chips, engineered proteins, molecular manufacturing and biomolecular electronic devices (Nicolini, 1995).

Waiting then for a new book on the very same subject to be done in the near future, this section intends to summarize the most recent pertinent to Nanobiotechnology and promising applications to industry and energy in the field of organic and biomolecular nanoelectronics.

4.1.1 *Nanosensors*

Nanosensors represent by far the most active area in the field of nanoelectronics with interesting and promising applications in the short term, which will be here summarized in key examples representative of organic and protein-based nanosensors (for a recent review see Nicolini *et al.*, 2006).

This subchapter summarizes with few key examples the significant potential of nanostructured organic matrices for developing intelligent sensors for gas and liquid having potential impact to health care and environment control (Nicolini *et al.*, 2001a; 2006).

4.1.1.1 *Protein-based nanosensors*

Light sensitive proteins and metallo-proteins are the most frequently used proteins in sensor technology in liquid and in air.

Starting from the effects of volatile anaesthetics on the structure of bacteriorhodopsin (bR) in the purple membrane in solution this work tries to investigate the interaction of ether and hydrocarbon type anesthetics vapors with self-assembled bR thin films (Figure 4.1). A dedicated constant flux chamber has been built to maintain the sample in a rather constant atmosphere of anaesthetics, during the absorption measurements. The kinetics of absorption and desorption have been determined. The results obtained have been compared with those of bR in solution (Maccioni et al., 1996).

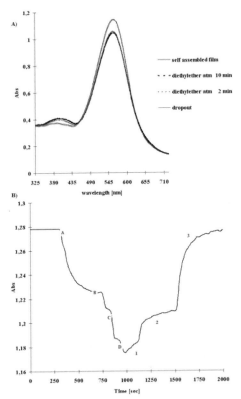

Figure 4.1. (a) The absorption spectra of the film pure and treated with diethylether during different periods of time and after relaxation. (b) The absorbance profile at 570 nm of a self-assembled film of bacteriorhodopsin upon exposition to vapors of diethyl ether by increasing pressure (A, B, C, D) and upon exposition to a compressed air stream for the desorption of the anesthetic (regions 1.2.3). (Reprinted with the permission from Maccioni et al., Bacteriorhodopsin thin film as a sensitive layer for an anaesthetic sensor, Thin Solid Films 284–285, pp. 898–900, © 1996, Elsevier).

The suspensions of PM have been obtained by dilution of the stock suspension (5 mg ml^{-1}) with distilled water. The concentration was checked spectrophotometrically, in particular the peak of proteins (at 280 nm) should not show the characteristic scattering signal. The concentration chosen for the experiments was 0.3 mg ml^{-1}. In the case of bR in a solution, the anaesthetic solution was just added to the bR solution and stirred for 10 min in order to equilibrate the system. In the case of chloroform, not mixable with water, the cuvette was shaken in order to promote the contact of PM with the solvent. The amount of the income anaesthetic vapours was qualitatively monitored injecting by a syringe different amounts proceeding from the heated vessel. In Figure 4.1a, a self-assembled film of bR in PM was treated with diethylether vapours: (a) the absorption spectra of the film pure and treated with diethylether during different periods of time is showed; (b) the decreasing of the absorbance at 570 nm was recorded during the injection of successive amounts of vapours of anaesthetic. In Figure 4.1a the absorbance maximum of the PM thick film appears at 563 nm and the intensity of absorbance decreases with the treatment with anaesthetic.

At the same time a broad band occurs at about 400 nm during the treatment and disappears after relaxation. It is difficult to give an interpretation of such a band, which cannot be precisely attributed to the anesthetic induced bR480, and bR380 forms further experiments will probably clarify the role of this band. Nevertheless this figure reveals either the functionality of the protein or the reversibility of the process qualitatively comparable with that obtained in solution.

Zones A, B, C, D are evident in Figure 4.1b, each of them corresponds to a progressive increase of the anaesthetic pres- sure: zone A is more wide than the others, this could be due to a higher pressure of the vapour injected in the cell or to an effect of saturation of the sample that makes more difficult the successive binding in the other zones. After the decrease of the absorbance of 0.1 units compressed air was injected in three steps (regions 1, 2, 3), obtaining almost the total recovering of the structure after 15 min. The work pointed out the possibility to use the self-assembled bR films as sensitive layers for optical biosensors for anesthetics. The important point is that the optical properties of the layer are reversible not only for ether type anesthetics but also for hydrocarbon type as chloroform. It allows one to consider such films as good candidates for biosensors of continuous monitoring of anesthetics.

4.1.1.1.1 P450's Langmuir films for clozapine, styrene and cholesterol

Recombinant cytochrome P450s consist of a large and highly diverse enzymes family that play a pivotal role in the metabolism of a wide variety of xenobiotics and drugs. For these attractive properties we have performed several studies to obtain nanostructures for device applications.

For each enzyme we have carried out structural and functional characterizations in order to optimize the immobilization process (Nicolini *et al.*, 2001; Ram *et al.*, 2001, Paternolli *et al.*, 2002a; Antonini *et al.*, 2003) sensing we have used several immobilization techniques including the layer-by-layer, the Langmuir-Blodgett, a gel-matrix and the self-assembly (solution spreading) and several forms of the given cytochromes including the wild type, the native recombinant and the GST-fused P4501A2, P4502B4 and P450scc cytochromes.

The conclusive new insights respect to all our previous papers published is the comparative undertaking capable to identify Langmuir-Blodgett among all various immobilization technologies as the one yielding the optimal sensing results as long recombinant P450 of high grade is utilized (Paternolli *et al.*, 2004, 2007).

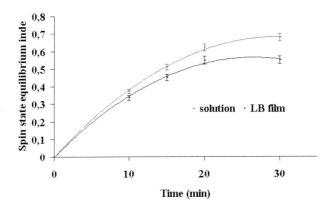

Figure 4.2. Graph of the spin state equilibrium index of cytochrome P4502B4 in solution and in LB film (30 layers). The data were acquired at different exposure times to styrene atmosphere, pointing to a shift to higher spin value (Reprinted with the permission from Paternolli *et al.*, Recombinant cytochrome P450 immobilization for biosensor applications, Langmuir 20, pp. 11706–11712, © 2004, American Chemical Society).

LB proves indeed superior to all other approaches being tested with the only exception of APA (see later). This has been generalized to all P450-based sensors being tested (Nicolini et al., 2001, 2006; Paternolli et al., 2002a, 2007), namely that the sensing can be optimized in terms of performance, stability, reusability and efficiency whenever LB immobilization and proper mutant are utilized.

Table 4.1. Substrates and primary products of the cytochrome P450s. (Reprinted with the permission from Paternolli et al., Recombinant cytochrome P450 immobilization for biosensor applications, Langmuir 20, pp. 11706–11712, © 2004, American Chemical Society).

Cytochrome	Reaction	Substrate	Product
P4501A2[a]	Demethylation	Clozapine	N-desmethylclozapine[d]
P4502B4[b]	Epoxydation	Styrene	Epoxystyrene
P450scc[c]	Hydroxylation	Cholesterol	Pregnenolone[e]

[a] Brosen, 1993; Brosen et al., 1993, Pirmohamed et al., 1995.
[b] Miller, 1988; Vaz et al., 1998.
[c] Nicolini et al., 2001; Ortiz de Montellano, 1986; Waterman and Simpson 1985.
[d] Cytochrome P4501A2 metabolizes the clozapine to produce N-desmethylclozapine and, secondarily, N-oxideclozapine.
[e] Cytochrome P450scc converts the cholesterol into pregnenolone and isocapraldehyde.

Spectroscopic biosensor for styrene is based on P4502B4 LB, whereby the characteristic Soret peak of the cytochrome P4502B4 free and complexed with its substrate was monitored in LB film.

In fact, it is known that the interaction between cytochrome and the substrate changes the protein spin state. The results are summarized in Figure 4.2.

The indexes of P4502B4 spin state equilibrium were calculated from the absorbance spectra acquired at different times of exposition to the styrene atmosphere. It was found that the exposure of cytochrome P4502B4 to a styrene saturated environment determines the progressive shift to higher spin value following the P4502B4 molecules binding to their substrates in both solution and film, as it is the case going from solution to LB film without styrene (Table 4.1).

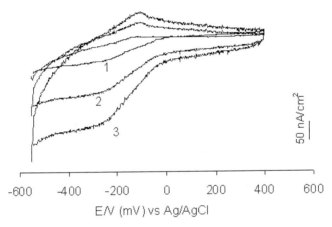

Figure 4.3. Cyclic voltammetry of cytochrome P450scc LB film with addition of cholesterol (Reprinted with the permission from Paternolli *et al.*, Recombinant cytochrome P450 immobilization for biosensor applications, Langmuir 20, pp. 11706–11712, © 2004, American Chemical Society).

The styrene detectable quantity found is limited by the home-made measurement system, in which the control of the working volume is difficult (Table 4.2).

Amperometric sensors for cholesterol based on P450scc in LB films. Protein studies utilizing electrochemical techniques such as cyclic voltammetry have provided insight into the functional properties of redox-active centers (Paternolli *et al.*, 2004). Electrochemical measurements are thereby performed in order to study the binding between cholesterol and cytochrome P450scc. The current values of

P450scc LB film are plotted as a function of cholesterol concentration additions in Figure 4.3.

Table 4.2. Physiological and minimum detectable concentrations of clozapine and cholesterol in the blood and of the styrene in the atmosphere. (Reprinted with the permission from Paternolli et al., Recombinant cytochrome P450 immobilization for biosensor applications, Langmuir 20, pp. 11706-11712, © 2004, American Chemical Society).

Detected metabolite	Physiological concentration range	Minimum detectable concentration (P450 μg in the LB per electrode
clozapine[a]	50–600 ng/mL	50 ng/mL
styrene[b]	0–85.2 mg/mc	500 mg/mc (4 μg P4502B4 in 40 LB
cholesterol[c]	100–220 mg/dL	11.6 mg/dL (5 μg P450scc in 40 LB)

[a] Haring et al., 1989; Buur-Rasmussen and Brosen, 1999; Rahden-Staron et al., 2001.
[b] American Conference of Governmental Industrial Hygienists, Inc., ACGIH, 1997.
[c] Report of the National Cholesterol Education and Treatment of High Blood Cholesterol, 1988.

Each addition consists of 50 μl of cholesterol solution (10 mM in TritonX-100) until reaching a final concentration of 100–750 μM. The time of the process is estimated to be about 3 minutes. The data show that the electrochemical process itself gives cytochrome P450scc electrons allowing it to react with cholesterol. The kinetics of the absorption and reduction process might be the result of the ion diffusion controlled process.

Moreover, the data plot suggests that the steady-state current is directly proportional to the cholesterol concentration until the beginning of the saturation trend. Amperometric sensor for clozapine based on P4501A2 was performed in gel matrix.

The electrochemical studies proved, however, that while the behavior of P4502B4 is comparable to P450scc in LB films (data not shown), the cytochrome P4501A2 LB films cause some problems with the cyclic voltammetry particularly in the study of the interaction between this enzyme and its substrate (clozapine), apparently due to the low pure grade (as shown in materials and methods). In fact, it is well known that molecular impurities on the electrodes may impede electron transfer and prevent enzyme-electrode electrical communication (Joseph et al., 2003).

For this reason we immobilized P4501A2 in a gel-matrix (Figure 4.4) as described in the materials and methods, and we employed chronoamperometry to verify the possibility of producing an amperometric sensor to detect clozapine. Aliquots of clozapine (40 µM in methanol) were added to the working mixture in order to obtain an amperometric response curve shows the resulting current as a consequence of constant potential.

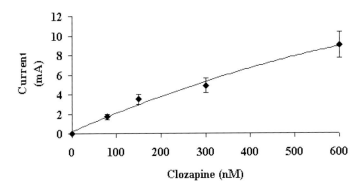

Figure 4.4. Current response of cytochrome P4501A2 gel-matrix as a function of the clozapine concentration by rhodium-graphite s.p.e. Working solution is 10 mM K-phosphate buffer, pH 7.4, The potential electrode was poised at -600 mV (Reprinted with the permission from Paternolli et al., Recombinant cytochrome P450 immobilization for biosensor applications, Langmuir 20, pp. 11706–11712, © 2004, American Chemical Society).

It can be observed that raising the clozapine concentration increased the response current, which appear stable at room temperature for over 30 days to 60% of its original value. Because the therapeutic range of clozapine in plasma is between 0.16 and 1.83 µM (50–600 ng/ml) (Haring et al., 1989; Buur-Rasmussen and Brøsen, 1999; Rahden-Staron et al., 2001) we conclude that the sensitivity extrapolated by our experiments is sufficient for routine measurements, even if the temporal stability of gel matrix appears quite less than that of the LB method (Table 4.2). By the combination of proper immobilization (LB), transducer and nanostructured mutants of high-grade stable and selective, P450-based sensors appear capable to detect the interaction with a wide range of organic substrates such as fatty acids, drugs, and toxic compounds. Only in the presence of low purity grade protein, as in the case of our preparation of P4501A2, is necessary to use a gel-matrix to warrant the optimal clozapine sensing (Figure 4.4).

4.1.1.1.2 Direct electron-transfer with gold nanoparticles

As shown in chapter 1, nanoparticles of metals have been investigated due to their novel material properties which differ from the bulk materials (Penn *et al.*, 2003; Han *et al.*, 2002). The gold nanoparticles display electronic, chemical and physical properties advantageous for application in bioelectrochemistry. Colloidal gold nanoparticles can adsorb proteins (Hu *et al.*, 2003).

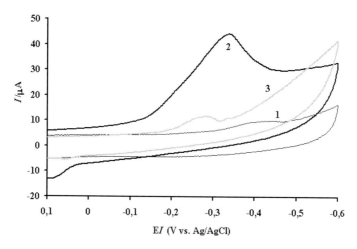

Figure 4.5. Cyclic voltammograms of screen-printed rhodium-graphite electrode with Au nanoparticles (1), with cytochrome P450scc and Au nanoparticles (2) and bare electrode with cytochrome P450scc (3). Experiments were performed in aerobic 100 mM phosphate buffer, 50 mM KCl, pH 7.4. Scan rate 50 mV/s^{-1}. Gold nanoparticles were prepared in the presence of dodecanthiol (Reprinted with the permission from Shumyantseva *et al.*, Direct electron transfer between cytochrome P450scc and gold nanoparticles on screen-printed rhodium-graphite electrodes, Biosensors & Bioeletronics 21, pp. 217–222, © 2005, Elsevier).

Bio-composites of gold nanoparticles and glucose oxidase were electrodeposited on indium tin oxide glass electrode for fabricating glucose biosensor (Bharathi and Nogami, 2001).

Gold nanoparticles was here utilized in the direct electron transfer from electrode to cytochromes P450scc, namely between cytochrome P450scc and Au colloid modified screen-printed rhodium graphite electrode (Shumyantseva et al., 2005). To construct a biosensor on the basis of cytochrome P450scc the amperometric response on the cholesterol addition has been measured (Figure 4.5).

The combination of bioelectrochemistry and nanobiotechnology has permitted thereby to construct a high sensitive amperometric biosensor for cholesterol measurements.

At vertical position of electrode in 1 ml electrochemical cell voltammograms were observed only for the first scan. Then enzyme was dissolved in solution. Vertical regime needs additional covalent binding of enzyme onto the electrode surface.

In planar regime only a small volume (20–60 µl) of buffer is needed. Integration of the reduction peak permits calculating the charge and thus the concentration of electro-active molecules on the surface of the Au-nanoparticles-electrode.

For the P450scc a value of 2.57 pmol (16.1 pmol/cm^2, when assuming a plane surface) was estimated, corresponding to 5.14% of the total amount of the loaded enzyme to be electro-active.

Electrodes with gold nanoparticles and P450scc were used for measurements of cholesterol concentration in solution. Figure 4.6 shows the steady-state current response of the cytochrome P450scc electrodes to cholesterol addition. In these experiments 100 µM cholesterol stock solution in 0.3% sodium cholate was used.

Aliquots of cholesterol were added to the analyzed solution. The sensitivity of this type of biosensor is 0.13 µA µM^{-1} and the detection limit is 70 µM of cholesterol in the presence of sodium cholate as detergent. This type of P450scc-electrode is quite sensitive and needs small volumes of analyzed solutions.

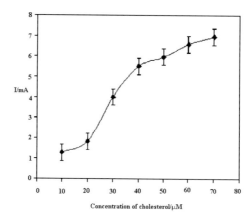

Figure 4.6. The amperometric response of screen-printed rhodium-graphite Au-P450scc electrode to increasing cholesterol concentration: 100 μM stock solution in 0.3 % sodium cholate, with 10 μM cholesterol being repeatedly added. Total volume of electrolyte was 60 μl. Current was measured at constant potential of -400mV (vs. Ag/AgCl) (Reprinted with the permission from Shumyantseva *et al.*, Direct electron transfer between cytochrome P450scc and gold nanoparticles on screen-printed rhodium-graphite electrodes, Biosensors & Bioeletronics 21, pp. 217–222, © 2005, Elsevier).

It is worth to notice that the solution casting nanostructured electrodes (Shumyantseva *et al.*, 2005) have an increased electron-transfer rate with respect to the clean rhodium-graphite electrodes; this fact may be due to the increased diffusion of the electrolyte in the nanostructured electrodes (Akiyama *et al.*, 2003), making this type of electrodes suitable for the detection of cholesterol in a small volume of electrolyte.

4.1.1.1.3 Improved mechanical stability and optimal performance with anodic porous alumina

To further improve the mechanical stability of electrodes based on P450scc for LDL-cholesterol detection and measure, anodic porous alumina (APA) was recently used (Stura *et al.*, 2007).

This inorganic matrix, which pores can be tuned in diameter modifying the synthesis parameters (see chapter 1), was realized with cavities 275 nm wide and 160 micron deep (as demonstrated with the AFM and SEM measurement shown in chapter 1), to allow the immobilization of P450scc macromolecules preserving their electronic sensitivity to its native substrate, cholesterol. Even if the sensitivity of the APA+P450scc system was slightly reduced with respect to the pure P450scc system, the readout was stable for a much longer period of time, and the measures remained reproducible inside a proper confidentiality band, as demonstrated with several cyclic voltammetry measures (Figure 4.7).

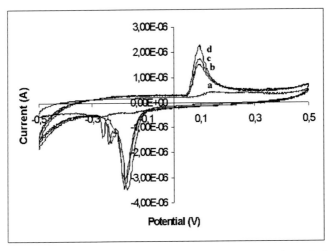

Figure 4.7. I–V current–voltage curves of s.p.e. of APA-P450scc. Electrode in presence of substrate LDL-cholesterol. The s.p.e. of APA-P450scc electrode was tested, after a month, in a 10mM K-phosphate buffer pH 7.4 in presence of LDL. (a) P450, (b) LDL 0.5 mg/ml, (c) LDL 1.1 mg/ml and (d) LDL 1.6 mg/ml. Results are representative of one of three similar experiments. (Reprinted with the permission from Stura *et al.*, Anodic porous alumina as mechanical stability enhancer for LDL-cholesterol sensitive electrodes, Biosensors & Bioeletronics 23, pp. 655–660, © 2005, Elsevier).

To optimize the adhesion of P450scc to APA, a layer of poly-l-lysine, a poly-cathion, was successfully implemented as intermediate organic structure (Figure 4.7). The functionality of a rhodium-graphite screen-printed electrode (s.p.e.) modified with anodic porous alumina, and the reported results prove the achievement of the optimal immobilization of cytochrome P450scc onto the modified electrode in both vertical and horizontal position, achieving far better results with respect to previously ones obtained in our laboratories (Shumyantseva *et al.*, 2005); and the optimization of the electron transfer's stability leading to the optimal detection of cholesterol in the clinical range concentration for longer times of use.

Table 4.3. Comparison among the results of the works about cholesterol detection using P450scc published by our group in the last years (Reprinted with the permission from Stura *et al.*, Anodic porous alumina as mechanical stability enhancer for LDL-cholesterol sensitive electrodes, Biosensors & Bioeletronics 23, pp. 655–660, © 2005, Elsevier).

P450scc	Response in current [nA μM^{-1}]	Time stability	Temperature stability	Range of cholesterol detectability [μM/cm^2]
In LS film (30 layers)[1]	0.4	90 days	Room temperature	50–750
With gold nanoparticles optimization[2]	130	2–3 days	4 °C	40–300
In Gel matrix[3]	400	10 days	Room temperature	50–1000
In APA+PLL matrix[4]	300	>5 months	Room temperature	50–1000

[1] Nicolini *et al.*, 2001
[2] Shumyantseva *et al.*, 2005
[3] Antonini *et al.*, 2004
[4] Stura *et al.*, 2007

In comparison with all other methods previously described (Table 4.3) the APA-P450scc electrode represents the most promising alternative for the existing amperometric biosensor, quite more stable and quite independent from its working position (either vertical or horizontal).

4.1.1.1.4 BR-based light adressable potentiometric sensor

Thin-film technologies (Ulman, 1991; Nicolini, 1997) allow the assembly of biological materials as bacteriorhodopsin (bR) in a 2D system for photocells (see later and Paternoli *et al.*, 2008; Nicolini *et al.*, 1999) and for Light Addressable Potentiometric Sensor (LAPS) development (Nicolini *et al.*, 1998; Fanigliulo *et al.*, 1996; Bousse *et al.*, 1994; Sartore *et al.*, 1992a,b,c).

For these devices, LB seems to be one of the most promising, due to its ability to form molecular systems having high packing degree and molecular order.

Therefore, the ability of bR to form thin films with excellent optical properties and the bR intrinsic properties themselves make it an outstanding candidate for use in optically coupled devices. bR thin layers have been widely studied (Hwang *et al.*, 1977a,b); Ikonen *et al.*, 1993; Shibata *et al.*, 1994; Méthot *et al.*, 1996; Sugiyama *et al.*, 1997) because they exhibit bi-stability in optical absorbance and they provide light-induced electron transport of protons through the membrane. Furthermore, their extremely high thermal and temporal stability also allows considering them as sensitive elements for electro-optical devices (Erokhin *et al.*, 1996; Miyasaka *et al.*, 1991). However, in order to use bR properties to provide photo-voltage and photocurrent, it is necessary to orient all the molecules in such a way that all the proton pathways are oriented in the same direction. The LB technique in its usual version does not allow this quite efficiently. When bR-containing membrane fragments are spread at the air/water interface, they orient themselves rather randomly in such a way that the proton pathway vectors are oriented in opposite directions in different fragments. Nevertheless, a technique of electrochemical sedimentation is known that allows the deposition of highly oriented bR layers. However, the layers deposited with this technique are rather thick and not well controllable in thickness. LB technique allows one to form a monolayer at the water surface and to transfer it to the surface of supports. Formation of the bR-containing membrane fragments monolayer at the air/water interface, however, is not a trivial task, for it exists in the form of membrane fragments. These fragments are rather hydrophilic and can easily penetrate the subphase volume.

In order to decrease the solubility, the subphase usually contains a concentrated salt solution. The efficiency of the film deposition by this approach (Sukhorukov *et al.*, 1992) was already shown. Nevertheless, it does not allow one to orient the membrane fragments. Because the hydrophilic properties of the membrane sides are practically the same, fragments are randomly oriented in opposite ways at the air/water interface. Such a film cannot be useful for this work, because the proton pumping in the transferred film will be automatically compensated; *i.e.*, the net proton flux from one side of the film to the other side is balanced by a statistically equal flux in the opposite direction. On the other hand, the technique of electrochemical sedimentation is known allowing the formation of rather thick bR films by orienting them in the electric field. Therefore, the following method was realized the results obtained scheme shown in Figure 4.8. A 1.5 M solution of KCl or NaCl (the effect of preventing bR solubility of these salts is practically the same) was used as a subphase. A platinum electrode was placed in the subphase. A flat metal electrode, with an area of about 70% of the open barrier area, was placed about 1.5–2.0 mm. above the subphase surface. A positive potential of 50–60 V was applied to this electrode with respect to the platinum one. Then bR solution was injected with a syringe into the water subphase in dark conditions.

The system was left in the same conditions for electric field-induced self-assembly of the membrane fragments for 1 h. After this, the monolayer was compressed to 25 mN/m surface pressure and transferred onto the substrate (porous membrane). The residual salt was washed with water. The water was removed with a nitrogen jet.

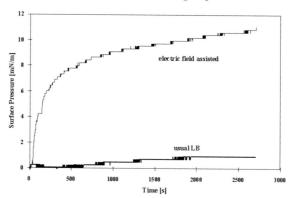

Figure 4.8. Surface pressure as a function of time in the electric-field assisted bR for the construction of a new kind of Light Addessable Potentiometric Sensor. The photosignal (see Table 4.4) has a similar dramatic enhancement with application of the electric field (Reprinted with the permission from Nicolini *et al.*, Towards light-addressable transducer bacteriorhodopsin based, Nanotechnology 9, pp. 223–227, © 1998, IOP Publishing Limited).

The dependence of the surface pressure upon the time with and without applied electric field is shown in Figure 4.8. It is clear that the electric field strongly improves the ability of the membrane fragments to form a monolayer at the water surface and thereby to enhance significantly the photosignal. X-ray measurements of the deposited multi-layers revealed practically the same structure in films prepared with the usual LB technique and electric field-assisted monolayer formation. This finding does not seem strange. In fact, an electric field only aligns the fragments at the air/water interface, providing equal orientation of the proton pathways. The layered structure in this case remains the same. X-ray curves from both types of samples revealed Bragg reflections corresponding to a spacing of 46 Å, which is in a good correspondence with the membrane thickness. In order to control the degree of bR orientation, photo-induced current was also measured. Photosignal was also measured, as a function of the illumination wavelength (Nicolini *et al.*, 1998). Moreover, one monolayer of bR was deposited onto the porous membrane.

Table 4.4. Photocurrent observed in a system using porous membranes covered with bR film deposited by the usual LB technique and electric field-assisted technique. A standard error of about 10% is observed over five independent positive measurements (Reprinted with the permission from Nicolini *et al.*, Towards light-addressable transducer bacteriorhodopsin based, Nanotechnology 9, pp. 223–227, © 1998, IOP Publishing Limited).

	Light-on current [pA]	Light-off current [pA]
Usual LB technique	15	10
Electric field-assisted monolayer formation	820	10

The results are summarized in Table 4.4. It is clear that the photoresponse in the case of electric field-assisted monolayer formation is much higher compared to that after a normal LB deposition (in the last case the signal value is comparable with the noise, indicating a mutually compensating orientation of the membrane fragments in the film).

The observed results allow the conclusion that the suggested method of electrically assisted monolayer formation is suitable for the formation of bR LB films, where the membrane fragments have preferential orientation. Since electric field-assisted monolayer formation at the air/water interface turned out to provide the possibility of highly oriented bR LB film formation, it was possible to suggest another application of bR films for transducing purposes.

The principles of device realization are described next. The scheme of the proposed device is presented in Figure 4.9. Porous membrane with deposited bR film separates two chambers with electrolytes and two platinum electrodes.

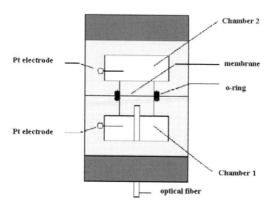

Figure 4.9. Schematic view of the measuring chamber used for the experiment with the bR membrane. Porous membrane with deposited bR film is separating two chambers with electrolytes. Light fibre is attached to the X-Y mover, which allow to illuminate the desirable parts of the membrane (Reprinted with the permission from Nicolini *et al.*, Towards light-addressable transducer bacteriorhodopsin based, Nanotechnology 9, pp. 223–227, © 1998, IOP Publishing Limited).

A light fiber is attached to the X-Y mover, which allows illuminating desirable parts of the membrane. Illumination of the membrane part will result in the proton pumping through it, carried out by bR. Therefore, a current between the electrodes will appear. This current must depend upon several factors, such as light intensity, pH of the electrolytes, and gradient of the pH on the membrane. One of the possible applications of the suggested device is mapping of 2D pH distribution in the measuring chamber, which can result from the working of enzymes immobilized in this chamber. By scanning the light over the membrane it will be possible to obtain the current proportional to the pH gradient at the illuminated point, and by maintaining the pH value fixed in the reference chamber it will be possible to calculate absolute pH values at different points over the whole membrane surface (Nicolini *et al.*, 1998; Fanigliulo *et al.*, 1996; Bousse *et al.*, 1994; Sartore *et al.*, 1992a,b,c). If different types of enzymes, producing or consuming protons during their functioning, are distributed over the area close to the membrane, the device will allow one to determine the presence of different substrates in the measured volume, performing, therefore, as a multiple enzymatic biosensor. Space resolution of the transducer is extremely high. Because each bR molecule performs proton pumping, it will be comparable with the protein size (about 2 nm). In practice, however, it will be limited by the possibility of focusing the light beam, which as shown separately can be now focused to nanosize range. But, in any case, it will be more than in existing transducers, subject to technological problems.

4.1.1.2 *Organic nanosensors*

Over the last decade the Biophysics Institute of the University of Genova and the Fondazione EL.B.A. in close cooperation with the Scientific and Technological Park of the Elba Island and with numerous leading international companies (STM, ABB, Edison, FIAT, Elsag-Bailey) have been quite active in developing neural sensors (Adami *et al.*, 1996a,b), based on organic materials (Ding *et al.*, 2001, 2002b; Ram *et al.*, 1997, 1997a, 1997b, 1997c; Bavastrello *et al.*, 2002, 2004, 2004a; Valentini *et al.*, 2004a) and utilizing a wide range of transducers, either amperometric (Antonini *et al.*, 2004; Shumyantseva *et al.*, 2004), potentiometric (Adami *et al.*, 1995a, 1996a,b, 2007), conductimetric (Valentini *et al.*, 2004a; Bavastrello *et al.*, 2004a), nanogravimetric (Nicolini *et al.*, 1997; Adami *et al.*, 1996a,b), spectrophotometric (Paternolli *et al.*, 2002a) and fluorometric (Paddeu *et al.*, 1995b). Several immobilizing techniques were also employed over the time, ranging from self-assembly (Shumyantseva *et al.*, 2004; Paternolli *et al.*, 2002a), to layer-by-layer (Ram *et al.*, 2001b; Shumyantseva *et al.*, 2001) and Langmuir-Blodgett (Valentini *et al.*, 2004b; Ghisellini *et al.*, 2004; Ding *et al.*, 2002a; Ram *et al.*, 2001a). We intend here to summarize the present state of the art with emphasis on the recent nanostructuring of sensing organic matrices and to critically assess the potential industrial relevance of emerging intelligent sensors for a wide variety of applications for health and environment.

4.1.1.2.1 Intelligent sensing for metals in the environment

In 1976 a new technique, derived from polarographic methods, was proposed, the Potentiometric Stripping Analysis (PSA).

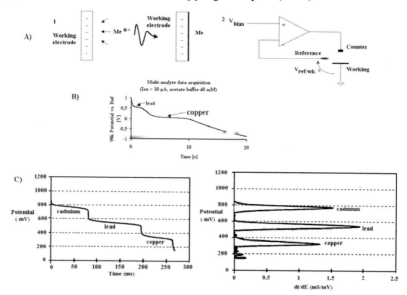

Figure 4.10. Schematic description of a PSA experiment. (A), part 1, the electrochemical process taking place during the first step is illustrated: the working electrode polarization causes the deposition of the metal ions on it; part 2 shows the circuital schematics to drive this phase. (B) describes, in the same manner, the second step of the analytical technique. (Figure 1 from Nicolini et al, 2006). (C) the data analysis software transforms and plot the acquired data E(t) in the inverse derivative form dt/dE as a function of the recorded potential, evidentiating the plateau of the stripping potentiogram as peaks in the derivative potentiogram, whose area is a linear function of the concentration of the metal in solution (A,B: Reprinted with the permission from Nicolini *et al.*, Nanostructured organic matrices and intelligent sensors, in Smart Biosensor Technology, (G. Knopf & Bassi A.S. eds), CRC Press, pp. 231–244, © 2006, Taylor & Francis Group LTD; C: Reprinted with the permission from Adami *et al.*, A potentiometric stripping analyzer for multianalyte screening, Electroanalysis 19, pp. 1288–1294, © 2007, Wiley-VCH Verlag GmbH & Co. KGaA).

This chemical method (Estela *et al.*, 1995) is very sensitive for the detection of metal ions in aqueous samples and consists of two steps: the ions are first electrolytically concentrated by deposition on a working electrode (preconcentration, or plating stage), then a metal stripping phase follows, during which no control is performed on the potential. The latter step is accomplished, usually, with a chemical oxidant in solution (Hg^{++} or dissolved oxygen).

After the preconcentration phase, the analytical signal (the potential of the working electrode) is recorded as a function of time and then utilized to obtain quantitative information about the metal ions in solution. A recent paper (Adami *et al.*, 2007) demonstrates the possibility to design and realize a simple and low-cost instrument for monitoring environmentally significant metals by implementing the PSA technique (Figure 4.10), connecting a cheap electronics to a computer.

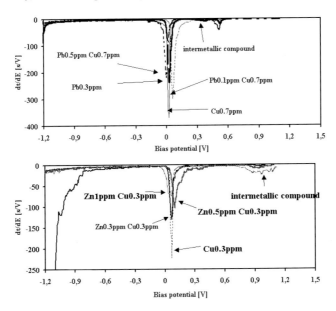

Figure 4.11. Intermetallic formation during the multianalyte determination of lead and copper (A) and zinc and copper (B). This effect, related both with the analytes and with the technique, affects the performance of the analyzer in terms of sensitivity and specificity (Reprinted with the permission from Adami *et al.*, A potentiometric stripping analyzer for multianalyte screening, Electroanalysis 19, pp. 1288–1294, © 2007, Wiley-VCH Verlag GmbH & Co. KGaA).

Such analyzer, suitable for heavy metals analyses is based on the Constant Current Potentiometric Stripping Analysis and on a very simple electrochemical cell. In the first step a potentiostat (Jagner *et al.*, 1993) drives the electrochemical cell through the counter electrode ensuring the proper biasing voltage $V_{ref/wk}$. In the second step, a constant oxidizing current (1–100 µA) is fed into the working electrode and the metal is forced to move again into the solution, producing a potential drop between reference and working electrodes. The system becomes a galvanostat and the potential drop $V_{ref/wk}$ is now recorded as a function of time.

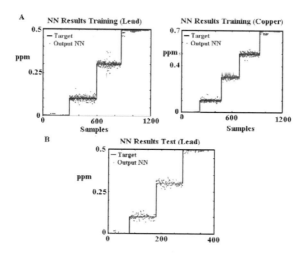

Figure. 4.12. A) NN response after the training phase. During this phase the net knows the final concentrations and is let running in order to detect the optimal node weights to obtain the desired result. Once finished, the net is well trained if the results (red) fit with the expected concentrations (—) for most samples. The figure shows both lead and copper results for the same set of samples, where there was simultaneous presence of both ions. B) NN response after the test phase. Here the net has fixed weights (those calculated and optimized in the previous step) and must detect the correct ion concentrations as the output of the net itself. (—) is again the correct ion concentration, while (red) are the detected values for all the samples (Reprinted with the permission from Adami *et al.*, A potentiometric stripping analyzer for multianalyte screening, Electroanalysis 19, pp. 1288–1294, © 2007, Wiley-VCH Verlag GmbH & Co. KGaA).

The software tools, user-friendly and easy to use, allow to detect the analytical signals and to extract the information data. With synthetic sample we obtained enough sensitivity to propose a first-screening apparatus but, when the sample is a real matrix, it can contain some different components, in particular some different metal ions, each one acting as a possible interferant (Figure 4.11). In order to overcome this problem (Adami *et al.*, 2007), we integrated in the proposed system a Neural Network Algorithm, the Multi Layer Perceptron (MLP), which appears able to reveal the real concentrations of the different metal ions, after a severe training (Figure 4.12).

From the results described it appears that sophisticated software implementation, such as those based on NN algorithm could be utilized for recovering an analytical signal derived from a multisensing application. Many examples were proposed for gas sensing but not so many for ions screening in solution. We are optimizing our NN for a better discrimination of metal traces, by changing some parameters and some strategies and we hope that a software solution will eliminate the need to introduce modifications in the proposed analyzer, such as the use of a modified working electrode (with a mercury layer or with some chemical agents, such as chelant compounds or resins) or of some sample treatments or of more sophisticated electronics.

4.1.1.2.2 Nanocomposites sensing for inorganic vapours

Various nanostructures have been investigated to determine their possible applications in biosensors. These structures include nanotubes, nanofibers, nanorods, nanoparticles and thin films (see chapter 1). A gas sensor, fabricated by selective growth of aligned carbon nanotubes (CNTs) by pulsed plasma on Si_3N_4/Si substrates patterned by metallic platinum, was recently presented for inorganic vapor detection at room temperature (Valentini *et al.*, 2004a).

Poly(o-anisidine) (POAS) deposition onto the CNTs device was shown to impart higher sensitivity to the sensor (Valentini *et al.*, 2004a). Upon exposure to HCl the variation of the CNTs sensitivity is less than 4%, while the POAS-coated CNTs devices over a higher sensitivity (*i.e.* 28%). The extended detection capability to inorganic vapors apparent in Figure 4.13 is attributed to direct charge transfer with electron hopping effects on intertube conductivity through physically adsorbed POAS between CNTs. The CNTs thin film was grown using a radio frequency pulsed plasma enhanced chemical vapor deposition (RF-PECVD) system (Gonzalez, 1992). Prior to the nanotube growth, a Si_3N_4/Si substrate was patterned with platinum film (60 nm thick) by vacuum deposition through shadow masks, containing rectangular stripes 30 μm wide and a back deposited thin film platinum heater commonly used in gas sensor applications. A thin film (3 nm) of Ni catalyst was deposited onto the Si_3N_4/Si substrates using thermal evaporation.

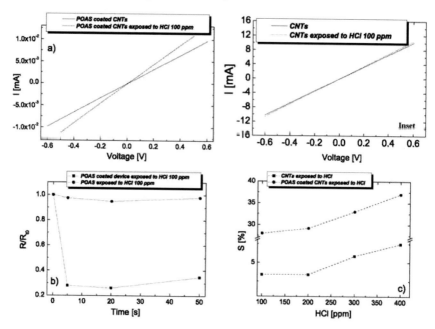

Figure 4.13. (a) *I-V* curves at room temperature of POAS coated devices to HCl vapor. The inset shows *I-V* curves of the CNTs device at room temperature to HCl 100 ppm. (b) The time-dependence change of the normalized resistance (R_{t0} is the initial resistance of the sample) of POAS coated device and CNT film at room temperature to HCl 100 ppm. (c) Sensitivity vs. HCl concentrations of POAS coated device and CNT .lm at room temperature (Reprinted with the permission from Valentini *et al.*, Sensors for inorganic vapor detection based on carbon nanotubes and poly(o-anisidine) nanocomposite material, Chemical Physics Letters 383, pp. 617–622, © 2004a, Elsevier).

Figure 4.13 shows two distinct groups of *I-V* curves resulting from nanotubes and polymer POAS coated nanotubes. Each group consists of results from several sensors. It can be seen (Valentini *et al.*, 2004a;

Nicolini et al., 2006) that the conductivity (slope) is almost the same for CNTs and POAS-coated CNTs, but it varies for the device when exposed to HCl. In particular, it shows that the conductance of nanotubes slightly increases), while the POAS coated device shows a significant conductance increment. It is interesting to note that the response transient of POAS coated device is a few seconds, while sensor based on resistance changes of POAS exhibits a poorer response time. The circumstance to maintain film resistance below 1 kΩ, which is significant lower with respect to that reported for POAS sensors (about 50 MΩ), makes POAS coated CNTs films integration in electronic circuitry easier and cheaper, since lower DC voltages are required to drive the sensor response. The doping process of poly-anilines is always associated to conformational modifications of the polymer chains, due to the local distortions created by the addition of H^+ ions to the basic sites and usually provides stable systems. It means that the conducting polymer in the doped form can be maintained in this state for long periods of time till the material reacts with basic reagents and strongly changes its chemical-physical properties. In other words, the reversibility of the process is not spontaneous. If we define sensor sensitivity *(S)* as the ratio *S =[(R_A − R_G)=R_A] x 100*, where R_A represents the resistance in air and R_G the resistance in vapor, the gas sensitivity increases from $S = 3:0\%$ to 27.9%. It reveals that by selecting proper polymer functionalization sensor sensitivity to HCl may be improved. In conclusion, CNTs thin films prepared by pulsed RF PECVD demonstrated their potentiality as a new class of materials for HCl detection for environmental applications, with polymer functionalization enhancing their sensitivity.

4.1.1.2.3 Organic gas sensing

Whenever a nanocomposite of multiwalled carbon nanotubes (MWNTs) embedded in poly(2,5-dimethylaniline) (PDMA) was synthesised by oxidative polymerisation, the nanocomposite (PDMA-MWNTs) showed a progressive spontaneous undoping process along the time associated to the instability of the doping agent, constituted by HCl, inside the polymeric matrix (Bavastrello et al., 2004a,b).

The study of the undoping process revealed that this phenomenon is related to the synergetic effect of the sterical hindrance of the substituents on the aromatic rings and the presence of MWNTs inside the polymeric matrix. The conducting properties connected to a doping-undoping equilibrium in the presence of the doping agent were also investigated. The instability of the doping process allowed us to fabricate a spontaneous reversible sensor for acid vapours by setting up a comparative potentiometric circuit and engineering the sensitive element directly on the circuit board. The fabricated devices were connected to the electrometer by means of silver wires and silver paint, as shown in the schematic of Figure 4.14.

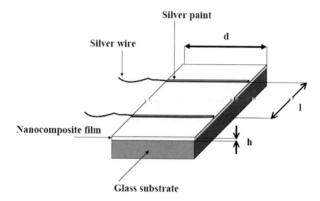

Figure 4.14. Schematic of devices employed for the determination of the specific resistance (Reprinted with the permission from Bavastrello *et al.*, Poly(2,5-dimethylaniline-MWNTs nanocomposite: a new material for conductometric acid vapors sensor, Sensors and Actuators B 98, pp. 247–253, © 2004b, Elsevier).

The instability of the doping process allowed us to fabricate a spontaneous reversible sensor for acid vapors (Bavastrello *et al.*, 2004b) by setting up a comparative potentiometric circuit and engineering the sensitive element directly on the circuit board. In order to test the response of the fabricated device after exposing to HCl vapors, an experimental set up was realized as following. The device was placed inside a container of 120 ml and thus sealed by means of a parafilm layer, letting the wires out through it to be connected to the electrometer. Volumes of 0.2 and 0.5 ml of saturated vapors of HCl were injected through the parafilm membrane. Current versus time measurements at a fixed bias of 10 V was then carried out. The relative results obtained from the experiments are shown in Figure 4.15, which illustrates the *V-I* characteristics of PDMA-MWNTs nanocomposite in the undoped and doped forms evidenced with curves 1 and 2, respectively.

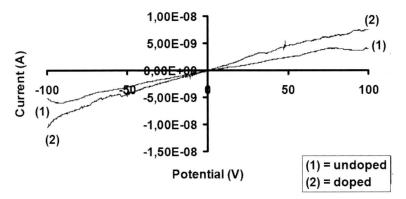

Figure 4.15. *V-I* characteristics of PDMA-MWNTs nanocomposite in both the undoped (curve 1) and doped (curve 2) forms. The material showed a quasi-linear behaviour. (Reprinted with the permission from Bavastrello *et al.*, Poly(2,5-dimethylaniline-MWNTs nanocomposite: a new material for conductometric acid vapors sensor, Sensors and Actuators B 98, pp. 247–253, © 2004b, Elsevier).

It can be observed that the system tends to return to the initial conditions of resistance after about 3 min since the injection of HCl vapors. This behavior is the consequence of the instability of EB/ES equilibrium, which can be shifted to the stable formation of the ES form only by means of a continuous dynamic inflation of acid vapors. Curve 1 represents the saturation level of the sensor device valuable in a concentration of 4200 ppm of HCl vapors, since higher levels of concentration gave the same results. Curve 2 shows the data obtained for an intermediate concentration between the absence and the saturation of vapors, and the device evidenced a response in function of HCl vapors concentration. The experiment was carried out by quickly injecting the acid vapors into the system containing the sensor device and the acquisition of data started immediately after the complete injection (Bavastrello *et al.*, 2004b). Other nanocomposites are now synthesized by using different monomers and concentrations of MWNTs.Ts inside the polymeric matrix for a wide range of gas sensor applications.

4.1.2 Passive elements

Electronic components are normally divided in three classes: passive electronic components (paragraph 4.1.2), active electronic components (paragraph 4.1.3) and energy electronic components (paragraph 4.2). The fist two classes consider the devices that don't generate (or store) electric power for supplying purposes; the third class contains all the elements dealing with generation or storage of electric energy.

Molecular electronics and nanostructuring of polymers for industrial applications are a challenge for many research groups across the world (Nicolini, 1996a,b). In particular, the development of discrete devices is a problem involving many difficulties, including the connection of the active elements and the circuit. In modern electronics, technical applications require capacitors of high value, so the aim of the research for passive electronic elements is to obtain a dispositive based on electrolyte with sufficient characteristics of breakdown voltage and capacity. An alternative way to silicon electronics (Figure 4.16) is the developing of circuital elements based on organic materials and low-cost technologies, allowing the production of lightweight, thin, flexible dispositives.

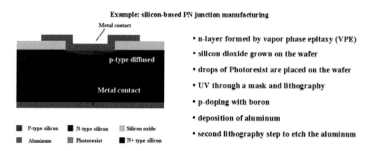

Figure 4.16. Silicon-based devices process steps, which for their difficulty provide a rationale for a new organic electronics.

International industries (NEC, Bayer and Kemet) are already developing the first commercial capacitors based on organic elements, reaching considerable results in capacity achievement but low applicable voltage. In the case of capacity typical problems of commercial organic capacitors are the excessive variability at different voltage ranges, and complicated and expensive oxidation of electrodes. The objective is to achieve a satisfying result with cheaper technology for producing both resistance and capacitance.

4.1.2.1 *Resistors*

Nanoparticles of CuS are typically formed by exposing the deposited LB films of copper stearate to the H_2S atmosphere for at least 12 hours (Erokhina *et al.*, 2003). The aggregation of the nanoparticles into thin layers was performed by washing the sample with chloroform after the reaction for removing stearic acid molecules. Thin layers of CuS were formed by aggregation of nanoparticles formed in LB precursor layer of copper stearate. Very thin layers are not uniform and can be represented as conductive aggregated "islands" with insulating gaps between them (Figure 4.17).

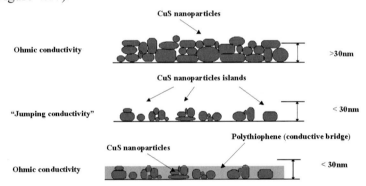

Composite based on polythiophene derivative and CuS nanoparticles in order to maintain linear conductivity in layers of semiconducting nanoparticles with thickness less than 30 nm.

Figure 4.17. Conducting nanostructure based on CuS islands nanoparticles bridged by conducting polymer.

Once the electrical properties of CuS layers of different thickness were examined, the dependence of an electrical conductivity upon the frequency was observed. The electrical conductivity appears strongly dependent upon the frequency when the thickness of the precursor copper stearate LB films is less than 30 bi-layers. For such thickness the resultant aggregated layer is not continuous one and can be considered as a film of "islands" separated one from the other. This fact is responsible for the increase of conductivity with frequency. When the thickness of the initial precursor layer is more than 30 bi-layers, the aggregated layer becomes uniform and displays ohmic conductance (Figure 4.17). Interestingly the heterostructure of less than 30 nm showed ohmic conductance only when is formed by CuS semiconducting nanoparticles and a conjugated polymer synthesized by oxidative copolymerization of 3-thiopheneacetic acid and 3-hexylthiophene leading to an amphiphilic polythiophene that allows the formation of a stable polymer layer at the air-water interface (Narizzano et al., 2005).

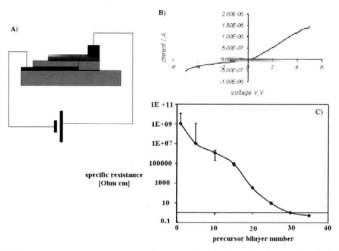

Figure 4.18. (A) The structure contains 30nm of CdS (red) and 30 nm of CuS (blue); (B); rectifying behavior of Cds-CuS heterostructure; (C) dependence of the specific resistance of aggregated CuS layers on the number of bilayers of copper arachidate LB precursor (Part C: Reprinted with the permission from Erokhina et al., Microstructure origin of the conductivity differences in aggregated CuS films of different thickness, Langmuir 19, pp. 776–771, © 2003, American Chemical Society).

Finally, when the thickness of the initial precursor LB layers was more than 25 bi-layers, the resulting aggregated films were uniform and their electrical conductivity was high (Erokhina et al., 2003). Linear voltage-current characteristics with clear rectifying behavior and with practically no dependence of the conductivity on frequency were measured for the films obtained from more than 30 bi-layers of the precursor (Figure 4.18 right below). Typical resistivity of such layers was less than 1.0.

Such low value of the specific resistance together with small thickness of the layers (21 nm) allows considering this material as very perspective for applications in electronics (Erokhina *et al.*, 2002). It is a well-known fact that for very thin metal films it is possible to observe a nonmetallic temperature dependence of the conductivity similar to that found in semiconductors, *i.e.*, an increase of conductivity with temperature. The structure of thin, aggregated layers of CuS nanoparticles, grown in Langmuir-Blodgett film precursors was investigated with atomic force microscopy (Erokhina *et al.*, 2003) along with the study of their electrical conductivity.

Very thin layers revealed an essentially insulating behavior. These layers were composed of isolated particle aggregates that had a mean thickness corresponding to the average particle diameter. The increase of the film thickness resulted in the formation of conducting pathways formed by the aggregates in the layer plane. Such samples revealed an increased conductivity. When the thickness of the initial precursor LB layers was more than 25 bi-layers, the resulting aggregated films were uniform and their electrical conductivity was high (Erokhina *et al.*, 2003). Finally, it is worth of notice an in-plane patterning process of aggregated nanoparticle thin layers of different inorganic conducting and semi-conducting capable to develop passive resistors produced in Langmuir-Blodgett precursors using film irradiation with an electron beam (Erokhin *et al.*, 2002a).

In conclusion, nanoparticles, formed in LB precursors, can be aggregated into thin inorganic layers to yield organic resistors of unique electrical properties at the nanoscale. When this thickness is low (one bi-layer of precursor), the particles form aggregates with the lateral size of about 70–80 nm and with a thickness of one individual particle diameter (2.3 nm). The layer is not uniform and homogeneous the film volume is only 8.5% filled by particle aggregates. Such "porosity" of the film determines its practically insulating behavior. Increased thickness of the precursor film results in significant changes of the aggregated layer structure and properties. Particles of the upper precursor layers tend to occupy empty spaces of the sub-layers during the aggregation process, resulting in the decrease of the height non-uniformity in the final film. Increased filling of the film volume by particle aggregates provides the formation of conducting pathways and a significant decrease of the film specific resistance. Further increase of the precursor film thickness results in the increase of the height non-uniformity of the aggregated film, providing a further increase of the filling of the film volume by particle aggregates, resulting in a pronounced decrease of the film specific resistance. The results obtained have demonstrated that effectively conducting thin inorganic layers can be prepared by the aggregation of CuS nanoparticles, produced in LB films, when the initial precursor thickness is more than 25 bi-layers. In this case, the layers are rather homogeneous. The variation in the film roughness is about 3 nm, corresponding to a variation of 15% of the average thickness, and electrical properties of these layers are stable and reproducible.

4.1.2.2 *Capacitors*

Electrolyte capacitor is a type of device offering intermediate characteristics between those of battery and simple dielectric capacitor. Usually, solid electrolyte capacitors are composed of an electrode (*e.g.* tantalum or aluminum), an electrolyte layer (*e.g.* manganese dioxide) and a very thin insulator oxide (0.01–1 Am) realized at or above approximately 275 °C.

This is a relatively violent reaction, which often damages the capacitor structure rendering it unuseful. Furthermore, deposition and pyrolysis have to be repeated several times since nitrates react with the oxide dielectric. In order to eliminate these drawbacks polymer electrolytic capacitors with a cathode of conducting polymer appears an adequate substitute for manganese dioxide (Figure 4.19). In this case, the production is quite simple: after the formation of a thin metal oxide by anodic oxidation, the tantalum or aluminum anode is immersed in the conducting polymer solution (Erokhin *et al.*, 2002). The use of a conducting polymer instead of manganese dioxide offers another advantage. The conductivity of manganese dioxide is quite low, thus, in order to increase the performance at high frequencies, it is better to use a polymer characterized by a higher conductivity than MnO_2. The rapid development of portable electronic devices has increased the demand for compact, lightweight, high capacity batteries. Polymers such as poly(ethylene oxide) (PEO) are widely studied due to their significant potential as "solid" polymer electrolytes (SPEs) in secondary (*i.e.*, rechargeable) lithium/polymer batteries. Its characteristics can be considered as an advantage for the realization of high-value electrolytic capacitors (Erokhin *et al.*, 2002). Al foil is used as a first electrode and since it presents a few nanometers thin native oxide layer, the manufacturing process is a very easy one since there is no electrochemical deposition of an additional oxide layer.

Recent work realized and tested such high value organic capacitors by using easy and low-cost fabrication techniques with solid solutions of Li salts in poly(ethylene oxide) (PEO), deposited by solution casting (Figure 4.19).

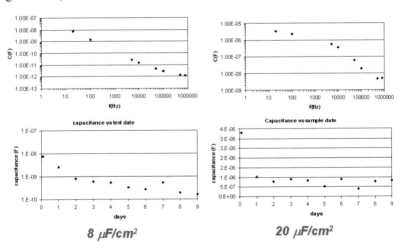

Figure 4.19. High value organic capacitors with poly(ethylene oxide) (PEO) deposited by solution casting, with LiCl salt content of 20% (right) and $LiClO_4$ salt content of 50% (left) at a frequency of 20 Hz.

With electrodes realized from different materials, the electrical tests of the elements were performed in order to check the excellent stability of properties in time and the high performance reproducibility with polymer electrolytic capacitors being formed by electrolytes of PEO/LiCl and PEO/LiClO$_4$ and by electrodes of Al and silver paint, without electrochemically deposited oxide layer (Erokhin et al., 2002). Since the solubility of LiCl in PEO solution turned out to be rather restricted, it was decided to study in a more detailed manner the element also with LiClO$_4$ as an ion source. With LiCl, the best capacity value of about 0.34 μF/cm^2 (specific value 8 μF/cm^2) was obtained with a salt content of 20% at a frequency of 20 Hz. With LiClO$_4$, the best capacity value of about 0.8 μF (specific value 20 μF/cm^2) was obtained with a salt content of 50% at a frequency of 20 Hz (Figure 4.19).

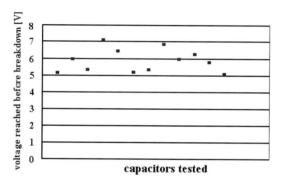

Figure 4.20. Breakdown voltage test on hybrid capacitor (Reprinted with the permission from Stura et al., Hybrid organic-inorganic electrolytic capacitors, IEEE Transaction on Nanobiosciences 1, pp. 141–145, © 2002, IEEE).

Concerning LiClO$_4$-based capacitors, there is a clear evidence of only a very low dependence of the capacity on the salt concentration, which is very interesting from an industrial point of view, as it can simplify the fabrication process. Capacity was found to be quite stable in time even if the measurements made day by day show small oscillations probably due to the environmental humidity variations. Both reproducibility and stability could be optimised by encapsulating the device with, for example, hydrophobic silicone elastomer coating. Finally, capacity measurements revealed a low-temperature dependency at a frequency of 50–60 Hz, which is also very important from the industrial applications point of view (Erokhin et al., 2002).

Organic capacitors are then among the most significant applications of conductive polymers, allowing the realization of lightweight and flexible dispositives. The main drawback of completely organic capacitors is the very low dielectric rigidity, due to irregular surface of the polymer-polymer interface. To overcome this, Stura *et al.* (2002) have used a hybrid solution using aluminum electrodes (both anode and cathode), with the very important characteristic of easy connection between contacts of the discrete dispositive and electrodes.

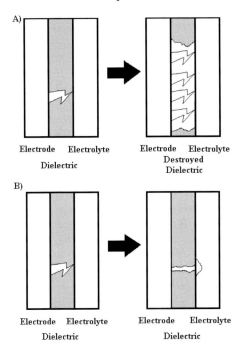

Figure 4.21. A) Standard capacitor breakdown; B) Hybrid capacitor self-healing (Reprinted with the permission from Stura *et al.*, Hybrid organic-inorganic electrolytic capacitors, IEEE Transaction on Nanobiosciences 1, pp. 141–145, © 2002, IEEE).

Chemical, technological, and functional tests have been conducted to estimate the effective values of the breakdown voltage of the hybrid capacitor (Figure 4.20), and its natural properties of self-healing and non-polarity with respect to standard capacitor has been investigated (Figure 4.21). Poly-ethylene oxide (PEO) deposited by solution casting is the matrix in which various salts have been solved, that separate in ions with the role of charge carriers. Our hybrid solution using aluminum electrodes appears to have the very important characteristic of easy connection between contacts of the discrete dispositive and electrodes.

4.1.2.3 Wires

Among the most challenging recent developments in molecular electronics is the construction of molecular wires of exceptional electronic properties from the viewpoint of both the chemical synthesis of conducting polymer (Reed *et al.*, 1997; Chen *et al.*, 1999; Collier *et al.*, 1999) and new physical means.

This latter approach has been strongly developed using scanning tunnelling microscopy that offers the possibility of both observation and manipulation of single atoms or molecules as shown by Bumm *et al.* (1996). Another approach has been developed by Reed *et al.* (1997) by simply connecting the benzene 1,4 dithiol self-assembled molecules onto two facing gold electrodes of mechanically controllable break junction to understand the charge transport through the molecules. Molecular logic gates were fabricated from an array of configurable switches from monolayer of redox-active rotaxanes sandwiched between metal electrodes as shown by Collier *et al.* (1999). It is necessary to have π-conjugated system organic molecule to the realization of the molecular wire and of many other potential applications.

Among the infinite possibilities of synthetic organic chemistry, linearly π-conjugated systems with nanometer dimensions can be synthesized as summarized by Ellenbogen and Love (2000). The molecular wire of chain molecule composed of repeating units bound together by conjugated π-bands in ethyl substituted 4,4-di(phenylene-ethynylene)benzenethiolate was already shown in literature (Pearson, 1994; Wu *et al.*, 1996). In this context, extended π-conjugated oligomers based on phenyleneethynylene and thiopheneethynylene have also been proposed as potential molecular wires (Reed, 1999). The introduction of alkyl chain in thiophene will allow us chain dimensions close to the present limits of the nanolithography (10 nm).

4.1.3 Active elements

A new generation of active electronic elements have been recently constructed by nanobiotechnology utilizing processes occurring at the nanoscale and based on LB-induced assembly and/or self-assembly of conductive polymers and biopolymers.

An example of the latter appears alternative to molecular beam epitaxy technology allowing the thickness resolution of 0.5 nm (Figure 4.22) whereby ultrathin semiconductor layers of different aggregated semiconductors are fabricated by a process of self-aggregation of different nanoparticles formed in organic layer.

Figure 4.22. SEM image of PbS-CdS-PbS superlattice. it is clearly possible to distinguish layers about 60 nm thick (Reprinted with the permission from Erokhin et al., Preparation of semiconductor superlattices from LB precursor, Thin Solid Films 327-329, pp. 503–505, © 1998a, Elsevier).

While molecular beam epitaxy is a complicated and expensive technique carried out at 10^{-10} vacuum with extra-pure materials, in this technique the layers are synthesized at normal conditions and fabrication of devices is based on these superlattices being a resonant tunneling diode with the quantum surrounded by two quantum barriers, or a semiconductor laser with the transitions of electrons through resonant levels within quantum wells in a semiconductor superlattice. The advantage is that in processes occurring at normal condition this technology is not expensive, without toxic waste and with low energy consumption. Today, the exploration of biopolymers and of neutral or pristine conjugated polymers for semiconductor device applications such as photovoltaic cells, field effect transistors, light-emitting diodes (LEDs), Schottky diodes and monoelectronic transistors have become the major point of interest.

4.1.3.1 *Schottky diode*

The Schottky diode has been fabricated using various classes of poly-aniline films. In this case a low work-function metal (Al, In, Sb etc.) is deposited on one side of poly-aniline film and the other side is vacuum deposited by high work-function metal electrode (Au, Ag etc.).

The possibility to utilize the PANI as Schottky diode was already established due to the semiconductor-like behavior of such polymers (Pandey *et al.*, 1997) summarizes the Schottky diode parameters realized on various types of poly aniline films. In later work we made an attempt to study the Schottky diode characteristics on a Langmuir-Blodgett monolayer film of poly(ortho-anisidine) (POAS) conducting polymer. The Schottky single junction was made by depositing POAS LB film on a flat graphite electrode, and subsequently approached by a second sharp electrode (Figure 4.23)

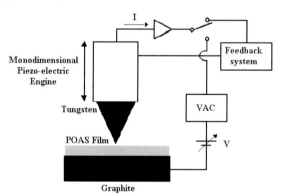

Figure 4.23. Experimental set-up for fabricating Schottky junctions. The positioning of a second point contact tungsten electrode on the POAS film, deposited onto a graphite flat electrode, is realised enabling a tunneling current to pass between the tungsten and the graphite electrodes. (Reprinted with the permission from Ram and Nicolini, Thin conducting polymeric films and molecular electronics, in Recent Research Development in Physical Chemistry 4, pp. 219–258, © 2000, Transworld Publishing).

Scanning tunnelling microscopy and spectroscopy on polymer films assured a device configuration similar to one for point-contact diodes that has been suggested to solve such problems (Pandey *et al.*, 1997). We made the point contact electrode positioned with a mono-dimensional piezo-mover until a tunnel current was achieved in order to avoid pine-hole between the flat electrode and the tip. Once realized, the Schottky junctions were tested in working devices (Figure 4.23).

The contact between the tungsten electrode and the POAS film was obtained with a feedback system and the piezomover was stopped, once reached to "a priori" imposed tunneling current between the graphite and the tungsten electrode.

The feedback system was switched off when the tunneling current was obtained in order to record possibly current-voltage (*I-V*) characteristics of the junctions. Figure 4.23 depicts *I-V* characteristics of Schottky junction when the tunneling current equals 0.9 nA set by the feed-back system at applied bias voltage of 4V.

The figure exhibits a pronounced rectifying effect from -0.3 V to +0.3 V. Potential increase above 4 V causes a change in current magnitude, giving also rise to high differential conductivity, indicating a good switching current passing through the barrier voltage.

Typically only 64% of the junctions have been found to be of the Schottky type while 36% of junctions characteristics differed somehow from Schottky junction behavior. Anomalous behavior can also be related to defects present in POAS films, like impurity or polymer break region or empty regions, or in the point contact electrode etching.

4.1.3.2 *Led*

Light-emitting diodes (LEDs) based on the electro-luminescent conjugated polymers (Figure 4.24) have attracted significant attention, both in academic research and industrial development, and are now on the stage of commercialization.

Polymer LEDs require properties such as shown in Table 4.5 and efforts have been made to design and synthesize electroluminescent polymers, tailor their properties, investigate the device physics, and engineer light-emitting diodes.

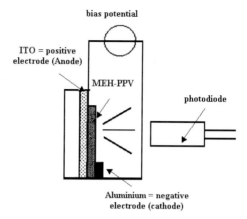

Figure 4.24. Schematic for producing polymer LEDs. The accomplishment of polymer LEDs scheme where, Al - 2000 Å, Emissive polymer layer =1500 Å, ITO work function value = 4.7 (positive electrode), Al work function value = 4.2 to 3.76 (negative electrode) (Reprinted with the permission from Nicolini et al., Supramolecular layer engineering for industrial nanotechnology, in Nano-surface chemistry, pp. 141 212 © 2001a, Marcel Dekker/Taylor & Francis Group LTD).

Among various conducting polymers, aromatic conducting polymers such as polythiophene, poly(p-phenylene vinylene), poly(pphenylene), poly(phenylene sulphide), and their derivatives have been used as the emission layer in polymer LED devices. A low voltage is applied over a thin film of the polymer. Subsequently the electrons and holes are injected from the electrodes; when a hole and an electron collide in the polymer, a local excited state can be formed that emits light (electro-luminescence), as shown in Figure 4.24. The basic requirements for the choice of electro-luminescent polymers are:
(1) the polymer should have good film-forming properties (smooth surfaces, no pinholes, minimum thickness of 50 nm);
(2) the polymer film should have good thermo-mechanical stability;
(3) the polymer films should be transparent;
(4) the polymer should be amorphous;
(5) the polymer should exhibit excellent heat, light, and environmental stability;
(6) special requirements for light-emitting polymers are light emission in the visible region and color tunability.

Table 4.5. Physical properties of interest for LED materials (Reprinted with the permission from Ram and Nicolini, Thin conducting polymeric films and molecular electronics, in Recent Research Development in Physical Chemistry 4, pp. 219–258, © 2000, Transworld Publishing).

Polymer	Properties	Notes	Dopant compounds
poly(phenylene vinylene) (PPV))	Bandgap = 2.5÷2.6 eV Intrinsic electrically conductivity 10^{-3} S/cm. Emission range 1.8–2.5 eV. Ionization potential = 5.1 eV Electron affinity = 2.6 eV	See forward and reverse potential	iodine, $FeCl_3$, K, Ca, Mg, or acids
poly(2-methoxy, 5-(2'-ethylhexyloxy)-1,4-phenylene vinylene) (MEH-PPV)	Emission peak = 605 nm	Structure: Es PANI electrode / MEH-PPV/Calcium Electrode	p- doping by sulfuric acid; n-type by sodium Iodine (I_2)= electron acceptor → oxidizing agent

To control the preceding parameters, Langmuir-Blodgett and layer-by-layer adsorption techniques for the preparation of PPV conjugated polymer films have recently been developed. These demonstrate the fabrication of optically transparent PPV-containing multi-layers of precursor PPV, which is converted to PPVs by thermal elimination. The PPV family of polymers serves as a prototypical conjugated polymer class for application as well as for fundamental understanding of the electronic processes in conjugated polymers. Ohmori *et al.* (1997) showed that other conjugated polymers like poly(3-alkylthiophene) are also useful as the emitting species in LEDs (Granstrom, 1997). Recent advances for both nonpolar, *i.e.*, "molecular device", and polymer LEDs indicate a great potential of organic-base diodes.

A survey of the preparation of electro-luminescence conjugated polymers evidences that the advances in synthetic methodologies for the preparation of thin films and fibers of PPVs enable them to be considered in various applications. The overall methodology can be roughly divided into three categories: precursor approach, side-chain derivatization and *in situ* polymerization. The sulfonium precursor route (SPR) to PPV is particularly well known and involves the polymerization of p-xylene bis-(tetrahydronium thiophenium chloride) or one of its analogues or derivatives, in the presence of water or methanol to give the corresponding sulphonium precursor polymer. Table 4.5 shows the common LED materials and their physical properties. The poly(2-methoxy-5-(2'-ethyl-hexyloxy) phenylene vinylene) (MEH-PPV), P-PPV, *etc.* have been synthesized by Gilch route. Furthermore, a modification of the Gilch route, namely the chlorine precursor route (CPR) was introduced in 1990 in order to avoid polymer precipitation. DP-PPVs have been prepared by this route in 1993. The overall impression is that CPR is very simple, general, versatile and reproducible and superior to the SPR and Gilch route for the preparation of PPV derivatives but SPR remains the best approach for the fabrication of PPV thin films. However, improvement in the luminance efficiency and durability of device remain unsolved. The processability of PPVs compounds has recently been increased for the fabrication of thin films for real device application. To increase the efficiency of devices, electron injection has also been significantly boosted. Green emission with a luminance of 100 Cd m^2 and an external quantum efficiency of 0.4 % at a driving voltage of 4 V has been obtained from a device using PPV. Poly(3-alkylthiophene)s provide a good example of how the colour of emission can be varied in polymer LEDs by modifying both the polymer structure and the design of the devices colour control by alteration of substituents and regio-regularity is also illustrated by the pyridine-based analogue of PPV and its derivatives.

The overall photoelectrical properties of PPV resemble closely those of aromatic dyes and pigments, which are the first generation of organic semiconductor materials. However, PPVs are much less stable than organic pigments, mainly because the vinylene groups are highly susceptible to photoxidation and it limits the lifetime of the device. However a different synthetic route may result in PPV with different π-conjugated chain lengths and chain configurations. These factors may affect the electronic and optoelectronic properties of the PPV. The one remaining obstacle to the widespread use of polymer LEDs is that, to date, their lifetimes remain lower than the best devices using inorganic or small-molecule organic compounds, although this gap is rapidly being closed. Singlet oxygen has been implicated in the cleavage of vinylene groups in MEH-PPV and bis-(cholestanyloxy)-PPV, leading to ester or aldehyde groups. A study of PPV-based LEDs revealed that exposure to air produced a drop in efficiency and an increase in threshold voltage, while degradation of the polymer is one of the primary processes involved in the breakdown of MEH-PPV-based LEDs. These results suggest that suitable protection from aerial oxidation may be a significant factor in improving the lifetime and efficiency of LED devices. Work is in progress about the implementation of this technology for LED operating in the infrared.

4.1.3.3 *Optical filtering and holography*

Optical information storage and processing utilized up to now only synthetic, photo-chromic and refractive materials as reversible recording media. However, the effect of side reactions on photochemical processes, the complicated crystal growth procedure (which is usually also expensive) impeded the realization of several important applications. The use of light-sensitive proteins, such as photosynthetic reaction centers and rhodopsins of various origins is the object of intensive studies in the recent years.

Bacteriorhodopsin, a retinal protein, the main source of which is *Halobacterium salinarium* (formerly *Halobacterium halobium*), is one of the most widely used in the experiments aimed at the application of proteins for the creation of alternative dynamic recording material instead of conventional synthetic ones (Hampp, 1993, 2000, Hampp and Zeisel, 2003; Hampp and Juchem, 2004). The extraordinary features, maturated during the long natural selection process, include, among others, very high stability, and reversibility of physicochemical processes, in the form of optically homogenous thin films. The possibility of gene engineering manipulations of bacterial strains that produce the protein permits the creation of mutants with optimal photochromic characteristics for specific applications such as optical filtering, re-writeable media and holography. According to the mechanism of functioning of bacteriorhodopsin, the absorption of a photon by a protein molecule leads to a fast (500 fs) trans-cis photoisomerization of retinylidene around the 13, 14 double bond with a quantum yield of 0.64. The subsequent steps involve thermal relaxation of several amino acids and of Schiff base linkage, involved in the proton transport, plus the backbone conformational changes. The states of bacteriorhodopsin are characterized by the retinylidene conformation and the initial state is the B-state with an all-trans configuration. In the dark conditions, there is an equilibrium between the D-state (dark-adapted, with 13-*cis* configuration) and the B-state. The proton adsorption converts the bR molecule into light-adapted B-state (all-*trans* retinal configuration). The subsequent conformational changes bring the protein into the L_{550}-state via the intermediate K_{590}-state by a thermal relaxation process. The subscript indices refer to characteristic wavelengths of the absorption spectra.

These features of photochromism are the most interesting ones in the determination of the applicability of bR to optical storage and information processing. However, an important point to be kept in mind is the operational lifetime of such media, evaluated in this case by the number of write/erase cycles that can be effected. In a holographic film the intensity and phase distribution of an object is recorded. The resulting hologram is a diffractive element which generates the original object wave in amplitude and phase when illuminated with the same reference beam as was employed during recording. The same happens with bR-films, but since they are reversible recording materials, these holograms can be either erased at any time or decay with the time constant of the M-lifetime (Hampp, 2000; Hampp and Juchem, 2004). The bR-film is used as short time storage for the reference hologram. With BR_{D96N}-films, a lifetime of the reference hologram of up to several minutes is obtained. After this time, a new reference hologram has to be recorded. The interference of the holographic image and the directly transmitted light can be described as a processing of two images that correspond to different times but originate from the same location.

The other example of bR-based image processing is holographic pattern recognition. This technique allows the correlation of two images that exist at the same time but are spatially separated. Common features of both patterns are detected. Since this technique is an analog computing method, signal-to-noise ratio is of central importance. Due to the polarisation recording properties of bR, an effective suppression of noise can be realized. Bacteriorhodopsin films counting the variant BR_{D96N}, which differs from the wild type bR_{WT} by a single amino acid exchange, Asp96→Asn, show significantly higher holographic diffraction efficiencies (η) than bR_{WT} films (Table 4.6).

Table 4.6. Holographic properties of bacteriorhodopsin films.

Resolution	≥ 5000 lines/mm
Optical density	1–20 (OD_{570})
Bleaching	90–100 percent (*e.g.*, BR_{D96N})
Index of refraction	1.47
Refraction index change	10^{-3}–10^{-2}
Diffraction efficiency	1–7 percent (type 2-3 percent)
Light sensitivity	1–80 mJ/cm^2 (B-type)
	30 mJ/cm^2 (M-type)
Polarization recording	possible
Reversibility	≥ 106 cycles
Thickness	10-500 µm (type 20 mm)
Speed	msec-sec
Aperture	Unlimited

A good correlation of the theoretically derived and experimentally measured values of the refractive index changes was found, indicating that the chromophore system of bacteriorhodopsin, which is formed by the retinal molecule, its Schiff base linkage to the protein moiety, and an inner shell of amino acids, behaves like an almost undisturbed chromophore with respect to the photo refractive properties at low actinic light intensities, despite the fact that all components of the chromophoric system are covalently linked to the amino acid matrix.

Bacteriorhodopsin films are however made with intact purple membranes, which contain about ten lipid molecules for one bacteriorhodopsin.

Advantage to use rhodopsins from bovine (Maxia *et al.*, 1995) or octopus (Paternolli *et al.*, 2008) to make thin films rather than bacteriorhodopsin is that rhodopsin can be extracted by detergent from the membrane without loosing its properties while isolated molecules of bacteriorhodopsin did no longer behave like a proton pump (Fisher and Oesterhelt, 1979).

We show furthermore that thermal stability is also a property of bovine rhodopsin in LB films, which can reach, in fact, a temperature up to 175 °C with a negligible loss of secondary structure (Maxia et al., 1995).

In a recent work (Paternolli et al., 2008) a new biomaterial resulting from the isolation of octopus rhodopsin starting from octopus photoreceptor membranes was obtained.

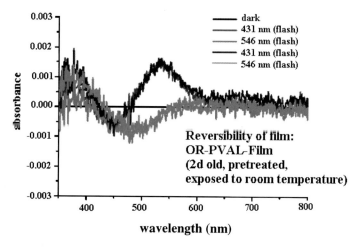

Figure 4.25. Reversibility of the octopus rhodopsin thin film (Reprinted with the permission from Paternolli et al., Photoreversibility and photostability in films of octopus rhodopsin isolated from octopus photoreceptor membranes, Journal of Biomedical Materials Research Part A, in press, © 2008, Wiley Periodicals, Inc., a Wiley Company).

Mass spectroscopic characterization was employed in order to verify the presence of rhodopsin in the extract, and photo reversibility and photo-chromic properties were investigated utilizing spectrophotometric measurements and pulsed light.

Thin films of octopus rhodopsin were realized, utilizing the gel-matrix entrapment method in polyvinyl alcohol solution. The results indicate that the photo-reversibility and the photo-stability of the octopus rhodopsin in gel-matrices are maintained in time and at room temperature up to few days (Figure 4.25) and after the exposure for several hours at room temperature.

4.1.3.4 *Displays*

Electrochromism is the property of a material or a system to change colour reversibly in response to an applied potential. There has been a growing interest in electrochromic materials for use in practical electrochromic displays. Oxides of W, Mo, V, Nb, Ti and a number of conducting polymers PPY, polythiophenes and PANIs are known to be electrochromic materials, the latter being considered as the most promising for electro-chromic displays (Figure 4.26).

The use of solvents plays an important role in the performance of the electro-chromic displays while problems associated with liquid electrolytes in such devices are due to their high degree of hydration, rapid proton insertion and stability against any chemical or electrochemical corrosion. A cell based on PANIs can be only used a thousand times due to the liquid solvent used, which is the disadvantage with respect to life-time of the electro-chromic cell. In practical applications it is preferable to employ solid-state materials for electro-chromic displays in order to minimize the problems of sealing using any hazardous liquids.

Electro-chromic displays are usually required to have thin layers configuration. The prospects for solid polymer electrolyte and conducting polymer look promising for the development of practical electro-chromic displays. The solid polymer electrolyte and conducting polymer form an ideal medium for a wide range of electrochemical processes, paving the way to easy fabrication processes. A solid polymer electrolyte (*i.e.* poly(ethylene oxide (PEO)) and its complexes) based on a conductive polymer would eliminate the use of toxic solvents. We have focused our attention on PEO, the use of which has been attempted as a solid electrolyte for the fabrication of electro-chromic displays, due to the recent synthesis of new PEO-complexes with improved low temperature electrical properties for the fabrication of electro-chromic displays based on conducting PANIs. The films of PANI, and its copolymer films of poly(aniline-co-o-toluidine) (PAOT) and poly(aniline-co-o-anisidine) (PAOA) have been electrochemically obtained on indium-tin-oxide (ITO) glass plates and silicon, respectively. The ultra thin films of substituted PANIs were fabricated by Langmuir-

Blodgett (LB) technique aiming towards their potential application in electrochromic cell. The current transients for coloration and decoloration have been analyzed in order to understand the charge transport for PANI and its conducting copolymer films of PAOT and PAOA. The electrochromic switching response time in different protonic acid media was studied on such poly(ortho-anisidine) POAS LB films.

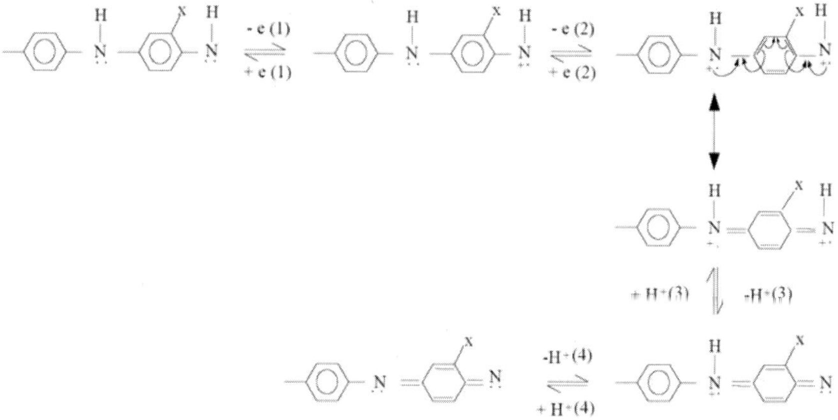

Figure 4.26. Reaction mechanism for PANI (X= H), poly(aniline –co-o-anisidine) (x = CH_3) (POAT) and poly(aniline-co-o-anisidine) (PAOA) (Reprinted with the permission from Ram and Nicolini, Thin conducting polymeric films and molecular electronics, in Recent Research Development in Physical Chemistry 4, pp. 219–258, © 2000, Transworld Publishing).

We have performed electrochemical studies for each electrochromic cell based on PANI, POAT and PAOA films. In evaluating the performance of the electro-chromic phenomenon of cells, we focused on the aspects mentioned, namely, switching time ($t_{1/2}$), the dependence of the switching time on the ionic diffusion process (D_o), the applied voltage required for the cell and the stability of the cell in the repeated cycling. In fact, the mechanism of PANI switching has an important bearing on these aspects. Bearing in mind that PANI system has a switching response; we applied a frequency from the functional generator coupled to electrochemical interface for recording the current versus time plot.

The electro-chromic cells have been studied at a frequency of 10^{-1} Hz obtained from the function generator. The applied potential for the switching reaction is listed in Table 4.7 for PANIs.

Table 4.7. Electrochromic parameters of conducting polymer and copolymeric films. (Reprinted with the permission from Ram and Nicolini, Thin conducting polymeric films and molecular electronics, in Recent Research Development in Physical Chemistry 4, pp. 219–258, © 2000, Transworld Publishing).

Conducting polymers	Applied voltage (V)	Response time (ms) $t_{1/2}$	Slope (n) reduction	Slope (n) oxidation	Diffusion coefficients $(D_0)\ 10^{-10}$ cm^2/sec	Process and life cycle
Polyaniline	-0.4–0.8	205	0.51	0.44	5.39	Diffusion 10^4
Poly(aniline -co-o-toluidine)	-1.0–1.2	299	0.13	0.37	0.40	Diffusion $10^4 <$
Poly(aniline -co-o-anisidine)	-0.6–1.0	143	0.07	0.06	18.95	Diffusion $>10^5$

The oxidation and reduction of the conducting polymer and copolymer films take place under steady-state conditions. The life-time of PAOA has been estimated to be 10^5 cycles.

4.1.3.5 Monoelectronic transistors

Inorganic semiconductor nanostructures formed inside fatty acid films, such as cadmium sulphide nanoparticles as small as 50 Å (see chapter 1), were found first by Smotkin *et al.* (1988) exposing cadmium arachidate LB films to atmosphere of H_2S, suggesting the idea that *V-I* characteristics out of them could display monoelectron behaviors (Devoret *et al.*, 1992) even at room temperature. During the reaction, the head groups of arachidic acid were protonated, and CdS was produced according to the following reaction:

$$[CH_3 (CH_2)_{18} COO]_2\ Cd + H_2S \rightarrow 2CH_3 (CH_2)_{18} COOH + CdS$$

Monoelectron junctions are of paramount importance in nanoelectronics because they could represent the basic elements for electronic chips exploiting an integration scale up to 10^4 folds the nowadays submicron limit, they could allow to implement a multi-state logic and function with very low power consumption.

Taking into account the actual particle size of CdS nanoparticles (Figure 4.27), it is indeed found monoelectron phenomena on CdS granules at room temperature (Facci et al., 1996).

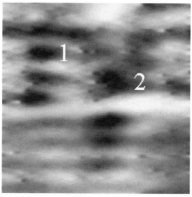

Figure 4.27. STM image of CdS granules inside a LB film of cadmium arachidate. Imaging parameters V_t=0.6 V (tip positive), I=1.5 nA, scanning speed 12Hz, image size 51.2 × 51.2 nm (Reprinted with the permission from Erokhin et al., Observation of room temperature mono-electron phenomena on nanometre-sized CdS particles, Journal of Physics D: Applied Physics, 28, pp. 2534–2538, © 1995a, IOP Publishing Limited).

Such kind of behavior, i.e. Coulomb Blockade (Devoret et al., 1992) was previously observed at low temperature on particles of larger size (Mullen et al., 1988), and by STM on metal granules even at room temperature (Shönenberger et al., 1992a,b).

This phenomenon, which can take place when a conductive or semiconductive granule is separated from two electrodes by two tunnelling junctions, consists in the quantified increase of the average number of electrons occupying the granule upon the bias voltage through the described structure (Devoret et al., 1992).

The presence of electrons in the granule provides an electric field, which prevents a further incoming electron to enter the granule until a suitable bias voltage is applied through the junction.

Each quantized increase of the average number of electrons takes place whenever the voltage through the structure changes of a quantity given by:

$$\Delta V = \frac{e}{C}$$

where e is the electron charge and C the capacity of the structure Coulomb Blockade appears when the following inequality (Averin and Likharev, 1986) holds true:

$$\frac{e^2}{2C} > kT$$

where e is the electron charge, C is the capacity of the structure, k is the Boltzmann constant and T the absolute temperature (*i.e.* when the effect of the thermal excitation is negligible with respect to electrostatic repulsion energy).

The classical approach to make the above inequality valid is to decrease the value of T, as there are technological limits in decreasing the value of C. CdS nanoparticles, however, seem to be right candidates for facing the problem of making the above formula valid by decreasing the value of C and, therefore, achieving high temperature Coulomb Blockade. Besides, as the sizes of CdS particles can be of the order of magnitude of tens of an Å, rough estimations foresee capacities as small as 10^{-18} F, which should display room temperature Coulomb Blockade (Facci *et al.*, 1996; Nicolini, 1996a).

The simplified scheme of the measuring set-up is shown in Figure 4.28, where arachidic acid monolayers were formed onto the surface of water containing 10^{-4} M of $CdCl_2$.

One bi-layer of cadmium arachidate was deposited with LB trough (MDT, Moscow) onto graphite substrate at the surface pressure of 28 mN m^{-1} by LB technique.

The samples were exposed to an atmosphere of H_2S for 15 minutes (Figure 4.28A) and the *V-I* characteristics measured on these samples (Figure 4.28B).

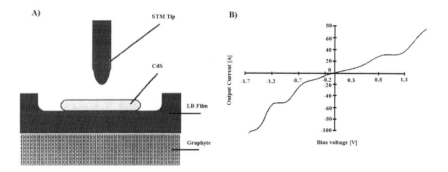

Figure 4.28. (A). The simplified scheme of the measuring set-up for tunnelling. (B). Voltage-current characteristics with single-electron conductivity. (A: reprinted with the permission from Erokhin et al., Observation of room temperature mono-electron phenomena on nanometre-sized CdS particles, Journal of Physics D: Applied Physics, 28, pp. 2534–2538, © 1995a, IOP Publishing Limited; B: reprinted with the permission from Nicolini et al., Supramolecular layer engineering for industrial nanotechnology, in Nano-surface chemistry, pp. 141–212 © 2001a, Marcel Dekker/Taylor & Francis Group LTD).

To achieve a deeper understanding of the process at issue and in order to give one more proof of the appearance of Coulomb Blockade, it was performed a theoretical simulation of the V-I characteristics expected from structures like those above described, by means of a semi empirical model (Nicolini, 1996a); the average electron occupation number in the island is estimated by means of the Boltzmann statistics:

$$<n> = \frac{\sum_{n=-\infty}^{+\infty} n e^{-\frac{E_n(V)}{kT}}}{\sum_{n=-\infty}^{+\infty} e^{-\frac{E_n(V)}{kT}}}$$

where n is the electron occupation number of the granule and $E_n(V)$ is the energy of the electron inside the granule. Within this frame the observed phenomena was corroborating the hypothesis that the behavior of V-I characteristics we observed on the wells inside the fatty acid film is really due to Coulomb Blockade at room temperature and that inside the wells are really present granules of nanometer sizes (Nicolini, 1996).

The experiment hereby described represents only the proof of principle and still many difficulties remain to overcome the possibility of exploiting phenomena such as the Coulomb Blockade from a technological point of view opening the way to a revolution in the concept of electronics. It was then comforting to find many years later unexpectedly similar mono-electronic transitions also in composite material (Figure 4.29) made by a multinational company on solution casted POAS for silver paint containing conductive nanoparticles of similar size as apparent by the associated atomic force microscopy.

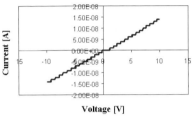 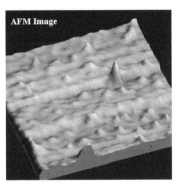

Process:
• CB particles distributed on solution casted POAS (liquid)
• Electrodes realized by silver paint

Results:
• Voltage step corresponds to 2.3x10^-19 F
• Particle size is 2 nm

Figure 4.29. Unexpected results displaying Coulomb staircase phenomenon in composite nanomaterials characterized electrically and morphologically (private communication, 2003).

Towards mono-electronic applications, we can say that these results represent only the first steps. Further steps toward the realization of stand-alone mono-electron devices involve the formation of a network of connections capable of addressing the single mono-electron junctions, which seems to be still far away.

4.1.4 *Quantum dots and quantum computing*

Quantum phenomenon appears already in a nanometric sized wire, where the conduction electrons meet ballistic transport and quantum nodes in the transverse direction (Figure 4.30).

This quantum conductivity originates in a nanometric sized wire due to the decreasing conductance with the decreasing size, resulting in the characteristic stair-like conductance decreasing which demonstrates that quantum conductivity is taking place. In comparison, in a normal size metallic wire the conduction electrons have multiple classical collisions through the wire.

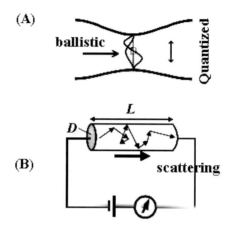

Figure 4.30. The quantum conductivity in a nanometric sized wire (A) is associated to a stair-like decreasing conductance (B), while in a classical metallic wire the conduction electrons have multiple collisions through the wire.

An other example of quantum phenomenon is exemplified by microtubule and tubulin (Hameroff *et al.*, 2002), where tubulin can undergo a conformational change from black to the white basis state depending on the localization of electrons in its hydrophobic pocket. A schematic representation of the superposed state is shown in (Figure 4.31).

Finally also single-electron phenomena were linked to the concept of quantum devices and of quantum dots (Glazmann and Shekhter, 1989). In particular, considering a ballistic model for the charge transport through a dot, it was possible to demonstrate that the current through it should be represented as a series of equidistant peaks whose positions correspond to the steps in the coulomb staircase. Moreover, the possibility of considering single-electron phenomena in a frame of a dot-based system theory allows consideration of even semiconductor nanoparticles as quantum dots, useful for single-electron junctions (Averin *et al.*, 1991).

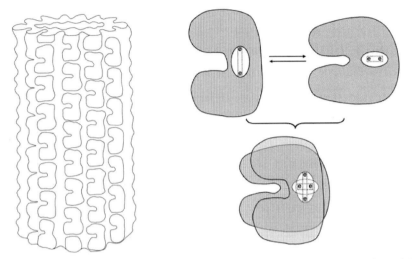

Figure 4.31. Left: microtubule, a cylindrical lattice of tubulin protein molecules. Right: Each tubulin molecule may occupy two classical conformations (top) or exist in quantum superposition of both conformational states (bottom), with each conformation coupled to position of a pair of electrons in an internal hydrophobic pocket. A tubulin may thus act as a classical bit (top) or as a quantum bit, or 'qubit'. The difference between the two conformations of tubulin, as well as the size of the hydrophobic pocket, have been exagerated for illustrative purposes.

The modeling of junctions based on these semiconductor quantum dots reveals that their behavior in terms of single-electron phenomena can result in current-voltage characteristics with differential negative resistance regions (Gritsenko and Lazarev, 1989). This fact was connected to the possibility of resonating tunneling through quantized energy levels inside the dot (Guinea and García, 1990; Beenakker, et al., 1991; Sumetskii, 1993; Groshev et al., 1991).

On the other hand, some work on the topic considers the presence of negative differential resistance in the current-voltage characteristics and the possibility of a coulomb staircase as different output of the very same phenomenon and, therefore, has tried to consider both of them in the very same conceptual frame (Beenakker, 1991; Stone et al., 1992; Prigodin et al., 1993).

This approach seems to be successful; in fact, it was possible to see models describing current-voltage curves presenting both stairs and negative resistance along them (He and Dassarma, 1993; Carrara *et al.*, 1996). The first stand-alone room-temperature single electron junction was made by depositing a semiconducting particle directly onto the tip of a very sharp electrode, avoiding in this case the use of an STM microscope, and it was possible to observe the coulomb staircase in such a system (Facci *et al.*, 1996).

In addition to the mainstream of element formation, several non traditional technological approaches were carried out for the formation of elements with nanometer sizes and their utilization for construction of single-electron elements (Wilkins *et al.*, 1989; Shónenberger *et al.*, 1992a; Dorogi *et al.*, 1995; Erokhin *et al.*, 1995a). Several possible applications of the phenomenon were discussed. The easiest one was to consider it an analog-digital transducer. In fact, continuous sweeping of the voltage applied to the junction results in the digital output of the current, providing, therefore, a fixed value of the current to the different voltage intervals.

Moreover, it is possible to vary the unit step of the digitization, taking a granule of different sizes. The next steps in the practical realization of such a transducer will be in a synthesis of the granule between preformed sharp metal electrodes. The electrodes can be prepared using the selective etching technique of the thin and narrow metal strips deposited onto insulating substrates. Several possible applications are proposed for systems, using three electrodes. In this case, two of them with a granule between them serve as analogs of the source and the drain in a field effect transistor.

The third one, the analog of the gate electrode, serves for the application of the electric field to the granule, which varies the character of the current flow between the source and the drain. Apart from transistor-like devices, single-electron junctions can also be useful for sensor applications.

The simplest one is the monitoring of H_2S^-. Since the formation of CdS nanogranules takes place when an initial cadmium arachidate layer is exposed to this gas, we can expect the appearance of single-electron conductivity only when it is present in the atmosphere.

4.2 Nanoenergetics Compatible with Environment

The recent upgrade in the new sources of renewable energy based on Nanobiotechnology represent the core of this subchapter.

Photovoltaic cells based on both polymers and biopolymers, organic batteries, hydrogen production and storage, and fuel cells represent most likely (considering the insurmountable problems of nuclear energy) the only possible solution to the dramatic consequences of Serra effect for the human race that in 200 years has burned nearly all fossils being accumulated in five millions years on the Earth planet.

4.2.1 *Photovoltaic cells*

Photovoltaic (PV) solar cells, which convert incident solar radiation directly into electrical energy, today represent the most common power source for Earth-orbiting spacecraft, such as the International Space Station, where a "photovoltaic engineering tested" (PET) was actually assembled on the express pallet.

The solid-state photovoltaic, based on gallium arsenide, indium phosphide, or silicon, proves to be capable, even if to different extents and with different performances, of operating in a reliable fashion at less than the 10-KW low-power range typical of the missions orbiting the Earth (Table 4.8); the electrical power generated over many orbital cycles supports both the electrical loads and the recharge of batteries.

Sunlight is practically an inexhaustible energy source, and increasing energy demand makes it a primary source of renewable energy. Sunlight possesses a very high energy potential and is an ecologically pure and easily accessible energy source.

The electrical power obtained from solar energy conversion is widely used in spacecraft power supply systems (the latest very important example is the International Space Station-ISS within the framework of Italian and European Space Agencies) and in terrestrial applications to supply autonomous customers with electrical power (portable equipment, houses, automatic meteostations, *etc.*).

Table 4.8. Photovoltaic parameters of various tested materials. (Reprinted with the permission from Nicolini and Pechkova, Nanostructured biofilms and biocrystals, Journal of Nanoscience and Nanotechnology 6, pp. 2209–2236, © 2006a, American Scientific Publishers, http://www.aspbs.com).

Cell	Efficiency	Notes
Inorganic-based cells Single junction thin-film polycrystalline cell	14%	Hydrogenated amorphous silicon (a-SM); cadmium Telluride (CdTe); copper indium diselenide (CuInSe$_2$)
Single-junction single crystal	30%	
Multijunction cells	> 30%	
Ga/As/CuInSe$_2$	21.3%	Year 1977/1988
GaAs/Si	31%	1988
AlGaAs/GaAs	24–28%	1988/1989
a-Si:WcuInSe$_2$	15.6%	1988
a-Si:H/a-Si:Ge:H	13.6%	1989
GaInP$_2$/GaAs	25%	1989
n-CdS/p-CdTe	15.8%	1993-heterojunction
Cr/chlorophyll-,affig junction	0.016% (monochrom. eff.)	λ = 745 run

The irregular incidence of sunlight on the Earth (daily and seasonal variations) represents one of its disadvantages, together with its low energy density. For these reasons, there is need to cover large areas with expensive semi-conducting solar cells, and consequently, costs are increased. Thus the electrical energy obtained in such a way is more expensive than that from conventional methods. Although reduction of pollutant emission is a key factor in the preservation of the environment and subsequently the quality of the life itself, the increase in costs retards the development of a large-scale solar power industry. Nowadays, solar cells are based on inorganic semi-conducting materials, namely, amorphous silicon (efficiency about 12%), multicrystalline silicon (18%), and CdTe (16%), and yield an average energy cost of about $5 per watt.

Given the present scenario, one can state that the emerging field of nanotechnology represents a new effort to exploit new materials as well as new technologies in the development of efficient and low-cost solar cells. In fact, the technological capabilities to manipulate matter under controlled conditions in order to assemble complex supramolecular structures within the range of 100 nm could lead to innovative devices (nano-devices) based on unconventional photovoltaic materials, namely, conducting polymers, fullerenes, biopolymers (photosensitive proteins), and related composites. Among such techniques, the most promising seems to be the Langmuir-Blodgett one.

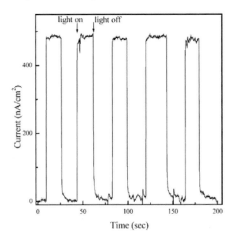

Figure 4.32. Photoelectrochemical response of ITO/(PDDA/CuTsPc-capped $TiO_2)_{15}$ thin films in an electrolyte of 0.1 mol dm^{-3} TBA TFB acetonitrile solution. The potential of the working electrode was set at 0.0 V versus the Pt counterelectrode (Reprinted with the permission from Ding *et al.*, Ultrathin films of tetrasulfonated copper phthalocyanine-capped titanium dioxide nanoparticles: fabrication, characterization and photovoltaic effect, Journal of Colloid and Interface Science 290, pp. 166–171, © 2005, Elsevier).

As far as organic materials for photovoltaic applications, such as conducting polymers (Figure 4.32), we note that present research is focused on understanding the physicochemical phenomena that underline the applicability of such new materials. Several research groups around the world are trying to develop new photovoltaic cells based on unconventional materials, particularly sunlight-converting solar cells based on conducting polymers and on composites. Photo-excitation, charge injection, and/or doping induce local electronic excitations necessary for charge transport. Among them, only the doping process (intercalation) is able to induce a permanent transition to a conductive state. As far as the photo-excitation process is concerned, we note that excited states produced by photon absorption, namely, excitons, have high relatively binding energies and do not dissociate themselves to give electrons and holes. Therefore, the entire process of exciton ionization is not a promising method for the development of an organic PV device. The correct energetic, which allows charge separation, can be provided either by the interfaces between molecular semiconductors or with electrodes.

4.2.1.1 Reaction centers-based

An alternative approach is the construction of photocell using photosynthetic bacterial membranes. The work presents experimental data, which allow making a conclusion about the possibility of constructing such type of cells using biological materials. To optimize performance we plan to try different types of LB deposition, namely orientation by pressure, electric field and different types of RC: from *Rhodobacter sphaeroides*, *Rhodopseudomonas viridis* and *Chromatium minutissimum*. Next step will be the use of proteins instead of membrane fragments to yield an increase in the efficiency of the light conversion.

The recent progress in research indicates that the disentanglement of photosynthesis will give strong impulse to the application of solar cells. Moreover, as far as the bioconversion of sunlight is concerned, it is known that photosynthesis starts with a charge separation process in the photosynthetic reaction centers (RCs): a photoactive bacterial protein. Therefore, several experiments have been carried out to test the possibility of the conversion of light to electrical energy by photochemical cells (or simply photocells) containing such photosynthetic bacterial proteins. The photocells developed have different geometries, and therefore it is very hard to classify the reliability of such devices. Nevertheless, the results indicated that the efficiency of quantum energy conversion is practically 100% (all the light energy was converted into the charge separation). Nevertheless, the energy conversion efficiency of the realized photosensitive units crucially depends on molecular orientation and is very difficult to estimate, because articles usually contain incomplete information.

Usually, such proteins are immobilized by thin-film fabrication techniques, such as self-assembly, electrical sedimentation, Langmuir-Blodgett, polymer matrix, or gel entrapment. Among such techniques, the most promising seems to be Langmuir-Blodgett. In fact, this technique allows the protein molecules to be organized in an ordered LB array. There are many publications on films of RCs. The films were dually characterized from different points of view. The discovery of heat and temporal stability had opened big possibilities in using the protein in devices (Nicolini *et al.*, 1993).

In the literature, Japanese authors (Yasuda *et al.*, 1993) suggested the making of a photo-device, using an RC LB film sandwiched between two electrodes (one of them transparent). Photo-voltage registered in this work was 4–5 mV for the film, which contained 44 layers. In our work (Facci *et al.*, 1998) we have shown that it is possible to adjust a tilt of RC molecules in the layer by controlling the surface pressure. On the other hand, for bR the possibility of increasing anisotropy by electric field application was shown. We can hope that by applying our technique of electric field-assisted deposition we will be able to increase this number (according to existing estimations, there is only about 12% of prevalent orientation with respect to the opposite one; with our technique, we can hope to increase this anisotropy). The other possible way of increasing the film anisotropy was suggested by Miyake *et al.* (1998) in LB8 conference. They voltage-biased the substrate during deposition with respect to the water subphase. He reported that even in this case the anisotropy of the film was improved. Thus, the films can be useful for the construction of devices for converting light to electron energy, working in the range of visible-near-IR spectrum (Nicolini, 1997).

4.2.1.2 Purple-membrane based

In recent years our understanding of the structure and function of several biological systems has grown rapidly. The study of bacteriorhodopsin (bR) protein and the elucidation of its function as a light-driven proton pump represent one of the most interesting examples (Oesterhelt *et al.*, 1991; Brauchle *et al.*, 1991; Birge, 1990).

BR is a light-transducing protein in the purple membrane (PM) of *Halobacterium Halobium.*

Its features allow one to identify and design several potential bioelectronics applications aimed at interfacing, integrating, or substituting for the silicon-based microelectronics systems, as well as developing molecular devices (Birge, 1992).

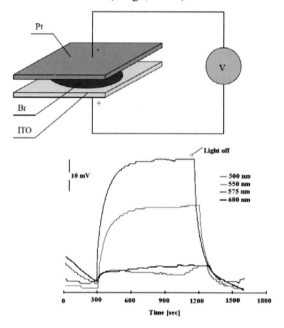

Figure 4.33. Photosignal measured for bR photo-induced current as function of time in the structure as shown above (Reprinted with the permission from Nicolini *et al.*, Supramolecular layer engineering for industrial nanotechnology, in Nano-surface chemistry, pp. 141–212 © 2001a, Marcel Dekker/Taylor & Francis Group LTD).

BR is a notable exception as compared to the usual biological molecules, being mechanically robust and chemically and functionally stable in extreme conditions, such as high temperatures (Hampp, 1993, Shen *et al.*, 1993; Zeisel and Hampp, 1996), which usually represents one of the key parameters of working conditions.

Furthermore, it possesses remarkable photonic and photovoltaic properties, which have been exploited for molecular device construction. For these reasons, bR has been adopted as a building block for a number of experimental prototypes previously discussed (Birge, 1990; Oesterhelt *et al.*, 1991; Fukuzawa *et al.*, 1996; Fukuzawa, 1994, Miyasaka *et al.*, 1991, Storrs *et al.*, 1996; Maccioni *et al.*, 1996; Chen and Birge, 1993), and particularly for photovoltaic cells (Bertoncello *et al.*, 2004; and Figure 4.33).

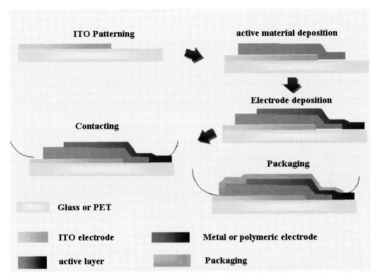

Figure 4.34. Photovoltaic cell fabrication process. (Reprinted with the permission from Bertoncello *et al.*, Bacteriorhodopsin-based Langmuir-Schaefer films for solar energy capture, IEEE Transactions on Nanobioscience 2, pp. 124–132 © 2003, IEEE).

The fabrication process was also optimized for the photovoltaic cell giving watt per area and per weight rather reproducible and utilizable for several application (mainly space), but in constant progress and further optimization using a combination of Gratzel cell and nanocomposite materials of inorganic, organic and biological origin and manufacturing (Stura *et al.*, in preparation).

4.2.2 Batteries

The batteries called "rocking-chair" systems are one of the most promising electrochemical energy storage systems, and they have a tremendous role in industrial applications of nanobiotechnology. Mounting concern regarding the environmental impact of throwaway technologies has caused a discernible shift away from primary batteries and toward rechargeable systems.

The secondary batteries ("rechargeable systems") have the advantage of being able to operate for many charge cycles without significant loss of performance. With technologies emerging today, an even higher demand for rechargeable batteries with high specific energy and power is expected.

Technological improvements, allowing to manipulate and investigate the properties of nanomaterials, are nowadays changing the approach to the energy storage and power supply vision. Modern nanoscale techniques led the market in the realization of nanostructured inorganic and organic materials increasing the efficiency of different devices, like lithium batteries, one of the most promising energy storage elements, obtaining everyday higher values of capacity, cyclability and environmental resistance. Each part of the battery, the anode, the cathode and the electrolyte, are here described analyzing the nanomaterials used for their realization.

Energy electronic components are elements designed in a power supply context, so oriented to the generation (conversion) or storage of energy in electrical form. Nanotechnology can help in the realization of materials with particular characteristics that make them suitable for the use in the energy field, particularly for the photovoltaic power generation and for the realization of parts of lithium-ion batteries, object of this review. The constant development of technological electronic devices is leading circuits to very small dimensions. The power supply must provide enough energy for the proper functionality of these structures, so it should be based on a material with high-energy storage/weight ratio. Energy storage/volume ratio is also very important to obtain valid small stand-alone devices. Lithium ion electrolytic cells for batteries are nowadays considered the best dealing with these two characteristics (Winter and Brodd, 2004).

A number of improvements of electrolytic lithium-ions cells for hybrid organic batteries were studied. Each part of the system is object of many studies, the cathode, the anode, and the electrolyte (Figure 4.35), that can be realized in liquid, gel or solid form. Another important characteristic to consider for developing a lithium ions battery is the cost of the materials involved, and it is one of the reasons for the use of polymers, spinels and other low cost materials.

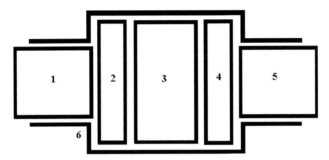

Figure 4.35. Physical layout of the hybrid organic battery. 1: Negative current collector (stainless steel). 2: Cathode 3: Electrolytic solution (LP30 pregnated porous membrane). 4: Anode (metallic lithium) 5: Positive current collector (stainless steel). 6: Hermetic plastic container (Reprinted with the permission from Stura and Nicolini, New nanomaterials for light weight lithium batteries, Analytica Chimica Acta 568, pp. 57–64 © 2006, Elsevier).

The overall effectiveness of different kinds of nanocomposite batteries is given in Table 4.9. The demand for high performance of lithium ion batteries generated strong incentives for the promotion of first-rate basic studies in different materials sciences, surface science, crystallography, spectroscopy, microscopy and electrochemistry.

Many worthy results have been achieved, among these the development of novel, high capacity negative electrodes based on amorphous silicon (Graetz *et al.*, 2003), different kind of tin alloys (Kim *et al.*, 2003), nanoparticles based on transition metal oxides (Nazar and Crosnier, 2004), inter-metallic compounds (Yin *et al.*, 2004) and new materials based on carbon (Li *et al.*, 2003). Intensive efforts are now in progress in order to develop new electrolyte solutions with non-flammable solvents (Zhang et al., 2003), for similar safety reason also salts such as Li-bi oxalato-borate (LiBOB) are topic of many worldwide researches (Xu et al., 2002) and, to assure a safe operation, over charge protection (Adachi et al., 1999).

Table 4.9. Overall effectiveness of different kinds of nanocomposite batteries (Reprinted with the permission from Stura and Nicolini, New nanomaterials for light weight lithium batteries, Analytica Chimica Acta 568, pp. 57–64 © 2006, Elsevier).

Battery type	Energy density: (Wh kg^{-1})	Energy density: (Wh l^{-1})	Maximum charge/discharge cycles	Nominal voltage
Ni-Cd	34	80	~1000	1.2
Ni-H	54	64	~5000	1.2
Pb acid	38	78	~500	2.0
Li-ion	152	305	~3000	3.3–4.0

Many efforts are spent to develop new materials for separators (Bohm, 1999) and new solid-state electrolytes, both polymeric (Scrosati, 2002) and ceramic (Birke and Weppner, 1999).

There also are highly intensive efforts developing new cathode materials with high redox potential (Wang et al., 2004), olivines (Pasquadi et al., 2004), and LiMn$_{2-x}$M$_x$O$_4$ spinel compounds.

The efficiency of different materials changes drastically between the raw form and nanostructured form, the same molecules organized in different spatial position (pellets, nanoparticles, nanotubes, high surface ratio elements) can change their macroscopical behaviour and electrical characteristics, because of their chemical and physical parameters (like surface energy).

To understand all the details regarding the structure of these materials and their surface chemistry other works are now in progress. The correlation between these parameters and the electrochemical behavior of electrode materials are still not completely defined.

The use of different analytical chemistry techniques such as neutron diffraction (Kim and Chung, 2004a), X-ray absorption near edge structure (Kim and Chung, 2004b), extended X-ray absorption fine structure (Okada et al., 2003) and *in situ* techniques like Raman spectroscopy (Dokko *et al.*, 2003), infrared spectroscopy (Santner *et al.*, 2004), various kind of atomic force and tunnelling microscopies (Aurbach *et al.*, 2002), X-ray diffraction (Hatchard and Dahn, 2004) and mass spectroscopy (Lanz and Novak, 2001) contribute to obtain clear relation between the chemical-morphological aspect and their behavior. The use of electrochemical impedance spectroscopy is also an important tool for the characterization of batteries and fuel cells. This technique yields quantitative information on a diverse range of processes including the analysis of state of charge, study of reaction mechanisms, film aestivation and corrosion processes.

4.2.2.1 *Lithium ion batteries elements*

The electrolytic cells used to test our electrodes (*i.e.*, the cathode) are typically assembled as shown in Figure 4.35.

An ABS plastic cylindrical box was used as container for the electrochemical elements, avoiding the dripping of the electrolytic solution. The anode is obtained from a foil of the material tested using a punch with 0.7 cm^2 circular head, the electrolyte is based on three disks of porous material with a surface of 1 cm^2 pregnated with the liquid electrolyte, if a solution is used, or a mass of plasticized gel if a polymeric electrolyte is used. The cathode is obtained using the same punch used for the anode, piercing a foil of the amalgamated materials. These three elements are put in close contact using the stainless steel cylindrical current collectors. The last elements also served as blocking objects to close the electrolytic cells.

4.2.2.2 *The cathode*

A considerable variety of cathode materials have been studied to develop efficient elements with high energy density: metallo-phosphate, lithium spinel materials, metallo-phosphates ($LiFePO_4$, $LiCoPO_4$, $LiMnPO_4$ and $LiNiPO_4$) with olivine-type structure attracted a noticeable attention as possible intercalation materials for the Li^+ ion (Padhi *et al.*, 1997; Garcìa-Moreno *et al.*, 2001; Yamada *et al.*, 2001).

This interest is focused mostly on the lithium iron phosphate (LiFePO$_4$) because of the high discharge voltage of 3.6V versus metallic lithium and the environmental compatibility of iron. Different approaches for enhancing the electrochemical performance of this compound were studied (Yamada et al., 2001; Huang et al., 2001; Franger et al., 2002) in order to get small particle size and good electronic contact between the particles of the sample. The mechanism of lithium extraction–insertion from LiFePO$_4$ is properly proved with ex-situ and in-situ X ray diffraction and electrochemical techniques (Padhi et al., 1997; Andersson et al., 2000; Takahashi et al., 2002). As the lithium extraction from LiFePO$_4$ proceeds for electrical field reasons, a new material, FePO$_4$, is formed. The dual phase characteristic of the electrochemical reaction seems to provoke low rate capability of this compound due to the hindered lithium diffusion through the interface between LiFePO$_4$ and FePO$_4$ (Padhi et al., 1997). Differently from what happens with FePO$_4$ with olivine-type structure, the existence of CoPO$_4$ and MnPO$_4$ crystallizing in the same space group is not proved yet. However, a two-phase mechanism of lithium deinsertion from LiMnPO$_4$ and LiCoPO$_4$ seems to take place also in this case (Li, et al., 2002; Delacourt et al., 2004; Amine et al., 2000; Okada et al., 2001). Delacourt et al. (2004) showed that the electrochemical deintercalation of the Li$^+$ ion from LiMnPO$_4$, prepared by a new precipitation mean, leads to the disappearance of LiMnPO$_4$ and to the formation of the new material with the same structure, but with lower cell volume, probably is MnPO$_4$. Bramnik et al. (2004) showed that LiCoPO$_4$ samples prepared by solid-state reaction at high temperature gives unsatisfactory electrochemical performance. The reversibility of lithium intercalation and deintercalation from LiCoPO$_4$ was improved by an artificial method based on the precursor NH$_4$CoPO$_4$·H$_2$O (Lloris et al., 2002), even if the cyclability of the material is not good. Diffraction patterns of Li$_x$CoPO$_4$ sample charged to $x = 0.05$ (from the electrochemical data) by refining a two-phase model were analyzed (Bramnik et al., 2004). The difference in the cell volumes for both iso-structural phases was found to be much lower in comparison with the data reported for manganese metallo-phosphate. Another interesting material for batteries cathodes is LiMn$_2$O$_4$, that attracted significant interest mostly because it has 4Vas voltage versus metallic lithium. Its reversible capacity (110–120mAh g^{-1}) and its characteristics are similar to the ones of the currently used $LiCoO_2$. The manganese spinel material is cheaper and more environmentally compatible, and undergoes a phase transition at 290 K, transforming itself from a cubic phase (at high temperature) to a orthorhombic phase (at low-temperature). This transition takes place for the critical concentration of Mn^{3+} ions (Jahn-Teller ions). This is due to the presence of Mn^{3+} ions, and this transition is probably responsible of the limited cyclability (Tarascon et al., 1994; Shimakawa et al., 1997).

$Li_xM_yMn_{2-y}O_4$ (where M is Al, Mg, Ti, V, Cr, Fe, Co, Ni, Cu or Zn) spinels are also currently investigated as possible cathode materials (de Koch et al., 1998; Shao-Horn et al., 2001; Kumagai et al., 1997; Iwata et al., 1999; Thirunakaran et al., 2001; Tarascon et al., 1991) and the results look encouraging. These studies, however, are limited to determining lattice parameters and electrochemical characteristics. For different dopants, the charging curve drastically changes its shape, suggesting a variation in the electronic structure of the manganese spinel. At this moment a very few research results on the electronic transport of doped manganese spinel are available, it is only known that its value of conductivity is low (10^{-4} S/cm) because of its small polaron mechanism (Marzec et al., 2002). Molenda et al. (1999, 2003) and Swierczek et al. 2003) noticed that the polaron mechanism of charge transport, in the manganese spinel is very stable.

Recent results in the cathode research introduced $LiNi_{0.5}Mn_{0.5}O_2$, providing over 30 cycles of charge/discharge with a capacity of 200 mAh g^{-1} (Makimura et al., 2003). Additional research has shown that the redox center is the divalent nickel, which becomes tetravalent due to the oxidation occurring during charging, and is then reduced to divalent nickel again during discharging at a 4V (Yoon et al., 2003; Nakano et al., 2003; Johnson et al., 2003). Manganese remains tetravalent during cycles at the 4V plateau, and the absence of trivalent manganese contributes to its structural stability. Another characteristic of $LiCo_{0.5}Mn_{0.5}O_2$ is its first charge and discharge capacities of respectively 220 mAh g^{-1} and 125 mAh g^{-1}, that makes it suitable for non rechargeable batteries too (Tsuda et al., 2005).

In the last ten years there is a particular attention to TiO_2 nanoparticles and their electronic properties (as shown earlier for photovoltaic cells). This material can attract interest for the realization of low voltage batteries, since its discharge voltage is 1.8 V versus Li/Li^+. However the insertion of 1 Li per TiO_2 unit corresponds to a capacity of 335 mAh g^{-1}. For the best results in the realization of cathodes for lithium batteries, 20 nm nanoparticles are normally used, since it has been reported that a sol-gel sample prepared starting with these nanoparticles undergoes an optimal partial phase transformation around 600 °C that generates a favourable material with good electrochemical characteristics. The insertion coefficient in anatase (TiO_2) is usually close to 0.5 (Kavan *et al.*, 1996), making these nanoparticles interesting materials for batteries designed for particular purposes.

4.2.2.3 *The anode*

Carbon-based materials are currently commercially used as anode materials due to the flat charge and discharge plateau and excellent cycling stability.

However, their theoretical maximum capacity is limited to 372 mAh g^{-1}, corresponding to the formation of LiC_6 (Fong and Sacker, 1990). Since the introduction of tin-based oxide composite by Fuji Photo Film Celltec in early 1997, great interest has been turned to metal, inter-metallic compounds, alloy anodes due to their extremely larger capacity compared to those of carbon-based materials. Also organic salts and polymers attracted interest lately, for the good mechanical and technological properties of organic matter. In the new battery technology there is a renewed interest in metal alloys and inter-metallic compounds for replacing graphitic carbon as the anode of choice in lithium-ion batteries.

The lithiation voltage of these materials is enough positive versus metallic lithium to minimize lithium-plating problems (occurring in case of overcharge). Furthermore, these materials offer a higher volumetric capacity than graphite. For example, the theoretical volumetric capacity of InSb is 1904 mAh cm^{-3} (assuming a constant electrode density of 5.6 g cm^{-3} throughout charge and discharge) compared with a theoretical 818 mAh cm^{-3} for graphite (ρ =2.2 g cm^{-3}). Two reactions can happen between inter-metallic compounds and lithium: insertion of lithium with no extrusion of metal atoms from the host structure, or insertion of lithium into the structure. For the first reaction typical materials are Cu6Sn5 and MnSb that, on lithiation, form Li$_2$CuSn (Kepler et al., 1999) and LiMnSb (Fransson et al., 2003), respectively. In the second case, the inter-metallic compound can be comprised entirely of elements that react with lithium, such as in SnSb (Rom et al., 2001), InSb (Tostmann et al., 2002) or Ag$_3$Sb (Vaughey et al., 2003) when discrete LixSn, LixSb, LixIn and LixAg phases are formed. In other cases, the intermetallic compound can consist of two elements, and only one of these reacts with lithium, as in FeSn$_2$ (Mao et al., 1999a), Cu$_2$Sb (Fransson et al., 2001) and CoSb$_3$ (Alcantara et al., 1999) in which case the Li$_x$Sn or Li$_x$Sb phases are cycled within electrochemically inactive Fe, Cu or Co metal matrixes (Sarakonsri et al., 2005).

Although alloys have much higher capacities than that of graphite, they undergo severe volume expansion/contraction and pulverization. The effect of these events limits the lifecycle of the anode and of the whole battery. The performance of an alloy anode can be improved if the active alloy is supported by inactive components, which provide structural stability during cycling (Mao et al., 1999b; Beaulieu et al., 2000).

Sn-Cu intermetallic compounds have been suggested as promising alternative anode materials (Thackeray et al., 2002; Wachtler et al., 2001; Winter and Besenhard, 1999). Kim et al. (2002) prepared nano-sized Cu$_6$Sn$_5$ electrodes by chemical reduction and found that their cyclability was significantly enhanced as compared with the same material prepared by sintering or mechanical alloying. Even though, the above preparation methods are not suitable for large-scale, low cost production of alloy electrode materials (Pu et al., 2005).

Among alloys, which can be potential anode materials for secondary Li-ion batteries, the Sb-based intermetallic compounds received much interest lately. Alcantara et al. (1999), first reported CoSb$_3$ as possible anode materials for Lithium ion batteries. Thereafter, many Sb-based intermetallic compounds, such as CrSb$_2$ (Fernandez-Madrigal et al., 2001), TiSb$_2$ (Larcher et al., 2000), Cu$_2$Sb (Fransson et al., 2001a),

Mn_2Sb (Fransson et al., 2003), $Co_{1-2y}Fe_yNi_ySb_3$ (Monconduit et al., 2002), SnSb (Besenhard et al., 1999) etc., were investigated. Although these compounds show slight lower capacity than pure Sb (660 mAh g^{-1}), they show improved cycling behavior. However, the long-term cyclability cannot meet the requirement for the practical application of these materials as anode in commercial Li-ion batteries. The main drawback of these compounds is the rapid capacity fade after repeated cycling, resulting from the large volume changes (expansion and compression) during the charge and discharge phases (Xie et al., 2005).

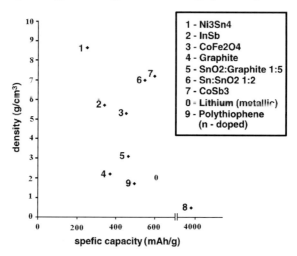

Figure 4.36. Comparison among different anode nanomaterials: density versus capacity (mAh g^{-1}) (Reprinted with the permission from Stura and Nicolini, New nanomaterials for light weight lithium batteries, Analytica Chimica Acta 568, pp. 57–64 © 2006, Elsevier).

The electrochemical application of organic ionic liquids (room temperature molten salts) that show attractive properties such as high ionic conductivity ($\sim 10^{-3}$ S cm^{-1}), wide electrochemical window (wider than 3.0 V), and good thermal and chemical stability, has been explored in the fields of photo-electrochemical cells (Nazeeruddin et al., 1993); Papageorgiou et al., 1996), lithium secondary batteries (Fuller et al., 1997,1998; Koch et al., 1996) and electrochemical capacitors (Nanjundiah et al., 1997). Among organic materials, dialkylimidazolium based ionic liquids with the weakly complexing anions (*e.g.* PF_6^-, BF_4^-, or $CF_3SO_3^-$, $(CF_3SO_2)_2N^-$) seem to be the most stable and conductive to date (Koel, 2000; Hagiwara and Ito, 2000; Bonhote *et al.*, 1996; McFarlane *et al.*, 2000). The 1-butyl-3-methylimidazole based ionic liquids with PF_6^- (Fuller *et al.*, 1998) and $(CF_3SO_2)_2N^-$ (Bonhote *et al.*, 1996) have high hydrophobicity besides other good properties (Fuller *et al.*, 1998). These materials are particularly useful for the realization of anodes for specific applications like lithium/seawater batteries (Zhang *et al.*, 2005). In Figure 4.36 is given the comparison among the anode nanomaterials: density versus capacity (mAh g^{-1}). This figure gives an idea of how certain materials, universally considered "good" in reality need large volumes to reach few milligrams, while others considered "mediocre" need instead small volumes to obtain large masses.

4.2.2.4 *The electrolyte*

Actually most industrial electrolytes are mixtures of ethylenecarbonate, diethylcarbonate and dimethylcarbonate with lithium hexafluorophosphate ($LiPF_6$) as salt. This type of electrolyte permits a great number of charge–discharge cycles without noticeable loss in capacity but the search of new electrolytes with higher thermal stability is of great importance (Botte *et al.*, 2001; Zhang *et al.*, 1998; von Sacken *et al.*, 1994; Dahn *et al.*, 1994; Du Pasquier *et al.*, 1998; Gee and Laman 1993; Hong *et al.*, 1998; Richard and Dahn, 1999; Ohta *et al.*, 1995; Roth, 1999; Kumai *et al.*, 1999).

Among polymeric electrolytes, poly(ethylene oxide)-lithium salt complexes are promising candidates as electrolytes for lithium polymer battery applications (Armand *et al.*, 1989; Gray, 1997; Gray and Armand, 2000; Lightfoot *et al.*, 1993; Vincent and Scrosati, 1993). Large research efforts were spent for the development of poly(ethylene oxide) electrolyte solutions allowing to combine high conductivity, good interfacial stability with lithium metal anode and good mechanical properties (Wieczorek *et al.*, 1989; Borghini *et al.*, 1995; Appetecchi *et al.*, 2000). A common approach is the use of a lithium salt having a very large counter-ion, able to interfere with the crystallization process of the polymer chains (Appetecchi *et al.*, 2001a; Rossi Albertini *et al.*, 1997), promoting amorphous regions and increasing the lithium ion transport across the polymeric electrolyte (Gray, 1997; Lascaud et al., 1994; Feuillade and Perche, 1975). Following this approach, Appetecchi et al. (2001b) have shown that the use of a lithium salt whit large anion, $N(SO_2CF_2CF_3)^{2-}$, enhances the conductivity of poly(ethylene oxide) based polymer electrolytes. In addition, these polymer electrolytes develop a very stable interface with lithium metal anode both under rest conditions and current flow (Appetecchi and Passerini, 2000).

Room temperature ionic liquids (RTIL) can also be used as safe electrolytes in electrochemical applications owing to their wide thermal stability, wide liquid-phase range, non-flammability and very low vapour pressure (Hu et al., 2004; Ngo et al., 2000). These RTIL are composed of a cation like quaternary ammonium (Sun et al., 1998), alkylpyridinium (Chum and Osteryoung, 1981), alkylpyrrolidinium (MacFarlane et al., 1999), alkylpyrazolium (Caja et al., 1999), alkyltriazolium (Vestergaard, et al., 1993), alkylphosphonium (Holbery and Seddon, 1999) and alkylimidazolium (Blanchard et al., 1999), combined with a variety of large anions having a delocalized charge (PF_6^-, BF_4^-). All RTILs show a high viscosity and therefore a relatively low conductivity. In order to decrease the viscosity and increase the conductivity, aprotic dipolar organic solvent may be added to RTIL, experimental results obtained by Chagnes et al. (2005) using butyrolactone confirm this.

Recently, to avoid liquid electrolytes and the related hydraulic insulation issues, some efforts were spent in the research of solid state or gel electrolytes. Ion-conducting polymer electrolytes have contributed to the development of lithium battery technology by replacing the liquid electrolyte and thereby enabling the fabrication of flexible, compact, and laminated solid-state structures free from leaks of the electrolyte (Croce, et al., 1998). Among these, solvent free polymer electrolytes formed by complexes of a lithium salt with a polyether such as poly(ethylene oxide) received considerable attention for their advantages in terms of the ease of fabrication, flexibility in dimensions, good mechanical properties, safety features, and good stability at the lithium interface (Ulrich et al., 2002; Gadjourova et al., 2001; Kim, 1998; Appetecchi et al., 1998). However, their low ionic conductivities have been the reason for them not being used in practical applications in rechargeable lithium batteries that require a value of above 10^{-4} S cm^{-1} at room temperature (Jeon et al., 2005). It is only above the melting temperature of crystalline poly(ehtylene oxide) – lithium salt complexes (~60 °C) that significative conductivity values ($\sigma >10^{-4}$ S/cm) are measured (Fauteux et al., 1995). Many efforts aimed to the lowering of operation temperatures of poly(ehtylene oxide) – lithium salt systems to the room temperature region have focused on the development of copolymerization (Fauteux et al., 1995; Soo et al., 1999; Allcock et al., 1986; Abraham et al., 1988; Tonge and Shriver, 1987; Xia and Smid, 1984; Cowie et al., 1985a,b; Gray et al., 1988) or cross-linking (Maccallum et al., 1984; Watanabe et al., 1986; Andrei et al., 1994; Killis et al., 1982; Cheradame et al., 1987) strategies and the use of suitable plasticizers (Gray, 1991; Abraham, 1993; Kelly et al., 1985) to create completely amorphous systems, therefore with enhanced conductivity. Incorporation of inorganic particles in the polymer matrix to obtain mechanical stability (Weston and Steele, 1982), and enhance interfacial properties (Capuano et al., 1991) and conductivity by suppressing crystallization of the PEO host has also been investigated (Croce et al., 1998, 1999; Wieczorek et al., 1995; Krawiec et al., 1995; Best et al., 1999; Capiglia et al., 1999).

An alternative strategy for creating polymer electrolyte systems with improved electrical and mechanical properties is through fabrication of polymer silicate nanocomposites. These materials are a class of compounds in which nanoscale clay particles are molecularly dispersed within a polymeric matrix (Yano et al., 1993; Messersmith and Stupp, 1992; Kojima et al., 1993; Krishnamoorti et al., 1996; Shi et al., 1996; Wang and Pinnavaia, 1998a,b). Recent commercial interest in these nanocomposites is derived from the fact that they show significant increases in tensile strength (Kojima et al., 1993), heat resistance (Messersmith and Stupp, 1992) and solvent resistance (Burnside and Giannelis, 1995) as well as decreases in gas permeability when compared with the bulk polymer (Messersmith and Stupp, 1992). These characteristics are useful also developing lithium batteries.

By now, lithium ions batteries are the most used devices in almost any kind of mobile devices, both for industrial and for commercial applications, overcoming most part of the issues typical of the previous technologies in batteries. In the last years, engineered polymers were implemented in the realization of lithium batteries, reducing the complexive weight of the power supply element providing acceptable values of capacity. Most industries included in their next future research plans studies of better nanomaterials for the application in lithium ion batteries.

Particular attention is oriented nowadays to organic cathodes, not for their capacity values but for the low cost of production, ease of synthesis and good technological properties like material modelling and solubility in organic solvents.

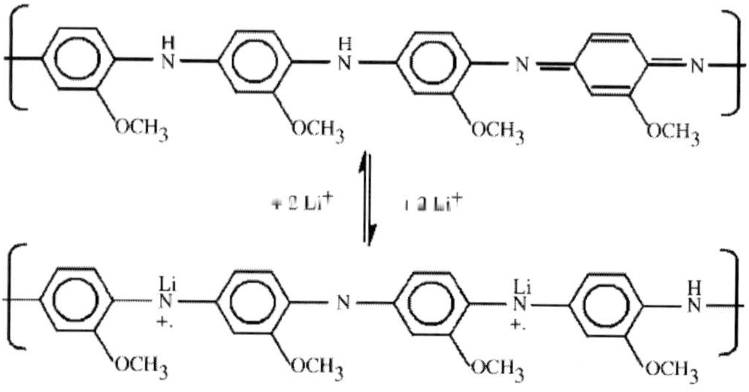

Figure 4.37. Chemical intercalation of the lithium ions in the polymer. (Reprinted with the permission from Stura and Nicolini, New nanomaterials for light weight lithium batteries, Analytica Chimica Acta 568, pp. 57–64 © 2006, Elsevier).

Our research group obtained acceptable values (130 mAh/g^{-1}) for a nanocomposite material based on poly(ortho-anisidine) and titanium dioxide nanoparticles, taking benefit from the optimization of the polymer chain occurring in presence of TiO$_2$ nanoparticles. This nanocomposite showed a particularly higher capacity value in the first charge and discharge cycle than in the following ones. When Li$^+$ ions reach the polymeric matrix, they dope the polymer taking the place usually occupied by the H$^+$, tied to the imines group. In this operation, a charge transfer occurs, and the difference of potential between the anode and the cathode changes.

The opposite process happens when the Li$^+$ ions leave the polymeric material and are released into the electrolytic solution, moving charge in the opposite direction and undoping the organic material (Stura *et al.*,

2004). Cathodes based on the above polymeric material and nanocomposite material (polyorthoanisidine with titanium dioxide nanoparticles) are typically intended for a low power and long life power supply, and tested in classical lithium-ion rechargeable batteries.

The results are promising considering that multiple cycles of charge-discharge tests result in sufficient performance of the synthesized materials that after 20 cycles tend reach a constant value of capacity, and can be readily explained in terms of the motion of lithium ions (Li^+) in the electrolyte, undergoing, when the electrical field is present (in the charge phase), an intercalation process in the polymer matrix, thereby doping the polymer similarly for what happens when the polymer is in presence of acid agents. A schematic of this process is presented in Figure 4.37. The present status of new materials by organic nanotechnology and their applications to molecular electronics has been recently overviewed (Nicolini *et al.*, 2005), with respect to the development of organic nanotechnology capable to yield new materials for a variety of technological applications. Particular emphasis has been placed on what has been accomplished in our laboratory in the last few years (Narizzano and Nicolini, 2005; Valentini *et al.*, 2004a; Carrara *et al.*, 2005; Stura *et al.*, 2002), whereby can be found in earlier papers the details on the supramolecular layer engineering and its application to industrial nanotechnology (Nicolini *et al.*, 2001a) and to molecular electronics (Nicolini, 1996b; Facci *et al.*, 1996). As previously pointed out (Nicolini *et al.*, 2005) the major drawback of the polymeric materials is the multi-step synthesis required for the functionalization and the stringent process requirements of the condensation polymerization. It was then our efforts in the last few years therefore to find shorter synthetic routes to process the polymers with predictable absorption wavelengths of light. Several types of polymer poly(p-phenylenevinylene) (PPV), derivatives have been synthesised in our laboratory, namely poly(2-methoxy-5-(2'-ethyl)hexyloxy-p-phenylenevinylene) (MEHPPV), whereby the Gilch route has been modified in order to increase the processability for specific device application.

The data here presented point to the successful engineering of organic nanotechnology-based materials and using polymer chemistry having potential industrial relevance in the area of lightweight lithium batteries. Figure 4.38 shows indeed the significant cyclability of present nanomaterials for lithium ion batteries, grouped in four large families.

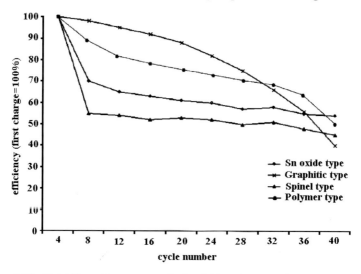

Figure 4.38. Ciclability of nanomaterials for lithium ion batteries (Reprinted with the permission from Stura and Nicolini, New nanomaterials for light weight lithium batteries, Analytica Chimica Acta 568, 57–64 © 2006, Elsevier).

We may then conclude that, although the work is still in progress in order to further optimize the parameters and to evaluate in more needed details, case by case, the optimal implementation of this technology within the required cost effectiveness, and the reproducibility within an highly competitive industrial context, it is conservative to conclude that light weight lithium methodology represents a promising general purpose tool for the design and production of new batteries.

4.2.3 *Hydrogen storage and fuel cells*

The main impediment to the use of hydrogen as a transportation fuel is the lack of a suitable storage system (Figure 4.39).

Compressed-gas storage is bulky and requires the use of high-strength containers. Liquid storage of hydrogen requires temperatures of 20 K and efficient insulation. Solid-state, storage offers the advantage of safer and more efficient handling of hydrogen, but promises at most 7% hydrogen by weight and more typically 2%.

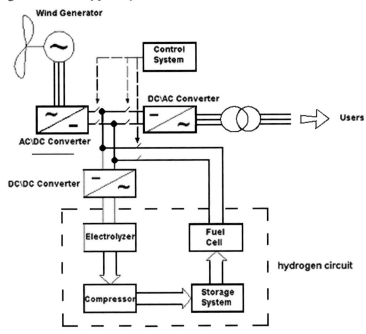

Figure 4.39. Hydrogen storage (Reprinted with the permission from Stura *et al.*, Hydrogen storage as stabilization for wind power: completely clean system for insulated power generation, Chemical Engineering Transactions 4, pp. 317–323 © 2004, AIDIC, http://www.aidic.it).

Various materials, such as palladium (Pd), palladium alloy, palladium-ruthenium alloys, nanocrystalline FeTi, mechanically alloyed amorphous Ni_{1-x}, Zr_x, alloys, carbon nanofibers, and carbon nanotubes, are employed for the storage of hydrogen. There have been reports that certain carbon graphite nanofibers are able to absorb and retain 67 wt% hydrogen gas at ambient temperature and moderate pressure, *i.e.*, up to 23 standard liters (2 grams of hydrogen per gram of carbon at 50–150 bars). The lowest hydrogen adsorption reported for any graphite fiber microstructure was shown to be 11 wt%. Approximately 90% of the adsorbed hydrogen can be desorbed at ambient temperature by reducing the pressure, while the balance is desorbed upon heating.

Such claims are especially noteworthy, given that up to this point the typical best value of hydrogen adsorption in carbon materials has been 4%, or 0.5 H/C. A large number of research Institutes and various companies are involved in the storage of hydrogen and production of full cells based on hydrogen: the Electric Power Research Institute, the American Gas Association, the Gas Research Institute, International Fuel Cells, Energy Partners, Ballard Power Systems, the Energy Research Corporation, MC Power, Westinghouse Electric Corp, Daimler-Benz, BMW, Volkswagen, Volvo, Renault, Peugeot, Siemens, Toyota, Honda, Toshiba, Mitsubishi, Fuji, and Sanyo. Fuel cell-powered cars (Figure 4.40B) and field emission by carbon nanotubes in spacecraft (Figure 4.40C) are being researched and tested.

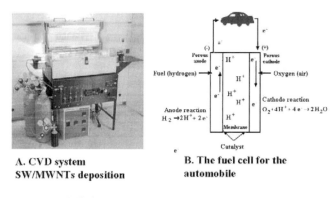

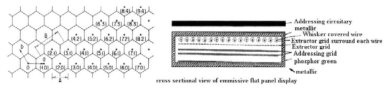

Figure 4.40. Chemical vapor deposition for carbon nanotubes manufacturing (A) in automotive fuel cell (B) and spacecraft field emitter (C).

Hydrogen is excellent for storage and would make certain sources more feasible. This would open doors to many alternative resources and begin to shift our use away from fossil fuels. Still, electrolysis and cryogenic cooling are both very expensive. Hydrogen storage is economically viable only when it is sent over very long distances, where piping hydrogen would be more efficient then sending electricity, or when a storage system is necessary, as in the case of solar or wind power (Stura et al., 2004). The use of Pd has revealed the restriction in storage capability due to the change in structures upon a few cycles of the adsorption-desorption process. The Pd becomes disordered after a few cycles of adsorption. The Pd-Ru structure remained almost unaltered after cycling, but the disadvantage could be that the efficiency of adsorption decreases during alloy formation. Several graphite nanostructures were prepared using Fe-Cu catalysts of different compositions, in order to generate a range of fiber sizes and morphologies. The hydrogen desorption measured from these materials was found to be less than the 0.01 R/C atom, compared to the other forms of carbons. The hydrogen exposed in the metal alloy $Ni_{4-n}Zr_n$, has shown that hydrogen resides in $Ni_{4-n}Zr_n$ (n = 4, 3, 2) tetrahedral interstitial sites, with a maximum hydrogen ratio of 1.9. Carbon adsorption techniques rely on the affinity of carbon and hydrogen atoms. Hydrogen is pumped into a container with a substrate of fine carbon particles, where molecular forces hold it. This method is about as efficient as metal hydride technology but is much improved at low temperatures, where the distinction between liquid hydrogen and chemical bonding needs to be considered. One of the most exciting advances recently has been the announcement of carbon nanofiber and carbon nanotube technologies. There is also the claim that up to 10 wt% was achieved for hydrogen storage in single-wall nanotubes. Owing to the potential importance of new materials with high hydrogen storage capacity for the worldwide energy economy, transportation systems, and interplanetary propulsion systems, carbon nanotubes could play an important role in hydrogen storage.

Iijima (1991) has focused much attention on both fundamental and applied research on carbon nanotubes (see CVD in Figure 4.40A) since the discovery of multiwall carbon nanotubes (MWNTs) in 1991. In particular, recent progress in research on the properties of single-wall carbon nanotubes (SWNTs), such as their atomic structure and electronic properties, hydrogen storage properties, mechanical properties, and property enhancement through nanotube modification, has been outstanding, due mainly to the availability of sufficient quantities of

SWNTs that can be obtained using the pulsed laser vaporization method and the electric are technique. It has been both predicted theoretically and demonstrated experimentally that SWNTs have many interesting properties. Pores of molecular dimensions can adsorb large quantities of gases, owing to the enhanced density of the adsorbed material inside the pores, a consequence of the attractive potential of the pore wall. Dillon *et al.* (1997) have shown that a gas can condense to high density inside narrow SWNTs. Simonyan *et al.* (1999) described the adsorption of molecular hydrogen gas onto charged single-wall nanotubes by grand canonical Monte Carlo computer simulation. The present availability of various fullerene structures points up a large gap in the intermediate size range between small, highly tangled ropes of nanotubes that are currently available in short lengths. Recently, laser vaporization and electric arc methods have best for even for obtaining a continuous process for SWNT production on a commercial scale. Therefore, from an applications standpoint, emphasis is given to the production of high-purity, high-yield, low-cost, large-scale, and easily handled SWNTs for the storage of hydrogen. Recently, a novel method for synthesizing SWNTs reported the catalytic hydrocarbon decomposition method, in which benzene is catalytically decomposed at 1100–1200°C, yielding SWNTs that are similar, on a nanometer scale, to those obtained by laser vaporization and electric-arc techniques. This growth method allows lower growth temperatures, permits semi-continuous or continuous preparation, and produces a large quantity of SWNTs at relatively high purity and low cost. However, subsequent experiments showed that the ends of the tubes remained open during the growth process, with highly reactive dangling bonds located around the tube ends.

In the near future, the possible synthesis of nanotubes with solid-gas potential will be more favorable to adsorption. The effect of hydrogen overpressure on the stability of adsorbed H_2 needs to be verified in the near future. The high-purity nanotube produced by laser vaporization, catalytic decomposition, or other techniques as chemical vapor deposition (Figure 4.40A) should be investigated. It is noteworthy that the synthesis of the SWNT with defined diameters and distances between the walls is difficult to perform at present, but future synthesis routes will allow more hydrogen adsorption in the SWNT. Some theoretical calculations, such as Monte Carlo simulation, were performed for the adsorption of hydrogen with carbon nanotubes, but the real mechanisms of adsorption and desorption are still unknown. Control of these parameters, coupled with improvements in production, purification, and alignment of SWNTs, may lead to a new technology for hydrogen storage.

4.3 Nanobiocatalysis

The Langmuir-Blodgett (LB) technique was successfully applied for the deposition of thin protein layers (Langmuir and Schaefer, 1938; Tiede, 1985); Lvov et al., 1990). LB organization of enzymes in film of different number of monolayers (see Figure 4.41 for the GST enzyme) not only preserved the structure and functionality of the molecules, but also resulted in the appearance of new, useful properties, such as enhanced thermal stability as shown earlier in this Volume and by Nicolini et al. (1993) and Erokhin et al. (1995).

The enzymatic activity of GST was evaluated spectrophotometrically following the conjugation of glutathione (GSH) thiol group to 1-chloro2,4-dinitrobenzene (CDNB) at a wavelength of 340 run (Habig et al., 1974) by a double-beam spectrophotometer (Jasco 7800). The GSH and CDNB concentrations were 2.5 mM and 0.5 mM, respectively. Ten covered spheres were placed into the cuvette. The diffusion effects (Antolini et al., 1995b) were avoided by carrying out the measurements under continuous stirring with a magnetic microstirrer (Bioblock scientific) at a speed of 600 rpm.

The deposition procedure described earlier allows one to obtain protein films chemically bound to the activated surface of spherical glass particles. Subsequent compression of preformed protein monolayer with these particles permitted to coverage of the particle area that initially has not come in contact with the monolayer. Even if such a procedure does not initially result in deposition of strictly one monolayer, this fact does not seem to be critical, because only the monolayer chemically attached to the surface remains after washing as can be seen from the rather constant functional activity remaining with increasing number of layers

(Figure 4.41). Indeed, whenever the single protein monolayer is properly formed by Langmuir-Blodgett technique at the saturating surface pressure (see earlier sections), the enzymatic functional activity appears nearly independent of the number of layer as proven for GST (Antolini *et al.*, 1995b) and alkaline phosphatase (Petrigliano *et al.*, 1996) enzymes.

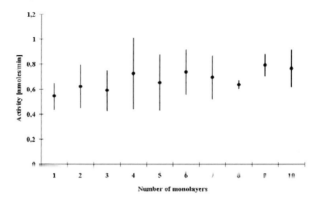

Figure 4.41. Dependence of GST activity after washing as function of the number of monolayers. For each point is given the error (for a confidential level of 95%). The reaction volume is 2 mL. (Reprinted with the permission from Antolini *et al.*, Heat-stable Langmuir-Blodgett film of glutathione-S-transferase, Langmuir 11, pp. 2719–2725 © 1995b, American Chemical Society).

Enhanced thermal stability enlarges the areas of application of protein films. In particular it might be possible to improve the yield of reactors in biotechnological processes based on enzymatic catalysis, by increasing the temperature of the reaction and using enzymes deposited by the LB technique. Nevertheless, a major technical difficulty is that enzyme films must be deposited on suitable supports, such as small spheres, in order to increase the number of enzyme molecules involved in the process, thus providing a better performance of the reactor. An increased surface-to-volume ratio in the case of spheres will increase the number of enzyme molecules in a fixed reactor volume. Moreover, since the major part of known enzymatic reactions is carried out in liquid phase, protein molecules must be attached chemically to the sphere surface in order to prevent their detachment during operation.

The aim of the work was the development of a technique to deposit enzyme LB films on the surface of small glass spheres and to test the enzymatic activity of such samples before and after thermal treatment.

The experiments were carried out with two enzymes, urease and glutathione-S-transferase (GST). Urease catalyses the hydrolysis of urea, while GST is an enzyme that catalyzes the reduction of compounds such as alkylants with a nucleophilic addition of the thiol of glutathione to electrophilic acceptors (*e.g.*, aryl and alkyl halides, quinones, organic peroxides) (Pickett and Lu, 1989). Both enzymes were chosen, since their activity can easily be tested by spectrophotometric measurements.

The LB technique was chosen for covering the spheres because it was shown to provide enhanced thermal stability of many types of proteins in deposited layers (Nicolini *et al.*, 1993; Erokhin *et al.*, 1995, 1995a), which no other technique is able to achieve. Since only the upper protein layer is involved in the catalytic activity, no special attention was paid to check whether the deposited layer is a monolayer or multilayer. However, the samples were thoroughly washed to remove protein molecules not bound covalently to the sphere surface, since during the functional test these molecules could contribute to the measured apparent catalytic activity. Borosilicate glass spheres with a diameter of 2 mm were used as substrates for the deposition.

The surface of the spheres was activated in the following way. Spheres were treated with boiling chloroform, rinsed on a glass filter, and dried under nitrogen, to be subsequently silanized with 3-glycidoxypropyl trimethoxysilane following the technique proposed by Malmquist and Olofsson (1989). Silanization of the spheres was performed in nitrogen flux in order to prevent reciprocal attachment of spheres and to activate their surface homogeneously.

The essential steps of the deposition procedure, which are the same for both enzymes, are illustrated in Figure 4.42. The protein solution was spread over the water subphase, and the monolayer was compressed up to 25 mN/m. Activated spheres were distributed over the monolayer in the following way: A plate with spheres over it was moved along the monolayer, while the weak nitrogen flow was used for transferring the spheres from the plate to the layer. It is important to have the plate in close vicinity to the water surface in order to keep all the particles floating over the monolayer. The spheres were kept at the surface for 30 minutes in order to provide chemical linking of the monolayer to activated surface. After this time the feedback system was switched off and the Wilhemy plate was removed from the water. The layer with particles was compressed until the minimum area (20 cm^2), which corresponds to the collapse of the monolayer, was reached. Even though this compression yields a multilayer film, such action seems to be necessary, since otherwise only half of the sphere surface was covered with protein monolayer, while compression induced the motion both of spheres and the monolayer, covering other regions of the spheres. The spheres were collected, washed with substrate buffer in order to remove parts of the monolayer not attached chemically to the sphere surface, and dried.

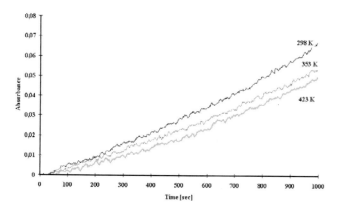

Figure 4.42. Urease activity test at different temperature (Reprinted with the permission from Nicolini, Heat-proof enzymes by Langmuir-Blodgett technique, Annals New York Academy of Science 799, pp. 297–311 © 1996, Blackwell Publishing).

The activity of enzymes in the film was estimated in the following way: In order to test the activity of urease, we utilized a calorimetric assay based on urea hydrolysis; the enzymatic reaction was followed at 590 nm, the suitable wavelength for bromcresol purple (Chandler *et al.*, 1982). Urea concentration was 1.67; ts 10^{-2} M. From the results of the urease activity test summarized it is clear that the deposition procedure preserved to a certain extent the enzyme catalytic activity. Heating the sample before testing decreased the enzyme in the film by about 30% but did not eliminate it completely. The results of the activity test of two samples are summarized in Table 4.10 together with reference values for a spontaneous reaction without enzyme.

Table 4.10. Slopes of activity for two LB samples of urease and for spontaneous reaction. (Reprinted with the permission from Nicolini *et al.*, Supramolecular layer engineering for industrial nanotechnology, in Nano-surface chemistry, pp. 141–212 © 2001a, Marcel Dekker/Taylor & Francis Group LTD).

Sample 1	7.01×10^{-4}
Sample 2	5.97×10^{-4}
Average	6.49×10^{-4}
Mean error	7.4×10^{-5}
Spontaneous reaction	2.49×10^{-6}

It is necessary to underline that enzymatic activity on spherical supports was higher than the respective value in "flat" films, which could indicate that apparent catalytic efficiency was improved due to an increased area-to-volume ratio.

As evident from results of the GST activity test presented in Figure 4.41, the enzyme is active in LB films. Comparison of these catalytic activity values with the balance values of GST LB film deposited onto flat silanized surfaces (Antolini *et al.*, 1995b), and presuming a linear dependence of activity on enzyme concentration, gives an effective amount of 2.44 pmol in the sample. Knowing the total area of spheres in the reaction medium (10 spheres with total area of about 125 mm^2) and taking into account that only the upper layer is involved in the reaction, we can estimate the surface density of the enzyme in the layer. The area per molecule was found to be 83 nm^2. On the other hand, the area per molecule can be estimated from strictly geometric considerations of protein sizes approximations taken from the RCSB Protein Data Bank (Berman *et al.*, 2000), and such calculations yield a value of 34 nm^2. Thus, comparing these values the conclusion can be reached that the amount of active enzyme in the film is about 41% of a maximum possible in the close-packed layer. The resultant activity is reported in Figure 4.31 together with the value for a spontaneous reaction.

The comparison of the results on enhanced thermal stability of proteins in LB films reported here and those already published again underlines that molecular close packing is a critical parameter responsible for this phenomenon.

The described procedure allows one to deposit protein, in particular, enzyme, LB films onto the surface of small spheres. Deposited multilayer film was washed in order to leave at the surface only a layer covalently attached to the activated surface.

The enzyme in the film preserves its catalytic activity and demonstrates highly increased thermal stability. The procedure can be useful for the fabrication of heat-proof active elements for bioreactors based on enzymatic catalysis.

4.3.1 Bioreactors

In recent time, Pastorino et al. (2004) showed the production of new and efficient catalytic biomaterials was analyzed also on lipase, investigating in toluene and comparing self-assembled lipase from *Candida rugosa*, *Mucor miehei* and *Rizhopus delemar*. Of these ones, *M. miehei* lipase resulted in the highest degradation yield and was used in further experiments.

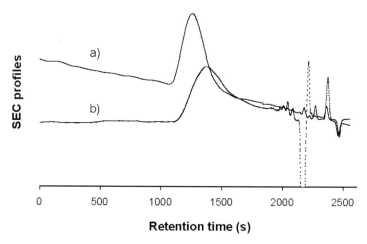

Figure 4.43. SEC profiles of (a) PCL and (b) degradation products of one-hour lipase-catalyzed PCL hydrolysis under the optimal conditions (Reprinted with the permission from Pastorino et al., Lipase-catalyzed degradation of poly(epsilon-caprolactone), Enzyme and Microbial Technology 35, pp. 321–326 © 2004, Elsevier).

Almost 74% of PCL (starting polyε-caprolactone, $M_n \cong 87000$ Da) was hydrolyzed at 40 °C by this lipase within 24 h, while a lower yield was observed with the other two lipases. Figure 4.30. illustrates the time course of the degradation of PCL catalyzed by *M. miehei* lipase under the optimal conditions (protocol A, 40°C). The peak due to PCL after one-hour reaction shifted to lower molecular weights and its area had increased, thus pointing out a wide distribution of the obtained molecular weights. Moreover the profile of the degradation products showed two peaks due to the simultaneous presence of both the dimer and the monomer. We found that the relative concentrations of reactants, the lipase source and the temperature all remarkably influenced the yield of the enzymatic hydrolysis of PCL. Moreover, the data obtained in this study provided indirect confirmation that the water content in the reaction mixture could dramatically influence the enzyme stability and activity. The stability of lipase in toluene at different temperatures proves good and the time course of the degradation reaction was relatively fast. It must be noticed that in water instead also for the lipase the optimal immobilization and biocatalytic process was achieved only with LS and protective plate nanostructured films.

Hans Kuhn was the first to design and make molecular machines on the basis of LB assembly (Kuhn, 1983). However, only recently (Troitsky *et al.*, 1996b, 2003) variations in the assembly composition and sequence of enzyme layer alternation were shown to influence the properties of the biocatalytic film, *i.e.*, its activity and structural stability. These studies have allowed us finally to produce biocatalysts with enhanced performance based on enzyme penicillin G-acylase (PGA). Penicillin acylases catalyse the hydrolysis of the side chain amide bond of penicillin (Figure 4.44). PGA preferentially hydrolyses penicillin G to give 6-aminopenicillanic acid (6-APA) and the side chain phenylacetic acid (Shewale and Sivaraman, 1989). 6-APA serves as a backbone for the synthesis of semisynthetic penicillins, providing a range of penicillin variants with differing antibiotic characteristics. The chemical methods for producing 6-APA are environmentally burdensome and require the use of hazardous chemicals. Current processes use penicillin acylases to remove the side chain from penicillin G/V, providing a green route to 6-APA (Parmar *et al.*, 2000). Immobilized PGA technology has had major success in making the enzymatic production of 6-APA economically viable. In those industrial process the immobilization is carried out in different ways; the most frequently used methods are based on membrane entrapment, chemical surface activation and physical absorption. However, these techniques do not provide control of the immobilization process or of the properties of the biocatalytic medium (Troitsky et al., 1996b).

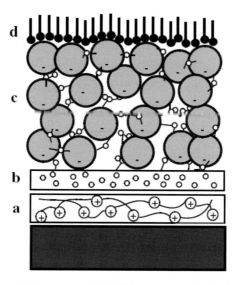

Figure 4.44. PGA model structure model of film structure obtained in the presence of an adsorbed layer of gluteraldehyde near the surface of the solid support. A frame-like film of PGA molecules cross-linked by GA (c) is formed over the sublayers of GA (b) and polymer p-DADMAC (a) and is protected by the monolayer of stearic acid (d). (Reprinted with the permission from Troitsky et al., A new approach to the deposition of nanostrructured biocatalytic films, Nanotechnology 14, pp. 597–602 © 2003, Elsevier).

Manipulation here described with LB monolayers and adsorbed layers without any lateral patterning is a promising methodology for the development of efficient biocatalytic films with a predetermined structure. Films of different structure and composition were investigated in order to demonstrate that the structure determines the functional properties of the biocatalytic film. The general approach can be described dividing the film into three hypothetical functional blocks (Figure 4.44). The functions of the bottom block are to provide high adhesion of the film to the surface of the solid support, to bind the enzyme molecules, and to orient them in a proper way. The function of the middle block is to catalyse the reaction while the top block provides protection for the enzyme layer. The purpose of depositing such a protective coating is to facilitate the storage of the biocatalyst and to increase the period of preservation of enzyme activity. The deposition procedure resulted in a considerable increase in the activity of the film per unit of the surface and the addition of a small amount of cross-linker near the surface where the adsorption of PGA should take place appeared to yield striking results. The enzyme activity, per unit of film surface, was immediately increased at least 20 times compared with the activity reported above for one closely packed monolayer. By further optimization of the film structure the best results were obtained for films with a poly(diallyldimethylammonium) chloride (p-DADMAC) sublayer in the bottom block (Figure 4.44).

4.3.1.1 *From lab scale to industrial scale*

For the purpose of industrial applications, the biocatalytic film has to be deposited on to supports of very large effective area to be utilized in bioreactors. For this reason, we are presently developing methods for technological realization of a new-patented procedure for large-scale nanobiocatalysis (Stura and Nicolini, 2006).

For practical applications it is indeed not enough to obtain just high efficiency and stability of the biocatalytic medium. It is also necessary to provide rather fast film deposition onto supports of very large effective area to be used in industrial bioreactors (Dubrovsky and Nicolini, 1994). Thus, we have performed a modification of the "protective plate" technique in order to carry out continuous deposition onto a flexible support (*e.g.* polymeric tape) of practically unlimited length. If the film containing the enzyme is deposited on to such tape, it is easy to make from the latter a compact roll with small gaps between the turns. The rolls with the immobilized biocatalyst can be used in bioreactors of different types. The idea behind the solution to the problem is to provide the possibility of transferring the flexible support from one compartment to another through the slits, which are located over the levels of liquids in the adjacent compartments, while holding the solutions inside the gap by capillary forces. The main time-consuming process is the deposition of LB monolayers.

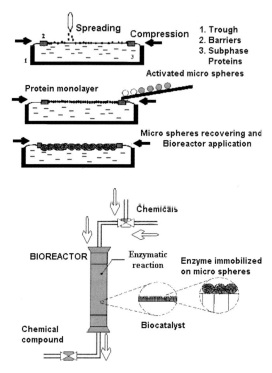

Figure 4.45. LB-based large scale bioreactor prototype (Reprinted with the permission from Nicolini C., Engineering of enzyme monolayer for industrial biocatalysis, Annals New York Academy of Sciences 864, pp. 435–441 © 1998b, Blackwell Publishing).

However, coating a 20–30 m long tape during the working day is quite possible (Nicolini, 1998b) (Figure 4.45). The other problems which arise when one tries to substitute small solid supports by long polymeric tape and to adopt a continuous process of deposition are the depletion of protein solution during the film deposition and transferring the cross-linker solution with the tape into the volume for protein adsorption. The calculations showed that either big volumes of protein solution should be used or a continuous supply of protein should be provided into the compartment, which should be of small volume. Transfer of cross-linker through the slit can be in principle be limited by careful elaboration of its design up to such a level that during the whole working day coagulation of protein in the volume will not occur. At present, development of the laboratory prototype of an apparatus based on this principle is in progress. Its design and the results of testing will be reported in a separate publication (Stura and Nicolini, 2006).

Figure 4.46. Prototype of industrial reactor, consiting of a rotating pilot reactor at laboratory level but displaying deposition industrially feasable (Reprinted with the permission from Nicolini and Pechkova, Nanostructured biofilms and biocrystals, Journal of Nanoscience and Nanotechnology 6, pp. 2209–2236 © 2006a, American Scientific Publishers, http://www.aspbs.com).

Based on the obtained results we can conclude that the methods of monolayer engineering are suitable for application in the field of biocatalysis. Although laboratory developments of the deposition technique do not prove the possibility of elaboration of the technology as well as its efficiency we do not see strong objections to this work. The main problem is working out a fast continuous technological process of biocatalyst deposition on to the surface of very long polymeric tape. At the same time, it seems that the problems of structural stability and preservation of enzymatic activity can be successfully solved. In conclusion, industrial applications, which already found their implementation in biocatalysis are gene engineering-based production of biologically essential macromolecules and the stereo-specific biosynthesis with the utilization of specific enzymes such as shown here for PGA and other enzymes (*i.e.*, urease, alcohol dehydrogenase, alkaline phosphatase, glutathione-S-transferase and DNA polymerase).

4.3.2 Bioactuators

The application of biological materials to different branches of modern technology has expanded enormously to bioactuators during the recent period. Examples of heat-proof and stable bioactuators are LB films based on the phenol oxidases and the laccase from animals which are specific for bioactive phenols (Tyr, DOPA, dopamine, adrenaline and noradrenaline), while enzymes from fungi and higher plants are active towards a wider spectrum of substrates (mono and polyphenols).

Laccase (p-diphenol: oxygen oxidoreductase) is a ubiquitous enzyme, which catalyzes:

$$\text{p-C}_6\text{H}_4(\text{OH})_2 + 1/2\, O_2 \rightarrow \text{p-C}_6\text{H}_4 O_2 + H_2O$$

while the enzyme cathechol oxidase (o-diphenol: oxygen oxidoreductase) catalyzes a similar reaction:

$$\text{o-C}_6\text{H}_4(\text{OH})_2 + 1/2\, O_2 \rightarrow \text{o-C}_6\text{H}_4 O_2 + H_2O$$

These two enzymes have been used:
- as bioelements in amperometric biosensors for the selective determination of neurotransmitters;
- for the construction of bioreactors for the enzymatic degradation of phenolic compounds in waste water.

Bibliography

Abraham, K. M. (1993). In: B. Scrosati (Ed.), *Applications of electroactive polymers*, Chapman & Hall, London, p. 75.

Abraham, K. M., Alamgir, M. and Perrotti, S. J. (1988). Rechargeable solid-state li batteries utilizing polyphosphazene-poly(ethylene oxide) mixed polymer electrolytes, *Journal of Electrochemistry Society*, 135, pp. 535–536.

Abraham, S., Vonderheid, E., Zietz, S., Kendall, F. M. and Nicolini, C. (1980). Reversible (G0) and nonreadily reversible (Q) noncycling cells in human peripheral blood. Immunological, structural, and biological characterization, *Cell Biophysics*, 2, pp. 353–371.

Adachi, M., Tanaka, K. and Sekai, K. (1999). Aromatic compounds as redox shuttle additives for 4 V class secondary lithium batteries, *Journal of the Electrochemical Society*, 146, pp. 1256–1261.

Adami, M., Sartore, M., Rapallo, A. and Nicolini, C. (1992). Possible developments of a potentiometric biosensor, *Sensors and Actuators B*, 7, pp. 343–346.

Adami, M., Alliata, D., Del Carlo, C., Martini, M., Piras, L., Sartore, M. and Nicolini, C. (1995). Characterization of silicon transducers with Si_3N_4 sensing surfaces by AFM and PAB systems, *Sensors and Actuators - B Chemical*, 24–25, pp. 889–893.

Adami, M., Piras, L., Lanzi, M., Fanigliulo, A., Vakula S. and Nicolini, C. (1994). Monitoring of enzymatic activity and quantitative measurements of substrates by means of a newly designed silicon–based potentiometric sensor, *Sensors and Actuators B*, 18–19, pp. 178–182.

Adami, M., Sartore, M. and Nicolini, C. (1995a). PAB: a newly designed and integrated potentiometric alternating biosensor, *Biosensors & Bioelectronics*, 10, pp. 155–167.

Adami, M., Sartore, M. and Nicolini, C. (1996a). Potentiometric and nanogravimetric biosensors for drug screening and pollutants detection, *Food Technology and Biotechnology*, 34, pp. 125–130.

Adami, M., Zolfino, I., Fenu, S., Nardelli, D. and Nicolini, C. (1996b). Potentiometric alternating biosensing toxicity tests on cell population, *Journal of Biochemical and Biophysical Methods*, 32, pp. 171–181.

Adami, M., Sartore, M. and Nicolini, C. (2007). A potentiometric stripping analyzer for multianalyte screening, *Electroanalysis*, 19, pp. 1288–1294.

Akiyama, T., Inoue, K., Kuwahara, Y., Terasaki, N., Niidome, Y. and Yamada, S. (2003). Particle-size effects on the photocurrent efficiency of nanostructured assemblies consisting of gold nanoparticles and a ruthenium complex–viologen linked thiol, *Journal of Electroanalytical Chemistry*, 303, pp. 550–551.

Alcantara, R., Fernandez-Madrigal, F. J., Lavela, P., Tirado, J. L., Jumas, J. C. and Olivier–Fourcade, J. (1999). Electrochemical reaction of lithium with the $CoSb_3$ skutterudite, *Journal of Materials Chemistry*, 9, pp. 2517–2521.

Allcock, H. R., Austin, P. E., Neenan, T. X., Sisko, J. T., Blonsky, P. M. and Shriver, D. F. (1988). Polyphosphazenes with etheric side groups: prospective biomedical and solid electrolyte polymers, *Macromolecules*, 19, pp. 1508–1512.

Amann, E., Ochs, B. and Abel, K. J. (1988). Tightly regulated tac promoter vectors useful for the expression of unfused and fused proteins in Escherichia coli. *Gene*, 69, pp. 301–315.

Amine, K., Yazuda, H. and Yamachi, M. (2000). Olivine $LiCoPO_4$ as 4.8 V electrode material for lithium batteries, *Electrochemical and Solid State Letters*, 3, pp. 178–179.

Anand, K., Pal, D., Hilgenfeld, R. (2002). An overview on 2-methyl-2,4-pentanediol in crystallization and in crystals of biological macromolecules, *Acta Crystallographica D*, 58, pp. 1722–1728.

Andersson, A., Kalska, B., Haggstrom, L. and Thomas, J. (2000). Lithium extraction/insertion in $LiFePO_4$: an X-ray diffraction and Mössbauer spectroscopy study, *Solid State Ionics*, 130, pp. 41–52.

André, J. M., Delhalle, J. and Brédas, J. L. (1991). *Quantum chemistry aided design of organic polymers*, World Scientific, Singapore.

Andrei, M., Marchese, L., Roggero, A. and Prosperi, P. (1994). Polymer electrolytes based on cross–linked silylated poly-vinyl-ether and lithium perchlorate, *Solid State Ionics*, 72, pp. 140–146.

Antolini, F., Trotta, M. and Nicolini, C. (1995a). Effect of temperature on optical properties of reaction centres organised in Langmuir–Blodgett films, *Thin Solid Films*, 254, pp. 252–256.

Antolini, F., Paddeu, S. and Nicolini, C. (1995b). Heat-stable Langmuir-Blodgett film of glutathione–S–transferase, *Langmuir*, 11, pp. 2719–2725.

Antonini, M., Ghisellini, P., Paternolli, C. and Nicolini, C. (2004). Electrochemical study of the engineerized cytochrome P450scc interaction with free and in lipoproteins cholesterol, *Talanta*, 62, pp. 945–950.

Antonini, M., Ghisellini, P., Pastorino, L., Paternolli, C. and Nicolini, C. (2003). Preliminary electrochemical characterizaton of cytochrome P4501A2–clozapine interaction. *IEE Procedings on Nanobiotechnology*, 150, pp. 31–34.

Appetecchi, G. B. and Passerini, S. (2000). PEO-carbon composite lithium polymer electrolyte, *Electrochemica Acta*, 45, pp. 2139–2145.

Appetecchi, G. B., Alessandrini, F., Duan, R. G., Arzu, A. and Passerini, S. (2001a). Electrochemical testing of industrially produced PEO-based polymer electrolytes, *Journal of Power Sources*, 1, pp. 42–46.

Appetecchi, G. B., Croce, F., Dautzenberg, G., Mastragostino, M., Ronci, F., Scrosati, B., Soavi, F., Zanelli, A., Alessandrini, F. and Prosini, P. P. (1998). Composite polymer electrolytes with improved lithium metal electrode interfacial properties, *Journal of Electrochemistry Society*, 145, pp. 4126–4132.

Appetecchi, G. B., Henderson, W., Villano, P., Berrettoni, M. and Passerini, S. (2001b). PEO–LiN(SO$_2$CF$_2$CF$_3$)$_2$ polymer electrolytes: I. xrd, dsc, and ionic conductivity characterization, *Journal of Electrochemistry Society*, 148, pp. A1171–A1178.

Appetecchi, G. B., Scaccia, S. and Passerini, S. (2000). Investigation on the stability of the lithium–polymer electrolyte interface, *Journal of Electrochemistry Society*, 147, pp. 4448–4452.

Argos, P., Rossman, M. G., Grau, U. M., Zuber, H., Frank, G., Tratschin, J. D. (1979). Thermal stability and protein structure, *Biochemistry*, 18, pp. 5698–5703.

Armand, M., Chabagno, J. M. and Duclot, M. J. (1989). In: P. Vashishita, J.N. Mundy, G.K. Shenoy (Eds.), Fast Ion Transport in Solid, Elsevier, New York.

Asherie, N. (2004). Protein crystallization and phase diagrams, *Methods*, 34, pp. 266–272.

Asherie, N., Pande, J., Pande, A., Zarutskie, J. A., Lomakin, J., Lomakin, A., Ogun, O., Stern, L. J., King, J. and Benedek, G. B. (2001). Enhanced crystallization of the Cys18 to Ser mutant of bovine gamma B crystalline, *Journal of Molecular Biology*, 314, pp. 663–669.

Aurbach, D., Koltypin, M. and Teller, H. (2002). In situ AFM imaging of surface phenomena on composite graphite electrodes during lithium insertion, *Langmuir*, 18, pp. 9000–9009.

Averin, D. V. and Likharev, K. K. (1986) Coulomb blockade of single-electron tunneling, and coherent oscillations in small tunnel junctions, *Journal of Low Temperature Physics*, 62, pp. 345–373.

Averin, D. V., Korotkov, A. N. and Likharev, K. K. (1991). Theory of single-electron charging of quantum–wells and dots, *Physical Review B*, 44, pp. 6199–6211.

Bains, W. and Smith, G. C. (1988). A novel method for nucleic acid sequence determination, *Journal of Theoretical Biology*, 135, pp. 303–307.

Baluchamy, S., Rajabi, H. N., Thimmapaya, R., Navaraj, A. and Thimmapaya, B. (2003). Repression of c-Myc and inhibition of G1 exit in cells conditionally overexpressing p300 that is not dependent on its histone acetyltransferase activity. *Proceedings of the National Academy of Science USA*, 100, pp. 9524–9529.

Bartolucci, S., Gagliardi, A., Pedone, E., De Pascale, D., Cannio, R., Camardella, L., Carratore, V., Rossi, M., Nicastro, G., De Chiara, C. and Nicolini, C. (1997). Thioredoxin from *Bacillus acidocaldarius*: Characterization, molecular modeling study and high–level expression in Escherichia coli, *Biochemical Journal*, 328, pp. 277–285.

Bartolucci, S., De Simone, G., Galdiero S., Improta, R., Menchise, V. and Pedone, C. (2003). An integrated structural and computational study of the thermostability of two thioredoxin mutants from Alicyclobacillus acidocaldarius, *Journal of Bacteriology*, 185, pp. 4285–4289.

Baserga, R. and Nicolini C. (1976). Chromatin structure and function in proliferating cells, *Biophysica et Biochemica Acta Reviews on Cancer*, 458, pp. 109–134.

Baud, F. and Karlin, S. (1999). Measures of residue density in protein structures, *Proceedings of the National Academy of Science USA*, 96, pp. 12494–12499.

Baughman, R. H., Zakhidov, A. A. and de Heer, W. A. (2002) Carbon Nanotubes - the Route Toward Applications, *Science*, 297, pp. 787–792.

Bavastrello, V., Ram, M. K. and Nicolini, C. (2002). Synthesis of multiwalled carbon nanotubes and poly(o–anisidine) nanocomposite material: fabrication and characterization of its Langmuir-Schaefer films, *Langmuir*, 18, pp. 1535–1541.

Bavastrello, V., Carrara, S., Ram, M. K. and Nicolini, C. (2004). Optical and electrochemical properties of poly(ortho-toluidine)-multi walled carbon nanotubes composite Langmuir–Schaefer films, *Langmuir*, 20, pp. 969–973.

Bavastrello, V., Erokhin, V., Carrara, S., Sbrana, F., Ricci, D. and Nicolini, C. (2004a). Morphology and conductivity in poly(ortho-anisidine)/carbon nanotubes nanocomposite films, *Thin Solid Films*, 468, pp. 17–22.

Bavastrello, V., Stura, E., Carrara, S., Erokhin, V. and Nicolini, C. (2004b). Poly(2,5-dimethylaniline-MWNTs nanocomposite: a new material for conductometric acid vapors sensor, *Sensors and Actuators B*, 98, pp. 247–253.

Beaulieu, L. Y., Larcher, D., Dunlap, R. A. and Dahn, J. R. (2000). Reaction of Li with grain-boundary atoms in nanostructured compounds, *Journal of Electrochemistry Society*, 147, pp. 3206–3212.

Beenakker, C. W. J. (1991). Theory of coulomb–blockade oscillations in the conductance of a quantum dot, *Physical Review B*, 44, pp. 1646–1656.

Beenakker, C. W. J., Vanhouten, H. and Staring, A. A. M. (1991). Influence of coulomb repulsion on the aharonov–bohm effect in a quantum dot, *Physical Review B*, 44, 1657–1662.

Bereznai, M., Pelsoczi, I., Toth, Z., Turzo, K., Radnai, M., Bor, Z. and Fazekas, A. (2003). Surface modifications induced by ns and sub–ps excimer laser pulses on titanium implant material. *Biomaterials*, 24, pp. 4197–4203.

Berman, H. M., Westbrook, J., Feng, Z., Gilliland, G., Bhat, T. N., Weissig, H., Shindyalov, I. N. and Bourne, P. E. (2000). The Protein Data Bank, *Nucleic Acids Research*, 28, pp. 235–242.

Berthet-Colominas, C., Monaco, S., Novelli, A., Sibai, G., Mallet, F. and Cusack, S. (1999). Head-to-tail dimers and interdomain flexibility revealed by the crystal structure of HIV-1 capsid protein (p24) complexed with a monoclonal antibody Fab, *EMBO Journal*, 18, pp. 1124–1136.

Bertoncello, P., Nicolini, D., Paternolli, C., Bavastrello, V. and Nicolini, C. (2003). Bacteriorhodopsin-based Langmuir-Schaefer films for solar energy capture, *IEEE Transactions on Nanobioscience*, 2, pp. 124–132.

Bertoncello, P., Notargiacomo, A. and Nicolini, C. (2004). Synthesis, fabrication and characterization of poly[3-3'(vinylcarbazole)] (PVK) Langmuir-Schaefer films, *Polymer*, 45, pp. 1659–1664.

Bertoncello, P., Notargiacomo, A., Erokhin, V. and Nicolini, C. (2006). Functionalization and photoelectrochemical characterization of poly[3-3'(vinylcarbazole)] multi-walled carbon nanotube (PVK–MWNT) Langmuir-Schaefer films, *Nanotechnology*, 17, pp. 699–705.

Bertoncello, P., Ram, M. K., Notargiacomo, A., Ugo, P. and Nicolini, C. (2002). Fabrication and physico–chemical properties of Nafion Langmuir-Schaefer films, *Physical Chemistry Chemical Physics*, 4, pp. 4036–4043.

Besenhard, J. O., Wachtler, M., Winter, M., Andreus, R., Rom, I. and Sitte, W. (1999). Kinetics of Li insertion into polycrystalline and nanocrystalline 'SnSb' alloys investigated by transient and steady state techniques, *Journal of Power Sources*, 81–82, pp. 268–272.

Best, A. S., Ferry, A., MacFarlane, D. R. and Forsyth, M. (1999). Conductivity in amorphous polyether nanocomposite materials, *Solid State Ionics*, 126, pp. 269–276.

Bharathi, S. and Nogami, M. (2001). A glucose biosensor based on electrodeposited biocomposites of gold nanoparticles and glucose oxidase enzyme, *Analyst*, 126, pp. 1919–1922.

Birge, R. R. (1990). Photophysics and molecular electronic applications of the rhodopsins, *Annual Review of Physical Chemistry*, 41, 683–733.

Birge, R. R. (1992). Protein-based optical computing and memories, *Computer*, 25, pp. 56–67.

Birke, P. and Weppner, W. (1999). Solid electrolytes, in: *Handbook of Battery Materials, Part III*, Besenhard J.O.B. (Ed.), Wiley/VCH, Weinheim, NY, Singapore, pp. 525–552.

Blanchard, L. A., Hancu, D., Beckman, E. J. and Brennecke, J. F. (1999). Green processing using ionic liquids and CO_2, *Nature*, 99, pp. 28–29.

Blodgett, K. B. (1934). Monomolecular films of fatty acids on glass, *Journal of the American Chemical Society*, 56, pp. 495–495.

Blodgett, K. B. (1935). Films built by depositing successive monomolecular layers on a solid surface, *Journal of the American Chemical Society*, 57, pp. 1007-1022.

Blodgett, K. B. and Langmuir, I. (1937). Built-up films of barium stearate ad their optical properties, *Physics Review*, 51, pp. 964–982.

Boeckmann, B., Bairoch, A., Apweiler, R., Blatter, M. C., Estreicher, A., Gasteiger, E., Martin, M. J., Michoud, K., O'Donovan, C., Phan, I., Pilbout, S. and Schneider, M. (2003). The Swiss-Prot Protein Knowledgebase and its supplement TrEMBL in 2003, *Nucleic Acids Research*, 31, pp. 365-370.

Bogozi, A., Lam, O., He, H., Li, C., Tao, N. J., Nagahara, L. A., Amlani, I. and Tsui, R., (2001), Molecular adsorption onto metallic quantum wires. *Journal of the American Chemical Society*, 123, pp. 4585–4590.

Bohm, H. (1999). in: J.O.B. Besenhard (Ed.), Handbook of Battery Materials, part III, Wiley/VCH, Weinheim, NY, Singapore, p. 565 (Chapter 11).

Bonhote, P., Dias, A. P., Papageorgiou, N., Kalyanasundaram, K. and Gratzel, M. (1996). Hydrophobic, highly conductive ambiente-temperature Molten salts, *Inorganic Chemistry*, 35, pp. 1168-1178.

Borghini, M. C., Mastragostino, M., Passerini, S. and Scrosati, B. (1995). Electrochemical properties of polyethylene oxide-Li[(CF$_3$SO$_2$)$_2$N]-Gamma-LiAlO$_2$ composite polymer electrolytes, *Journal of Electrochemistry Society*, 142, pp. 2118–2121.

Botte, G. G., White, R. E. and Zhang, Z. (2001). Thermal stability of LiPF$_6$-EC:EMC electrolyte for lithium ion batteries, *Journal of Power Sources*, 97–98, pp. 570–575.

Boussaad, S., Dziri, L., Arechabaleta, R., Tao, N. J. and Leblanc, R. M. (1998), Electron-transfer properties of cytochrome c Langmuir-Blodgett films and interactions of cytochrome c with lipids, *Langmuir*, 14, pp. 6215–6219.

Bousse, L., Mostarshed, S., Hafeman, D., Sartore, M., Adami, M. and Nicolini, C. (1994), Investigation of carrier transport through silicon–wafers by photocurrent measurements, *Journal of Applied Physics*, 8, pp. 4000–4008.

Bradford, M. M. (1976). A rapid sensitive method for the quantitation of microgram quantities of protein utilizing the principle of protein-dye binding, *Analytical Biochemistry*, 72, pp. 248–254.

Bramanti, E., Benedetti, E., Nicolini, C., Berzina, T. S., Erokhin, V., D'Alessio A. and Benedetti, E. (1997). Qualitative and quantitative analysis of the secondary structure of cytochrome C Langmuir-Blodgett films, *Biopolymers*, 42, pp.227–237.

Bramnik, N. N., Bramnik, K. G., Buhrmester, T., Baehtz, C., Ehrenberg, H. and Fuess, H. (2004). Electrochemical and structural study of LiCoPO$_4$–based electrodes, *Journal of Solid State Electrochemistry*, 8, pp. 558–564.

Brauchle, C., Hampp, N. and Oesterhelt, D (1991), Optical applications of bacteriorhodopsin and its mutated variants, *Advanced Materials*, 3, pp. 420–428.

Braud, C., Baeten, D., Giral, M., Pallier, A., Ashton-Chess, J., Braudeau, C., Chevalier, C., Lebars, A., Léger, J., Moreau, A., Pechkova, E., Nicolini, C., Soulillou, J. P. and Brouard, S. (2008). Immunosuppressive drug–free operational immune tolerance in human kidney transplants recipients: I. Blood gene expression statistical analysis. *Journal of Cellular Biochemistry*, 103, pp. 1681–16902.

Brige, A., Leys, D., Meyer, T. E., Cusanovich, M. A. and Van Beeumen, J. J. (2002). The 1.25 angstrom resolution structure of the diheme NapB subunit of soluble nitrate reductase reveals a novel cytochrome c fold with a stacked heme arrangement, *Biochemistry*, 41, pp. 4827–4836.

Brøsen, K. (1993). The pharmacogenetics of the selective serotonin reuptake inhibitors. *Clinical Investigator*, 71, pp. 1002–1009.
Brøsen, K., Skjelbo, E., Rasmussen, B. B., Poulsen, H. E. and Loft, S. (1993). Fluvoxamine is a potent inhibitor of cytochrome P4501A2, *Biochemical Pharmacology*, 45, pp. 1211–1214.
Bucher, H, Kuhn, H., Sperling, W., Tillmann, P. and Wiegand, J. (1967). Controlled transfer of excitation through. thin layers *Mol. Cryst.*, 2, pp. 199–230.
Bumm, L. A., Arnold, J. J., Cygan, M. T., Dunbar, T. D., Burgin, T. P., Jones, L., Allara, D. L., Tour, J. M. and Weiss, P. S. (1996). Are single molecular wires conducting?, *Science*, 271, pp. 1705–1707.
Burmeister, W. P. (2000). Structural changes in a cryo–cooled protein crystal owing to radiation damage, *Acta Crystallographica Section D: Biological Crystallography*, 56, pp. 328–341.
Burnside, S. D. and Giannelis, E. P. (1995). Synthesis and properties of new poly(dimethylsiloxane) nanocomposites, *Chemistry of Materials*, 7, pp. 1597–1600.
Butte, A. (2002). The use and analysis of microarray data, *Nature Reviews Drug Discovery*, 1, pp. 951–960.
Buur-Rasmussen, B. and Brøsen, K. (1999). Cytochrome P450 and therapeutic drug monitoring with respect to clozapine, *European Neuropsychopharmacology*, 9, pp. 453–459.
Bykov, V. A. (1996). Langmuir-Blodgett films and nanotechnology, *Biosensors and Bioelectronics*, 11, pp. 923–932.
Caffrey, M. (2003). Membrane protein crystallization, *Journal of Structural Biology*, 142, pp. 108–132.
Caja, J., Dunstan, T. D. J., Ryan, D. M. and Katovic, V. (1999). The Electrochemical Society and The Electrochmical Society of Japan Meeting Abstracts, vol. 99–2, pp. 2252.
Cantor, C. R., Mirzabekov, A. and Southern, E. (1992). Report on the sequencing by hybridisation workshop, *Genomics*, 13, pp. 1378–1383.
Cantrell, D. (2002). Protein kinase B (Akt) regulation and function in T lymphocytes. *Seminars in Immunology*, 14, pp. 19–26.
Capiglia, C., Mustarelli, P., Quartarone, E., Tomasi, C. and Magistris, A., (1999). Effects of nanoscale SiO_2 on the thermal and transport properties of solvent-free, poly(ethylene oxide) (PEO)-based polymer electrolytes, *Solid State Ionics*, 118, pp. 73–79.
Capuano, F., Croce, F. and Scrosati, B. (1991). Composite polymer electrolytes, *Journal of the Electrochemical Society*, 138, pp. 1918–1922.
Carrara, S., Bavastrello, V., Ram, M. K. and Nicolini, C. (2006). Nanometer sized polymer based Schottky junction. *Thin Solid Films* 510, pp. 229–234.

Carrara, S., Erokhin, V., Facci, P. and Nicolini, C (1996). On the role of nanoparticle sizes in monoelectron conductivity, in: *Nanoparticles in Solids and Solutions*, J Fendler, 1 Décáni (eds.), Kluwer, Netherlands, vol. 18, pp. 497–503.

Carrara, S., Gussoni, A., Erokhin, V. and Nicolini, C. (1995). On the degradation of conducting Langmuir–Blodgett films, *Journal of Materials Science: Materials in Electronics*, 6, pp. 79–83.

Carrara, S., Riley, D. J., Bavastrello, V., Stura, E. and Nicolini, C. (2005). Methods to fabricate nanocontacts for electrical addressing of single molecules, *Sensors and Actuators B*, 105, pp. 542–548.

Caruthers, M. H. (1985). Gene synthesis machines: DNA chemistry and its uses, *Science*, 230, pp. 281–285.

Céspedes O., Bari M. A., Dennis C., Versluijs J. J., Jan G., O'Sullivan J., Gregg J. F. D. and Coey J. M. (2002). Fabrication and characterisation of Ni nanocontacts, *Journal of Magnetism and Magnetic Materials*, 242–245, pp. 492–494.

Chagnes, A., Diaw, M., Carre, B., Willmann, P. and Lemordant, D. (2005). Imidazolium-organic solvent mixtures as electrolytes for lithium batteries, *Journal of Power Sources*, 145, pp. 82–88.

Chandler, H. M. Cox, J. C., Healey, K., MacGregor A., Premier, A. A. and Hurrell, J. G. R. (1982), An investigation of the use of urease-antibody conjugates in enzyme immunoassays, *Journal of Immunological Methods*, 53, pp. 187–194.

Chang, B. L, Zheng, S. L., Isaacs, S. D., Wiley, K. E., Turner, A., Li, G., Walsh, P. C., Meyers, D. A., Isaacs, W. B. and Xu, J. (2004). A polymorphism in the CDKN1B gene is associated with increased risk of hereditary prostate cancer, *Cancer Research*, 64, pp. 1997–1999.

Chayen, N. E., (2005). Methods for separating nucleation and growth in protein crystallisation, *Progress in Biophysics and Molecular Biology*, 88, pp. 329–337.

Chen, J., Reed, M. A., Rawlett, A. M. and Tour, J. M. (1999). Large on-off ratios and negative differential resistance in a molecular electronic device, *Science*, 286, pp. 1550–1552.

Chen, Z. P. and Birge, R. R. (1993). Protein-based artificial retinas, *Trends in Biotechnology*, 11, pp. 292–300.

Cheradame, H., LeNest, J. F. (1987). In: J.R. Maccallum, C.A. Vincent (Eds.), Polymer Electrolyte Reviews, vol. 1, Elsevier, London, p. 103.

Chernov, A. A. (1997). Crystals built of biological macromolecules, *Physics Reports - Review Section of Physics Letters*, 288, pp. 61–75.

Chou, K. C., Carpacci, L., Maggiora, G. M., Parodi, L. A. and Schulz, M. W. (1992). An energy-based approach to packing the 7-helix bundle of bacteriorhodopsin, *Protein Science*, 1, pp. 810-827.

Chum, H. L. and Osteryoung, R. A. (1981). In: D. Inman, D. Lovering (Eds.), Ionic Liquids, Plenum Press, New York, 1981.

Ciric Marjanovic, G. and Mentus, S. (1998). Charge-discharge characteristics of polythiophene as a cathode active material in a rechargeable battery, *Journal of Applied Ectrochemistry*, 28, pp. 103–106.

Collier, C. P., Wong, E. W., Belohradský, M., Raymo, F. M., Stoddart, J. F., Kuekes, P. J., Williams, R. S. and Heath, J. R. (1999). Electronically configurable molecular-based logic gates, *Science*, 285, pp. 391–394.

Costa-Krämer, J. L., García, N., García–Mochales, P., Serena, P. A., Marqués, M. I. and Correia, A. (1997). Conductance quantization in nanowires formed between micro and macroscopic metallic electrodes, *Physical Review B* 55, pp. 5416–5424.

Covani, U., Marconcini, S., Giacomelli, L., Sivozhelezov, V., Barone, A. and Nicolini, C. (2008). Bioinformatic prediction of leader genes in human periodontitis, *Journal of Periodonthology*, in press.

Cowie, J. M. G. and Ferguson, R. (1985). Glass and subglass transitions in a series of poly(itaconate ester)s with methyl-terminated poly(ethylene oxide) side chains, *Journal of Polymer Science: Polymer Physics Edition*, 23, pp. 2181–2191.

Cowie, J. M. G. and Martin, A. C. S. (1985). Ionic conductivity of poly(diethoxy(3)methyl itaconate) containing lithium perchlorate *Polymer Communications*, 26, pp. 298–303.

Croce, F., Appetecchi, G. B., Persi, L. and Scrosati, B. (1998). Nanocomposite polymer electrolytes for lithium batteries, *Nature*, 394, pp. 456–458.

Croce, F., Curini, R., Martinelli, A., Persi, L., Ronci, F., Scrosati, B. and Caminiti, R. (1999). Physical and chemical properties of nanocomposite polymer electrolytes, *Journal of Physical Chemistry B*, 103, pp. 10632–10638.

Cui, X. D., Primak, A., Zarate, X., Tomfohr, J., Sankey, O. F., Moore, A. L., Moore, T. A., Gust, D., Harris, G. and Lindsay S. M. (2001). Reproducible measurement of single–molecule conductivity, *Science,* 294, pp. 571–574.

Cusack, S., Belrhali, H., Bram, A., Burghammer, M., Perrakis, A. and Riekel, C. (1998). Small is beautiful: protein micro–crystallography, *Nature Structural Biology* 5, pp. 634–637S.

Dahn, J. R., Fuller, E. W., Obravae, M. and von Sacken, U. (1994). Thermal stability of Li_xCoO_2, $LixNiO_2$ and λ-MnO_2 and consequences for the safety of Li-ion cells, *Solid State Ionics*, 69, pp. 265-270.

Dale, G. E., Oefner, C. and D'Arcy, A. (2003). The protein as a variable in protein crystallization. *Journal of Structural Biology*, 142, pp. 88–97.

Dante, S., DeRosa, M., Francescagli, O., Nicolini, C., Rustichelli, F. and Troitsky, V. I. (1996). Supramolecular ordering of bipolar lipids from Archaea in Langmuir-Blodgett films by low–angle X–ray diffraction, *Thin Solid Films,* 285, pp. 459–463.

D'Arcy, A., Stihle, M., Kostrewa, D. and Dale, G. (1999). Crystal engineering: a case study using the 24 kDa fragment of the DNA gyrase B subunit from Escherichia coli, *Acta Crystallographica Section D: Biological Crystallography*, 55, 1623–1625.

Darkrim, F. and Levesque, D. (1998). Monte Carlo simulations of hydrogen adsorption in single–walled carbon nanotubes, *Journal of Chemical Physics*, 109, pp. 4981–4984.

Dasgupta, S., Iyer, G. H., Bryant, S. H., Lawrence, C. E. and Bell, J. A. (1997). Extent and nature of contacts between protein molecules in crystal lattices and between subunits of protein oligomers, *Proteins*, 28, pp. 494–514.

Davies, A., Gowen, B. E., Krebs, A. M., Schertler, G. F. and Saibil, H. R. (2001). Three-dimensional structure of an invertebrate rhodopsin and basis for ordered alignment in the photoreceptor membrane. *Journal of Molecular Biology*, 314, pp. 455–463.

De Heer, W. A., Chatelain, A. and Ugarte, D. (1995). A carbon nanotube field–emission electron source. *Science*, 270, pp. 1179–1180.

De Koch, A., Ferg, E. and Gummow, R. J. (1998). The effect of multivalent cation dopants on lithium manganese spinel cathodes, *Journal of Power Sources*, 70, pp. 247–252.

De Rosa, M., Gambacorta, A. and Gliozzi, A. (1986). Structure, biosynthesis, and physicochemical properties of archaebacterial lipids, *Microbiological Reviews*, 50, pp. 70-80.

De Rosa, M., Gambacorta, A., Nicolaus, B., Chappe, B. and Albrecht, P. (1983). Isoprenoid ethers: backbone of complex lipids of the archaebacterium Sulfolobus solfataricus, *Biochimica et Biophysica Acta – Lipids and Lipid Metabolism*, 753, pp. 249-256.

Decher, G. R., (1996). Comprehensive supramolecular chemistry, In: *Templating, self-assembly and self-organization*, J.–P. Sauvage and M. W. Hosseini, Eds. Pergamon Press, Oxford 507, p. 9.

Decher, G. (1997). Fuzzy nanoassemblies: towards layered polymeric multicomposites. *Science*, 277, pp. 1232–1237.

Decher, G., Hong, J. D. and Schmitt (1992). Build-up of ultrathin multilayer films by a self–assembly process .3. Consecutively alternating adsorption of anionic and cationic polyelectrolytes on charged surfaces. *Thin Sold Films*, 210, pp. 831–835.

Delacourt, C., Poisot, P., Morcrette, M., Tarascon, J. M. and Masquelier, C. (2004). One–step low–temperature route for the preparation of electrochemically active LiMnPO4 powders. *Chemistry of Materials*, 16, pp. 93–99.

Delvaux, M. and Demoustier–Champagne, S. (2003). Immobilisation of glucose oxidase within metallic nanotubes arrays for application to enzyme biosensor. *Biosensors and Bioelectronics*, 18, pp. 943–951.

Devoret, M. H., Esteve, D. and Urbina, C. (1992). Single-electron transfer in metallic nanostructures, *Nature*, 360, pp. 547–553.

Diaspro, A., Bertolotto, M., Vergani, L. and Nicolini, C. (1991). Polarized light scattering of nucleosomes and polynucleosomes. In situ and in vitro studies, *IEEE Transaction on Biomedical Engineering*, 38, pp. 670–678.

Diaspro, A., Radicchi, G. and Nicolini, C. (1995). Polarized light scattering: a biophysical method for studying Bacterial cells, *IEEE Transactions on Biomedical Engineering*, 42, pp. 1038–1043.

Dillon, A. C., Jones, K. M., Bekkedahl, T. A., Kiang, C. H., Bethune, D. S. and Heben, M. J. (1997). Storage of hydrogen in single-walled carbon nanotubes, *Nature*, 386, pp. 377–379.

Ding, H. M., Ram, M. K. and Nicolini, C. (2001). Nanofabrication of organic/inorganic hybrids of TiO_2 with substituted phthalocyanine or polythiophene. *Journal of Nanoscience and Nanotechnology*, 1, pp. 207–213.

Ding, H., Bertoncello, P., Ram, M. K. and Nicolini, C. (2002a). Electrochemical investigations on MEH-PPV/C_{60} nanocomposite Langmuir-Schaefer films, *Electrochemistry Communications*, 4, pp. 503–505.

Ding, H., Ram, M. K. and Nicolini, C. (2002b). Construction of organic-inorganic hybrid ultrathin films self-assembled from poly(thiophene-3-acetic acid) and TiO_2. *Journal of Materials Chemistry*, 12, pp. 3585–3590.

Ding, H., Zhang, X., Ram, M. K. and Nicolini, C. (2005). Ultrathin films of tetrasulfonated copper phthalocyanine-capped titanium dioxide nanoparticles: fabrication, characterization and photovoltaic effect, *Journal of Colloid and Interface Science*, 290, pp. 166–171.

Dokko, K., Shi, Q. F., Stefan, I. C. and Scherson, D. A. (2003). In situ Raman spectroscopy of single microparticle Li+ intercalation electrodes, *Journal of Physical Chemistry B*, 107, pp. 12549–12554.

Dorogi, M., Gomez, J., Osifchin, R., Andres, R. P. and Reifenberger, R. (1995). Room–temperature coulomb–blockade from a self-assembled molecular nanostructure, *Physical Review B*, 52, pp. 9071–9077.

Dortbudak, O., Haas, R., Bernhart, T. and Mailath-Pokorny, G. (2001). Lethal photosensitization for decontamination of implant surfaces in the treatment of periimplantitis. *Clinical Oral Implants Research*, 12, pp. 104–108.

Dotan, N., Cohen, N., Kalid, O. and Freeman, A. (2001). Supramolecular assemblies made of biological macromolecules. In: *Nanosurface chemistry* (editor M. Rossof), Marcel Dekker, New York, pp. 461–471.

Drexhage, K. H. and Kuhn, H. (1966). Optical and electrical phenomena on monomolecular layers. In: *Basic problems in thin film physics* (Niemaer, R. and Mayer, II, eds.), p. 339. Göttingen, Vandenhoeck and Ruprecht.

Drmanac, S., Stavropoulos, N. A., Labat, I., Vonau, J., Hauser, B., Soares, M. B. and Drmanac, R. (1996). Gene–representing cDNA clusters defined by hybridization of 57,419 clones from infant brain libraries with short oligonucleotides probes, *Genomics*, 37, pp. 29–40.

Drude, P., (1902). The theoryof optics. New York, Dover Pubblications Inc., pp. 287–292.

Dubrovsky, T. and Nicolini, C. (1994). Preparation and immobilization of Langmuir Blodgett films of antibodies conjugated to enzyme for potentiometric sensor application, *Sensors Actuators*, 22, pp. 69–73.

Ducruix, A. and Giege, R. (1999). *Crystallization of nucleic acids and proteins. A practical approach*, Eds. Ducruix, R. and Giege R., Oxford University Press, New York (USA).

Du Pasquier, A., Disma, F., Bowmer, T., Gozdz, A. S., Amatucci, G. and Tarascon, J. M. (1998). Differential scanning calorimetry study of the reactivity of carbon anodes in plastic Li–ion batteries, *Journal of the Electrochemical Society*, 145, pp. 472–477.

Dyda, F., Hickman, A. B., Jenkins, T. M., Engelman, A., Craigie, R. and Davies, D. R., (1994). Crystal structure of the catalytic domain of HIV-1 integrase: similarity to other polynucleotidyl transferases, *Science*, 266, pp. 1981–1996.

Ekström, F., Stier, G. and Sauer, U. H. (2003). Crystallization of the actin-binding domain of human actinin: analysis of microcrystals of SeMet-labelled protein, *Acta Crystallographica Section D–Biological Crystallography*, 59, pp. 724–726.

Ellenbogen, J. C. and Love, J. C. (2000). Architectures for molecular electronic computers: 1. Logic structures and an adder designed from molecular electronic diodes, *Proceedings of the IEEE*, 88, pp. 386–426.

Ellington, A.D. and Szostak, J. W. (1990). In-vitro selection of Rna molecules that bind specific ligands, *Nature*, 346, pp. 818–822.

Epstein, A. J. and MacDiarmid, A. G. (1991). Novel concepts in electronic polymers - polyaniline and its derivatives, *Makromolekulare Chemie–Macromolecular Symposia*, 51, pp. 217–234.

Erokhin, V., Facci, P. and Nicolini, C. (1995). Two-dimensional order and protein thermal stability: high temperature preservation of structure and function, *Biosensors and Bioelectronics*, 10, pp. 25–34.

Erokhin, V., Facci, P., Carrara, S. and Nicolini, C. (1995a). Observation of room temperature mono-electron phenomena on nanometer-sized CdS particles, *Journal of Physics D: Applied Physics*, 28, pp. 2534–2538.

Erokhin, V., Facci, P., Carrara, S. and Nicolini, C. (1996). Monoelectron phenomena in nanometer scale particles formed in LB films, *Thin Solid Films*, 284–285, pp. 891–893.

Erokhin, V., Facci, P., Kononenko, A., Radicchi, G. and Nicolini, C. (1996a). On the role of molecular close packing on the protein thermal stability, *Thin Solid Films*, 284–285, pp. 805–808.

Erokhin, V., Facci, P., Carrara, S., Nicolini, C. (1997). Fatty acid based monoelectronic device, *Biosensors & Bioelectronics*, 12, pp. 601–606.

Erokhin, V., Feigin, L., Ivakin, G., Klechkovskaya, V., Lvov, Y. and Stiopina, N. (1991). Formation and X-ray and electron-diffraction study of Cds and Pbs particles inside fatty-acid matrix, *Makromol. Chem.: Macromol. Symp.*, 46, pp. 359-363.

Erokhin, V., Kayushina, R., Lvov, Yu. and Feigin, L. (1990). Langmuir-Blodgett films of immunoglobulins as sensing elements, *Il Nuovo Cimento*, 120, pp. 1253–1258.

Erokhin, V., Carrara, S., Amenitch, H., Bernstorff, S. and Nicolini, C. (1998). Semiconductor nanoparticles for quantum devices, *Nanotechnology*, 9, pp. 158–161.

Erokhin, V., Facci, P., Gobbi, L., Dante, S., Rustichelli, F. and Nicolini, C. (1998a). Preparation of semiconductor superlattices from LB precursor, *Thin Solid Films*, 327–329, pp. 503–505.

Erokhin, V., Raviele, G., Glatz–Reichenbach, J., Narizzano, R., Stagni, S. and Nicolini, C. (2002). High value organic capacitors, *Mat. Sci. Eng. C*, 22, pp. 381–385.

Erokhin, V., Troitsky, V., Erokhina, V., Mascetti, G. and Nicolini, C. (2002a). In-plane patterning of aggregated nanoparticle layers, *Langmuir*, 18, pp. 3185–3190.

Erokhina S, Erokhin V, Nicolini C, (2002) Electrical properties of thin copper sulfide films produced by the aggregation of nanoparticles formed in LB precursor, *Colloid and Surface*, A198–200, 645–650.

Erokhina, S., Erokhin, V., Nicolini, C., Sbrana, F., Ricci, D. and Di Zitti, E. (2003). Microstructure origin of the conductivity differences in aggregated CuS films of different thickness, *Langmuir*, 19, pp. 766–771.

Estela, J. M., Tomás, C., Cladera, A. and Cerdà, V. (1995). Potentiometric stripping analysis: a review, *Critical Reviews in Analytical Chemistry*, 25, pp. 91–141.

Ewing, F., Fortsythe, E. and Pusey, M. (1994). Orthorhombic lysozyme solubility, *Acta Crystallographica Section D–Biological Crystallography*, 50, pp. 424–428.

Facci, P., Erokhin, V. and Nicolini, C. (1993), Nanogravimetric gauge for surface density measurements and deposition analysis of LB films, *Thin Solid Films*, 230, pp. 86–89.

Facci, P., Erokhin, V., Tronin, A. and Nicolini, C. (1994). Formation of ultrathin semiconductor films by CdS nanostructure aggregation, *Journal of Physical Chemistry*, 98, pp. 13323–13327.

Facci, P., Erokhin, V. and Nicolini, C. (1994a) Scanning tunnelling microscopy of a monolayer of reaction centers, *Thin Solid Films*, 243 pp. 403–406.

Facci, P., Erokhin, V., Antolini, F. and Nicolini, C. (1994b). Chemically induced anisotropy in antibody Langmuir Blodgett films, *Thin Solid Films*, 237, pp. 19–21.

Facci, P., Radicchi, G., Erokhin, V. and Nicolini, C. (1995). On the mobility of Immunoglobulines G in Langmuir–Blodgett films, *Thin Solid Films*, 269, pp. 85–89.

Facci, P., Erokhin, V., Carrara, S. and Nicolini, C. (1996). Room-temperature single-electron junction, *Proceedings of the National Academy of Sciences USA*, 93, pp. 10556–10559.

Facci, P., Erokhin, V., Paddeu, S. and Nicolini, C. (1998). Surface pressure induced structural effects in photosynthetic reaction center Langmuir-Blodgett films, *Langmuir*, 14, pp. 193–198.

Fanigliulo, A., Accossato, P., Adami, M., Lanzi, M., Martinoia, S., Paddeu, S., Parodi, M. T., Rossi, A., Sartore, M., Grattarola, M. and Nicolini, C. (1996). Comparison between a LAPS and a FET-based sensor for cell-metabolism detection, *Sensors and Actuators B Chemical*, 32, pp. 41–48.

Faulon, J. L., Sale, K. and Young, M. (2003). Exploring the conformational space of membrane protein folds matching distance constraints. *Protein Science*, 12, pp. 1750–1761.

Fauteux, D., Massucco, A., McLin, M., Vanburen, M. and Shi, J. (1995). Lithium polymer electrolyte rechargeable battery, *Electrochimica Acta*, 40, pp. 2185–2190.

Fernandez–Madrigal, F. J., Lavela, P., Perez–Vicente, C. and Tirado, J. L. (2001). Electrochemical reactions of polycrystalline $CrSb_2$ in lithium batteries, *Journal of the Electroanalytical Chemistry*, 501, pp. 205–209.

Ferreira, M. and Rubner, M. F. (1995). Molecular–level processing of conjugated polymers .1. Layer-by-layer manipulation of conjugated polyions, *Macromolecules* 28, pp. 7107–7114.

Feuillade, G. and Perche, P. (1975) Ion-conductive macromolecular gels and membranes for solid lithium cells, *Journal of Applied Electrochemistry*, 5, pp. 63–69.

Fischer, C. W., Caudle, D. L., Wixtrom, C. M., Quattrochi, L. C., Tuckey, R. H., Waterman, M. R. and Estabrook, R. W. (1992). High-level expression of functional human cytochrome P4501A2 in Escherichia coli, *The FASEB Journal*, 6, pp. 759–764.

Fischer, T. and Hampp, N. A. (2004). Encapsulation of purple membrane patches into polymeric nanofibers by electrospinning, *IEEE Transactions on Nanobioscience*, 3, pp. 118–120.

Fischer, T., Neebe, M., Juchem, T. and Hampp, N. A. (2003). Biomolecular optical data storage and data encryption, *IEEE Transaction on Nanobioscience*, 2, pp. 1–5.

Fischer, U. and Oesterhelt, D. (1979). Chromophore equilibria in bacteriorhodopsin, *Biophysical Journal*, 28, pp. 211–230.

Fodor, S. P. A., Rava, R. P., Huang, X. C., Pease, A. C., Holmes, C. P. and Adams, C. L. (1993). Multiplexed biochemical assays with biological chips, *Nature*, 364, pp. 555–556.

Fong, R., Vonsacken, U. and Dahn, J. R. (1990). Studies of lithium intercalation into carbons using nonaqueous electrochemical–cells, *Journal of the Electrochemistry Society*, 137, pp. 2009–2013.

Foreman, T. M., Khalil, M., Meier, P., Brainard, J. R., Vanderberg, L. A. and Sauer, N. N. (2001). Effects of charged water-soluble polymers on the stability and activity of yeast alcohol dehydrogenase and subtilisin Carlsberg, *Biotechnology and Bioengineering*, 76, pp. 241–246.

Foresi, J. S., Villeneuve, P. R., Ferrera, J., Thoen, E. R., Steinmeyer, G., Fan, S., Joannopoulos, J. D., Kimerling L. C., Smith. H. I. and Ippen E. P. (1997). Photonic-bandgap microcavities in optical waveguides, *Nature*, 390, pp. 143-145.

Franger, S., Le Gras, F., Bourbon, C. and Rouault, H. (2002). LiFePO$_4$ synthesis routes for enhanced electrochemical performance, *Electrochemical and Solid State Letters*, 5, pp. A231–A233.

Fransson, L. M. L., Vaughey, J. T., Benedek, R., Edstrom, K., Thomas, J. O. and Thackeray, M. M. (2001). Phase transitions in lithiated Cu2Sb anodes for lithium batteries: an in situ X–ray diffraction study, *Electrochemistry Communications*, 3, pp. 317–323.

Fransson, L. M. L., Vaughey, J. T., Edstrom, K. and Thackeray, M. M. (2003). Structural transformations in intermetallic electrodes for lithium batteries – An in situ x–ray diffraction study of lithiated MnSb and Mn2Sb, *Journal of the Electrochemistry Society*, 150, pp. A86–A91.

Fu, A., Gu, W., Larabell, C. and Alivisatos, A. P., (2005). Semiconductor nanocrystals for biological imaging, *Current Opinion Neurobiology*, 15, pp. 568–575.

Fukuzawa, K. (1994). Motion-sensitive position sensor using bacteriorhodopsin, *Applied Optics*, 33, pp. 7489–7495.

Fukuzawa, K., Yanagisawa, L. and Kuwano, H. (1996). Photoelectrical cell utilizing bacteriorhodopsin on a hole array fabricated by micromachining techniques, *Sensors Actuators B*, 30, pp. 121–126.

Fuller, J., Breda, A. C. and Carlin, R. T. (1998). Ionic liquid-polymer gel electrolytes from hydrophilic and hydrophobic ionic liquids, *Journal of Electroanalytical Chemistry*, 459, pp. 29–34.

Fuller, J., Carlin, R. T. and Osteryoung, R. A. (1997). The room temperature ionic liquid 1-Ethyl-3-methylimidazolium tetrafluoroborate: electrochemical couples and physical properties, *Journal of Electrochemistry Society*, 144, pp. 3881–3886.

Gadjourova, Z., Andrew, Y. G., Tunstall, D. P. and Bruce, P. G. (2001). Ionic conductivity in crystalline polymer electrolytes, *Nature*, 412, pp. 520–523.

Galkin, O. and Vekilov, P. (2000). Control of protein crystal nucleation around the metastable liquid–liquid phase boundary, *Proceedings of the National Academy of Science USA*, 97, pp. 6277–6281.

Gambacorta, A., Trincone, A., Nicolaus, B., Lama, L. and de Rosa. M. (1994). Unique features of lipids of archaea, *Syst. Appl. Microbiol.*, 16, pp. 518–527.

Gao, M., Richter, B. and Kirstein, S. (1997). White-light electroluminescence from self-assembled Q-CdSe/PPV multilayer structures, *Advanced Materials*, 9, pp. 802–805.

García, N., Muñoz, M. and Zhao, Y. W. (1999). Magnetoresistance in excess of 200% in Ballistic Ni Nanocontacts at Room Temperature and 100 Oe, *Physics Review Letters*, 89, pp. 2923–2926.

García, N., Muñoz, M., Osipov, V. V., Ponizovskaya, E. V., Qian, G. G., Saveliev, I. G. and Zhao Y. W. (2002). Ballistic magnetoresistance in different nanocontact configurations: a basis for future magnetoresistance sensors, *Journal of Magnetism and Magnetic Materials*, 240, pp. 92–99.

Garcia-Moreno, O., Alvarez-Vega, M., Garcia–Alvarado, F., Garcia-Jaca, J., Gallardo-Amores, J. M., Sanjuan, M. L. and Amador, U. (2001). Influence of the structure on the electrochemical performance of lithium transition metal phosphates as cathodic materials in rechargeable lithium batteries: A new high-pressure form of LiMPO$_4$ (M = Fe and Ni), *Chemistry of Materials*, 13, pp. 1570–1576.

Garibaldi, S., Brunelli, C., Bavastrello, V., Ghigliotti, G. and Nicolini, C. (2006). Carbon nanotube biocompatibility with cardiac muscle cells, *Nanotechnology*, 17, 391–397.

Garman, E. and Nave, C. (2002). Radiation damage to crystalline biological molecules: current views, *Journal of Synchrotron Radiation*, 9, pp. 327–328.

Gee, M. A. and Laman, F. C. (1993). Thermal stability study of LiAsF$_6$ electrolytes using accelerating rate calorimetry, *Journal of Electrochemistry Society*, 140, pp. L53–L55.

Gehrke, R. (1992). An ultrasmall angle scattering instrument for the DORIS-III bypass, *Review of Scientific Instruments*, 63, pp. 455–458.

Genies, E. M., Boyle, A., Lapkowski, M. and Tsintavis, C. (1990). Polyaniline - a historical survey, *Synthetic Metals*, 36, pp. 139–182.

Ghisellini, P., Paternolli, C., Antonini, M. and Nicolini, C. (2004). P450scc mutant nanostructuring for optimal assembly, *IEEE Transactions on Nanobioscience*, 3, pp. 121–128.

Ghisellini, P., Paternolli, C., Chiossone, I. and Nicolini C. (2002). Spin state transitions in Langmuir-Blodgett films of recombinant cytochrome P450scc and adrenodoxin, *Colloid and Surface B*, 23, pp. 313–318.

Giacomelli, L. and Nicolini, C. (2006). Gene expression of human T lymphocytes cell cycle: Experimental and bioinformatic analysis, *Journal of Cellular Biochemistry*, 99, pp. 1326–1333.

Glazmann, L. I. and Shekhter, R. I. (1989). Coulomb oscillations of the conductance in a laterally confined heterostructure, *Journal of Physics: Condensed Matter*, 1, pp. 5811–5816.

Gonzalez, F. J. (1992). Human cytochromes P450: problems and prospects, *Trends in Pharmacological Sciences*, 13, pp. 346–352.

Göpel, W. (1998). Bioelectronics and Nanotechnologies, *Biosensors and Bioelectronics*, 13, pp. 723–728.

Gorschlüter, A., Sundermeier, C., Roß, B. and Knoll, M. (2002). Microparticle detector for biosensor application, *Sensors and Actuators B–Chemical*, 85, pp. 158–165.

Gourley, P. L. (2005). Brief overview of BioMicroNano technologies, *Biotechnology Progress*, 21, pp. 2–10.

Graetz, J., Ahn, C. C., Yazami, R. and Fultz, B. (2003). Highly reversible lithium storage in nanostructured silicon, *Electrochemical and Solid State Letters*, 6, pp. A194–A197.

Graham, D. L., Ferreira, H. A., Freitas, P. P. and Cabral, J. M. S. (2003). High sensitivity detection of molecular recognition using magnetically labelled biomolecules and magnetoresistive sensors, *Biosensors and Bioelectronics*, 18, pp. 483–488.

Granstrom, M. (1997). Novel polymers light-emitting diode designs using poly(thiophenes), *Polymers for Advanced Technologies*, 8, pp. 424–430.

Grasso, V., Lambertini, V., Ghisellini, P., Valerio, F., Stura, E., Perlo, P. and Nicolini, C. (2006). Nanostructuring of a porous alumina matrix for a biomolecular microarray, *Nanotechnology*, 17, pp. 795–798.

Gray, F. M. (1991). *Solid polymer electrolytes: fundamentals and technological applications*, VCH Publishers, Inc., New York.

Gray, F. M. (1997). *Polymer Electrolytes*, Royal Society of Chemistry Monographs, Cambridge.

Gray, F. M. and Armand, M. (2000). In: T. Osaka, M. Datta (Eds.), Energy Storage System for Electronics, Gordon and Breach, Amsterdam.

Gray, F. M., Maccallum, J. R., Vincent, C. A. and Giles, J. R. M. (1988). Novel polymer electrolytes based on ABA block copolymers, *Macromolecules*, 21, pp. 392–397.

Grigorieff, N., Ceska, T. A., Downing, K. H., Baldwin, J. M. and Henderson, R. (1996). Electron-crystallographic refinement of the structure of bacteriorhodopsin. *Journal of Molecular Biology*, 259, pp. 393–421.

Gritsenko, O. V. and Lazarev, P. I. (1989). in *Molecular Electronics* (Hong F.T., Ed.), Plenum Press, New York, pp. 277.

Groshev, A., Ivanov, T. and Valtchinov, V. (1991). Charging effects of a single quantum level in a box, *Physical Review Letters*, 66, pp. 1082–1085.

Guinea, F. and García, N. (1990). Scanning tunneling microscopy, resonant tunneling, and counting electrons - a quantum standard of current, *Physical Review Letters*, 65, pp. 281–284.

Gulik, A., Luzzati, V., De Rosa, M. and Gambacorta, A. (1985). Structure and polymorphism of bipolar isopranyl ether lipids from archaebacteria, *Journal of Molecular Biology*, 182, pp. 131–149.

Guryev, O., Dubrovsky, T., Chernogolov, A., Dubrovskaya, S., Usanov, S. and Nicolini, C. (1997). Orientation of cytochrome P450scc in Langmuir-Blodgett monolayers, *Langmuir*, 13, pp. 299–304.

Guryev, O., Erokhin, V., Usanov, V. and Nicolini, C. (1996). Cytochrome P450scc spin state transitions in ther thin solid films, *Biochemistry and Molecular Biology International*, 39, pp. 205–214.

Habig, W. H., Pabst, M. J. and Jakoby, W. B. (1974). Glutathione S-transferases. The first enzymatic step in mercapturic acid formation, *Journal of Biological Chemistry*, 249, pp. 7130–7139.

Hagiwara, R. and Ito, Y. (2000). Room temperature ionic liquids of alkylimidazolium cations and fluoroanions, *Journal of Fluorine Chemistry*, 105, pp. 221–227.

Hameroff, S., Nip, A., Porter, M. and Tuszynski, J. (2002). Conduction pathways in microtubules, biological quantum computation, and consciousness. *Biosystems* 64, pp. 149–168.

Hampp, N. (1993). Optical-materials - heat-proof proteins, *Nature*, 366, pp. 12–12.

Hampp, N. (2000). Bacteriorhodopsin as a photochromic retinal protein for optical memories, *Chemical Reviews*, 100, pp. 1755–1776.

Hampp, N. and Brauchle, C. (2003). Bacteriorhodopsin and its functional variants: potential applications in modern optics, *Photochromism: Molecules and Systems*, pp. 954-975.

Hampp, N. and Juchem, T. (2004). Improvement of the diffraction efficiency and kinetics of holographic gratings in photochromic media by auxiliary light. *Optics Letters*, 29, pp. 2911–2913.

Hampp, N. and Zeisel, D. (1994). Mutated bacteriorhodopsins - versatile media in optical image processing, *IEEE Eng. Med. Biol.*, 13, pp. 67–74.

Han, X., Chen, W., Zhang, Z., Dong, S. and Wang, E. (2002). Direct electron transfer between hemoglobin and a glassy carbon electrode facilitated by lipid–protected gold nanoparticles, *Biochimica et Biophysica Acta – Bioenergetics*, 1556, pp. 273–277.

Hancock, B. C. and Zografi, G. J. (1997). Characteristics and significance of the amorphous state in pharmaceutical systems, *Journal of Pharmaceutical Sciences*, 86, pp. 1–12.

Hann, R. A. (1990). Molecular structure and monolayer properties. In *Langmuir Blodgett Films*. G. Roberts, editor. Plenum Press, New York, pp. 19–22.

Hanzal–Bayer, M., Renault, L., Roversi, P., Wittinghofer, A. and Hillig, R. C. (2002). The complex of Arl2-GTP and PDEdelta: From structure to function, *EMBO Journal*, 21, pp. 2095–2106.

Haring, C., Barnas, C., Saria, A., Humpel, C. and Fleischhacker, W. W. (1989). Dose-related plasma levels of clozapine, *Journal of Clinical Psychopharmacology*, 9, pp. 71–72.

Hatchard, T. D. and Dahn, J. R. (2004). In situ XRD and electrochemical study of the reaction of lithium with amorphous silicon, *Journal of the Electrochemical Society*, 151, pp. A838–A842.

He, H. X., Boussaad, S., Xu, B. Q., Li, C. Z., Tao, N. J. (2002). Electrochemical fabrication of atomically thin metallic wires and electrodes separated with molecular-scale gaps, *Journal of Electroanalytical Chemistry*, 522, pp. 167–172.

He, S. and Dassarma, S. (1993). Quantum electron–transport through narrow constrictions in semiconductor nanostructures, *Physical Review B*, 48, pp. 4629–4635.

Henderson, R. (1990). Cryoprotection of protein crystals against radiation-damage in electron and X-ray diffraction, *Proceedings of The Royal Society of London Series B–Biological Sciences*, 241, pp. 6–8.

Hendrickson, W. A. and Wüthrich, K. (1997). Eds., *Macromolecular Structures* (Current Biology, London).
Hillebrecht, J. R., Wise, K. J., Koscielecki, J. F. and Birge, R. R. (2004). Directed evolution of bacteriorhodopsin for device applications, *Methods Enzymology*, 388, pp. 333–347.
Hipps, K. W. (2001). It's all about contacts, *Science*, 294, pp. 536–537.
Hirai, T. and Subramaniam, S. (2003). Structural insights into the mechanism of proton pumping by bacteriorhodopsin, *FEBS Letters*, 545, pp. 2–8.
Holbery, J. D. and Seddon, K. R. (1999). Ionic liquids, *Clean Products and Processes*, 1, pp. 223–236.
Holm, L. and Sander, C., (1996). The FSSP database: fold classification based on structure-structure alignment of proteins, *Nucleic Acids Research*, 24, pp. 206–209.
Hong, J. S., Maleki, H., Al Hallaj, S., Redey, L. and Selman, J. R. (1998). Electrochemical-calorimetric studies of lithium-ion cells, *Journal of Electrochemistry Society*, 145, pp. 1489–1501.
Hoölzel, R., Gajovic–Eichelmann, N., Bier, F. F. (2003). Oriented and vectorial immobilization of linear M13 dsDNA between interdigitated electrode-towards single molecule DNA nanostructures, *Biosensors and Bioelectronics*, 18, pp. 555–564.
Hsie, A. W. and Puck, T. T. (1971). Morphological transformation of Chinese hamster cells by dibutyryl adenosine cyclic 30:50-monophospate and testosterone, *Proc. Nat. Acad. Sci. USA*, 68, pp. 358–361.
Hu, S. Q., Xie, J. W., Xu, Q. H., Rong, K. T., Shen, G. L. and Yu, R. Q. (2003). A label–free electrochemical immunosensor based on gold nanoparticles for detection of paraoxon, *Talanta*, 61, pp. 769–777.
Hu, Y., Li, H., Huang, X. and Chen, L. (2004). Novel room temperature molten salt electrolyte based on LiTFSI and acetamide for lithium batteries, *Electrochemistry Communications*, 6, pp. 28–32.
Huang, H., Yin, S. C. and Nazar, L. F. (2001). Approaching theoretical capacity of $LiFePO_4$ at room temperature at high rates, *Electrochemical and Solid State Letters*, 4, pp. A170–A172.
Hunte, C. and Michel, H. (2003). Membrane protein crystallization. In: Hunte, C., von Jagow, G, Schagger, H. (Eds.), *Membrane Protein Purification And Crystallization: A Practical Guide*. San Diego: Academic Press, pp. 143–160.
Hwang, S. B., Korenbrot, J. and Stoeckenius, W. (1977a). Proton transport by bacteriorhodopsin through an interface film, *Journal of Membrane Biology*, 36, pp. 137–158.
Hwang, S. B., Korenbrot, J. I. and Stoeckenius, W. (1977b). Structural and spectroscopic characteristics of bacteriorhodopsin in air-water interface films, *Journal Membrane Biology*, 36, pp. 115–135.
Iijima, S. (1991). Helical microtubules of graphitic carbon, *Nature*, 354, pp. 56–58.

Ikonen, M, Sharonov, A. Y., Tkachenko, N. V. and Lenunetyinen, H. (1993). The kinetics of charges in dry bacteriorhodopsin Langmuir-Blodgett films – an analysis and comparison of electrical and optical signals, *Advanced Materials for Optics and Electronics*, 2, pp. 211–220.

Inacker, O., Kuhn, H., Mobius, D., and Debuch, G. (1976). Manipulation in molecular dimensions, *Zeitschrift Fur Physikalische Chemie-Frankfurt*, 101, pp. 337–360.

Isakov, N. and Altman, A. (2002). Protein kinase C(theta) in T cell activation, *Annual Review of Immunology*, 20, pp. 761–794.

Itoh, K. and Adelstein, R. S. (1995). Neuronal cell expression of inserted isoforms of vertebrate nonmuscle myosin heavy chain II-B, *Journal of Biological Chemistry*, 270, pp. 14533–14540.

Iwata, E., Takahashi, K., Maeda, K. and Mouri, T. (1999). Capacity failure on cycling or storage of lithium-ion batteries with Li-Mn-O ternary phases having spinel-framework structure and its possible solution, *Journal of Power Sources*, 81, pp. 430–433.

Jacobs, J. W. and Fodor, P. A. (1994), Combinatorial chemistry - applications of light-directed chemical synthesis, *Trends in Biotechnology*, 12, pp. 19–26.

Jagner, D., Sahlin, E., Axelsson, B. and Ratana-Ohpas, R. (1993). Rapid method for the determination of copper(II) and lead(II) in tap water using a portable potentiometric stripping analyzer, *Analytica Chimica Acta*, 278, pp. 237–242.

Jen, A. and Merkle, H. P. (2001). Diamonds in the rough: Protein crystals from a formulation perspective, *Pharmaceutical Research*, 18, pp. 1483–1488.

Jeng, M. F., Campbell, A. P., Begley, T., Holmgren, A., Case, D. A., Wright, P. E. and Dyson, H. J. (1994). High-resolution solution structures of oxidized and reduced Escherichia coli thioredoxin, *Structure*, 2, pp. 853–868.

Jeon, J. D., Cho, B. W. and Kwak, S. Y. (2005). Solvent–free polymer electrolytes based on thermally annealed porous P(VdF–HFP)/P(EO–EC) membranes, *Journal of Power Sources*, 143, pp. 219–226.

Jerry, D. J., Dickinson, E. S., Roberts, A. L. and Said, T. K. (2002). Regulation of apoptosis during mammary involution by the p53 tumor suppressor gene, *Journal of Dairy Science*, 85, pp. 1103–1110.

Jessensky, O., Muller, F. and Gosele, U. (1998). Self-organized formation of hexagonal pore arrays in anodic alumina, *Applied Physics Letters*, 72, pp. 1173–1176.

Johnson, C. S., Kim, J. S., Kropf, A. J., Kahaian, A. J., Vaughey, J. T., Fransson, L. M. L., Edstrom, K. and Thackeray, M. M. (2003). Structural characterization of layered LixNi0.5Mn0.5O2 ($0 < x <= 2$) oxide electrodes for Li batteries, *Chemistry of Materials*, 15, pp.2313–2322.

Joseph, S., Rusling, J. F., Lvov, Y. M., Friedberg, T., and Fuhr, U. (2003). An amperometric biosensor with human CYP3A4 as a novel drug screening tool, *Biochemical Pharmacology*, 65, pp. 1817–1826.

Jost, O., Gorbunov, A., Liu, X., Pompe, W. and Fink, J. (2004) Single–walled carbon nanotube diameter. *Journal of Nanoscience and Nanotechnology*, 4, pp. 433–440.

Kandler, O. (1992). Where next with archaebacteria, *Biochemical Society Symposium*, 58, pp. 195-207.

Karacs, A., Fancsaly, A. J., Divinyi, T., Peto, G. and Kovach, G. (2003). Morphological and animal study of titanium dental implant surface induced by blasting and high intensity pulsed Nd–glass laser, *Materials Science Enginnering C*, 23, pp. 431–435.

Karan, D., Kelly, D. L., Rizzino, A., Lin, M. F. and Batra, S. K. (2002). Expression profile of differentially-regulated genes during progression of androgen-independent growth in human prostate cancer cells, *Carcinogenesis*, 23, pp. 967–975.

Kashchiev D., (2000). *Nucleation: Basic Theory with Applications*, Butterworth–Heinemann, Oxford.

Kavan, L., Gratzel, M., Rathousky, J. and Zukal, A. (1996). Nanocrystalline TiO_2 (anatase) electrodes: Surface morphology, adsorption, and electrochemical properties, *Journal of the Electrochemistry Society*, 143, pp. 394–400.

Kawabe, T., Suganuma, M., Ando, T., Rimura, M., Hori, H. and Okamoto, T. (2002). Cdc25C interacts with PCNA at G2/M transition, *Oncogene*, 21, pp. 1717–1726.

Kelly, I. E., Owen, J. R. and Steele, B. C. H. (1985). Poly(ethylene oxide) electrolytes for operation at near room temperature, *Journal of Power Sources*, 14, pp 13–21.

Kendall, F., Swenson, R., Borun, T., Rowinski, J. and Nicolini, C. (1977). Nuclear morphometry during the cell cycle, *Science*, 196, pp. 1106–1109.

Kepler, K. D., Vaughey, J. T. and Thackeray, M. M. (1999). $Li_xCu_6Sn_5$ ($0 < x < 13$): An intermetallic insertion electrode for rechargeable lithium batteries, *Electrochemical and Solid State Letters*, 2, pp. 307–309.

Khrapko, K. R., Lysov, Y. P., Khorlyn, A. A., Shick V. V., Florentiev, V. L. and Mirzabekov, A. D. (1989). An oligonucleotide hybridisation approach to DNA sequencing, *FEBS Letters*, 256, pp. 118–122.

Killis, A., Lenest, J. F., Cheradame, H. and Gandini, A. (1982). Ionic conductivity of polyether-polyurethane networks containing $NaBPh_4$: a free volume analysis, *Die Makromolekulare Chemie*, 183, pp. 2835–2845.

Kim, D. G., Kim, H., Sohn, H. J. and Kang, T. (2002). Nanosized Sn-Cu-B alloy anode prepared by chemical reduction for secondary lithium batteries, *Journal of Power Sources*, 104, pp. 221–225.

Kim, D. W. (1998). Composite gel electrolyte for rechargeable lithium batteries, *Journal of Power Sources*, 55, pp. 7–10.

Kim, H., Garavito R. M. and Lal, R. (2000). Atomic force microscopy of the three-dimensional crystal of membrane protein, OmpC porin, *Colloids Surf B Biointerfaces*, 19, pp. 347–355.

Kim, J. H., Jeong, G. J., Kim, Y. W., Sohn, H. J., Park, C. W. and Lee, C. K. (2003). Tin-based oxides as anode materials for lithium secondary batteries, *Journal of the Electrochemical Society*, 150, pp. A1544–A1547.

Kim, J. M. and Chung, H. T. (2004a). Role of transition metals in layered Li[Ni,Co,Mn]O-2 under electrochemical operation, *Electrochimica Acta*, 49, pp. 3573–3580.

Kim, J. M. and Chung, H. T. (2004b). The first cycle characteristics of Li[Ni1/3Co1/3Mn1/3]O-2 charged up to 4.7 V, *Electrochimica Acta*, 49, pp. 937–944.

Kimura, Y. and Bianco, P. R. (2006). Single molecole studies of DNA binding proteins using optical tweezers, *Analyst*, 131, pp. 868–874.

Kiselyova, O. I., Guryev, O. L., Krivosheev, A. V., Usanov, S. A. and Yaminsky, I. V. (1999). Atomic force microscopy studies of Langmuir-Blodgett films of cytochrome P450scc: Hemeprotein aggregation states and interaction with lipids, *Langmuir*, 15, pp. 1353–1359.

Koch, V. R., Dominey, L. A., Nanjundiah, C. and Ondrechen, M. J. (1996). The intrinsic anodic stability of several anions comprising solvent-free ionic liquids, *Journal of Electrochemistry Society*, 143, pp. 798–803.

Koel, M. (2000). Physical and chemical properties of ionic liquids based on the dialkylimidazolium cation, *Proc. Estonian Acad. Sci. Chem.*, 49, pp. 145–155.

Kojima, Y., Usuki, A., Kawasumi, M., Okada, A., Kurauchi, T. and Kamigaito, O. (1993). Synthesis of nylon-6-clay hybrid by montmorillonite intercalated with epsilon-caprolactam, *Journal of Applied Polymer Science Part A: Polymer Chemistry*, 31, pp. 983–986.

Krätschmer, W., Lamb. L. D., Fostiropoulos, K. and Huffman, D. R. (1990). Solid C_{60}: a new form of carbon, *Nature*, 347, pp. 354-358.

Krawiec, W., Scanlon, L. G., Fellner, J. P., Vaia, R. A. and Giannelis, E. P. (1995). Polymer nanocomposites – a new strategy for synthesizing solid electrolytes for rechargeable lithium batteries, *Journal of Power Sources*, 54, pp. 310–315.

Kreisler, M., Kohnen, W., Marinello, C., Gotz, H., Duschner, H., Jansen, B. and D'Hoedt, B. (2002). Bactericidal effect of the Er: Yag laser on dental implant surfaces: An in vitro study. *Journal of Periodontology*, 73, pp. 1292–1298.

Kreisler, M., Kohnen, W., Marinello, C., Schoof, J., Langnau, E., Jansen, B. and D'Hoedt, B. (2003). Antimicrobial efficacy of semiconductor laser irradiation on implant surfaces, *International Journal of Oral Maxillofacial Implants*, 18, pp. 706–711.

Kuhn, H. (1965). *Pure and Applied Chemistry*, 11, pp. 345.

Kuhn, H. (1981). Information, electron and energy transfer in surface layers, *Pure and Applied Chemistry*, 53, pp. 2105-2122.

Kuhn, H. (1983). Functionalized monolayer assembly manipulation, *Thin Solid Films*, 99, pp. 1–16.

Kumagai, N., Ooto, H. and Kumagai, N. (1997). Preparation and electrochemical characteristics of quaternary Li–Mn–V–O spinel as the positive materials for rechargeable lithium batteries, *Journal of Power Sources*, 68, pp. 600–603.

Kumai, K., Miyashiro, H., Kobayashi, Y., Takei, K. and Ishikawa, R. (1999). Gas generation mechanism due to electrolyte decomposition in commercial lithium-ion cell, *Journal of Power Sources*, 81–82, pp. 715–719.

Kumar, S. and Nussinov, R. (2001). How do thermophilic proteins deal with heat ?, *Cellular and Molecular Life Science*, 58, pp. 1216–1233.

Kumar, S. and Nussinov, R. (2002). Close-range electrostatic interactions in proteins. *Chembiochemistry*, 3, pp. 604–617.

Kuznetsov, Y. G., Malkin, A. J. and McPherson, A. (2001). The liquid protein phase in crystallization: a case study – intact immunoglobulins, *Journal of Crystal Growth*, 232, pp. 30–39.

LaBaer, J. and Ramachandran, N. (2005). Protein microarrays as tools for functional proteomics. *Current Opinion in Chemical Biology*, 9, pp. 14–19.

LaBaer, J. (2006). Functional proteomics for biomarker and target discovery, *Molecular and Cellular Proteomics*, 5, pp. S140-S140.

Laemmli, U. K. (1970). Cleavage of structural proteins during the assembly of the head of bacteriophage, T4. *Nature*, 227, pp. 680–685.

Langmuir, I. and Schaefer, V. J. (1938). Activities of urease and pepsin monolayers, *Journal of the American Chemical Society*, 60, pp. 1351–1360.

Langmuir, I. and Schaefer, V. J. (1939). Properties and structure of protein monolayers, *Chemical Reviews*, 24, pp. 181–202.

Lanz, M. and Novak, P. (2001). DEMS study of gas evolution at thick graphite electrodes for lithium-ion batteries: the effect of gamma–butyrolactone, *Journal of Power Sources*, 102, pp. 277–282.

Lascaud, S., Perrier, M., Valle, A., Besner, C., Prud'homme, J. and Armand, M. (1994). Phase–diagrams and conductivity behavior of poly(ethylene oxide) molten-salt rubbery electrolytes, *Macromolecules*, 27, pp. 7469–7477.

Lazzari, R. (2002). IsGISAXS: a program for grazing-incidence small-angle X-ray scattering analysis of supported islands. *Journal of Applied Crystallography*, 35, pp. 406–421.

Leonhardt, A., Berglundh, T., Ericsson, I. and Dahlen, G. (1992). Putative periodontal pathogens on titanium implants and teeth in experimental gingivitis and periodontitis in beagle dogs, *Clinical Oral Implant Research*, 3, pp. 112–119.

Li, C. Z., He, H. X. and Tao, N. J. (2000). Quantized tunneling current in the metallic nanogaps formed by electrodeposition and etching, *Applied Physics Letters*, 77, pp. 3995–3997.

Li, G. H., Azuma, H. and Tohda, M. (2002). LiMnPO$_4$ as the cathode for lithium batteries, *Electrochemical and Solid State Letters*, 5, pp. A135–A137.

Li, H., Nadarajah, A. and Pusey, M. L. (1999). Determining the molecular–growth mechanisms of protein crystal faces by atomic force microscopy, *Acta Crystallographica D: Biological Crystallography*, 55, pp. 1036–1045.

Li, N. C., Mitchell, D. T., Lee, K. P. and Martin, C. R. (2003). A nanostructured honeycomb carbon anode, *Journal of the Electrochemical Society*, 150, pp. A979–A984.

Lightfoot, P., Mehta, M. A. and Bruce, P. G. (1993). Crystal structure of the polymer electrolyte poly(ethylene oxide)$_3$:LiCF$_3$SO$_3$, *Science*, 262, pp. 883–885.

Lindhe, J., Berglundh, T., Ericsson, B., Lijienberg, B. and Marinello, C. (1992). Experimental breakdown of periimplant and periodontal tissues. A study in the dog, *Clinical Oral Implants Research*, 3, pp. 9–16.

Lindino, C. A. and Bulhoes, L. O. S. (1996). The potentiometric response of chemically modified electrodes, *Analytica Chimica Acta*, 334, pp. 317-322

Lloris, J. M., Perez Vicente, C. and Tirado, J. L. (2002). Improvement of the electrochemical performance of LiCoPO(4)5Vmaterial using a novel synthesis procedure, *Electrochemical and Solid State Letters*, 5, pp. A234–A237.

Luecke, H., Schobert, B., Lanyi, J. K., Spudich, E. N. and Spudich, J. L. (2001). Crystal structure of sensory rhodopsin II at 2.4 Angstroms: Insights into color tuning and transducer interaction, *Science* 293, pp. 1499–1503.

Luger, K., Rechsteiner, T., Flaus, A. J., Waye, M. M. Y. and Richmond, T. J. (1997a) Characterization of nucleosome core particles containing histone proteins made in bacteria, *Journal of Molecular Biology*, 272, pp. 301–311.

Luger, K., Mäder, A. W., Richmond, R. K., Sargent, D. F. and Richmond, T. J. (1997b). Crystal Structure of the nucleosome core particle at 2.8 Å resolution, *Nature*, 389, pp. 251–260.Luzzati, V., Gambacorta, A., De Rosa, M., Gulik, A. (1987). Polar lipids of thermophilic prokaryotic organisms: chemical and physical structure, *Annual Review of Biophysics and Biophysical Chemistry*, 16, pp. 25–47.

Lvov, Y. M., Erokhin, V. V. and Zaitsev, S. Y. (1990). Protein Langmuir-Blodgett films, *Biologicheskie Membrany*, 7, pp. 917–937.

Maccallum, J. R., Smith, M. J. and Vincent, C. A. (1984). The effect of radiation–induce crosslinking on the conductance of LiClO$_4$PEO electrolytes, *Solid State Ionics*, 11, pp. 307–312.

Maccioni, E., Radicchi, G., Erokhin, V., Paddeu, S., Facci, P. and Nicolini, C. (1996). Bacteriorhodopsin thin film as a sensitive layer for an anaesthetic sensor, *Thin Solid Films*, 284–285, 898–900.

MacFarlane, D. R., Meakin, P., Sun, J., Amini, N. and Forsyth, M. (1999). Pyrrolidinium imides: A new family of molten salts and conductive plastic crystal phases, *Journal of Physical Chemistry B*, 103, pp. 4164–4170.

Majka, J. and Burgers, P. M. (2004). The PCNA–RFC families of DNA clamps and clamp loaders. *Prog. Nucleic Acid Res. Mol. Biol.*, 78, pp. 227–260.

Majka, J., Chung, B. Y. and Burgers, P. M. (2004). Requirement for ATP by the DNA damage checkpoint clamp loader, *Journal of Biological Chemistry*, 279, pp. 20921–20926.

Makimura, Y. and Ohzuku, T. (2003). Lithium insertion material of LiNi1/2Mn1/2O2 for advanced lithium-ion batteries, *Journal of Power Sources*, 119, pp. 156–160.

Malmquist, M. and Olofsson, G. (1989). Methods of silanization surfaces, US Patent 4,833,093.
Mao, O., Dunlap, R.A. and Dahn, J. R. (1999a). Mechanically alloyed Sn-Fe(-C) powders as anode materials for Li-ion batteries – I. The Sn_2Fe-C system, *Journal of the Electrochemistry Society*, 146, pp. 405–413.
Mao, O., Turner, R. L., Courtney, I. A., Fredericksen, B. D., Buckett, M. I., Krause, L. J. and Dahnl, J. R. (1999b). Active/inactive nanocomposites as anodes for Li-ion batteries, *Electrochemical and Solid State Letters*, 2, pp. 3–5.
Marconcini, L., Giacomelli, L., Barone, A., Covani, U., Nicolini, C. (2007). Leader Genes in Osteogenesis, *Bone*, in preparation
Margolin, A. L. and Navia, M. A. (2001). Protein crystals as novel catalytic materials, *Angewandte Chemie–International Edition*, 40, pp. 2205–2222.
Marzec, J., Swierczek, K., Przewoznik, J., Molenda, J., Simon, D. R., Kelder, E. M. and Schoonman, J. (2002). Conduction mechanism in operating a $LiMn_2O_4$ cathode, *Solid State Ionics*, 146, pp. 225–237.
Masuda, H. and Fukuda, K. (1995). Ordered metal nanohole arrays made by a two-step replication of honeycomb structures of anodic alumina, *Science*, 268, pp. 1466–1468.
Masuda, H. and Satoh, M. (1996). Fabrication of gold nanodot array using anodic porous alumina as an evaporation mask, *Jpn. J. Appl. Phys.*, 35, pp. L126–L129.
Masuda, H., Haseqwa F. and Ono, S. (1997). Self-ordering of cell arrangement of anodic porous alumina formed in sulfuric acid solution, *Journal of the Electrochemistry Society*, 144, pp. L127–130.
Matsui, Y., Sakai, K., Murakami, M., Shiro, Y., Adachi, S., Okumura, H. and Kouyama, T. (2002). Specific damage induced by X–ray radiation and structural changes in the primary .photoreaction of bacteriorhodopsin, *Journal of Molecular Biology*, 324, pp. 469–481.
Matysik, J., Alia Bhalu, B. and Mohanty, P. (2002). Molecular mechanisms of quenching of reactive oxygen species by proline under stress in plants, *Current Science*, 82, 525–532.
Maxia, L., Radicchi, G., Pepe, I. M. and Nicolini, C. (1995). Characterization of Langmuir-Blodgett films of rhodopsin – thermal-stability studies, *Biophysical Journal*, 69, pp. 1440–1446.
McFarlane, D. R., Sun, J., Golding, J., Meakin, P. and Forsyth, M. (2000). High conductivity molten salts based on the imide ion, *Electrochimica Acta*, 45, pp. 1271–1278.
McMillan, R. A., Paavola, C. D., Howard, J., Chan, S. L., Zaluzec, N. J. and Trent, J. D. (2002). Ordered nanoparticle arrays formed on engineered chaperonin protein templates, *Nature Materials*, 1, pp. 247–252.
McPherson, A. (1999). *Crystallization of Biological Macromolecules*, CSHL Press, Cold Spring Harbor.

McPherson, A., Malkin, A. J. and Kuznetsov, Y. G. (2000). Atomic force microscopy in the study of macromolecular crystal growth, *Annual Review of Biophysics and Biomolecular Structure*, 29, pp. 361–410.

Meiney, P., Fischer, J. E., McGhie, A. R., Romanow, W., Denenstein, A. M., McCauley, J. P. and Smith. A. B. (1991). Orientational ordering transition in solid C_{60}, *Physical Review Letters*, 66, pp. 2911-2914.

Messana, I., Cabras, T., Inzitari, R., Lupi, A., Zuppi, C., Olmi, C., Fadda, M.B., Cordaro, M., Giardina, B. and Castagnola, M. (2004). Characterization of the human salivary basic proline-rich protein complex by a proteomic approach. *Journal of Proteome Research*, 3, pp. 792–800.

Messersmith, P. B. and Stupp, S. I. (1992). Synthesis of nanocomposites – organoceramics, *Journal of Materials Research*, 7, pp.2599–2611.

Méthot, M., Boucher, F., Salesse, C., Subirade, M. and Pézolet, M. (1996). Determination of bacteriorhodopsin orientation in monolayers by infrared spectroscopy, *Thin Solid Films*, 285, pp. 627–630.

Miedlich, S. U., Gama, L., Seuwen, K., Wolf, R. M. and Breitwieser, G. E. (2004). Homology modeling of the transmembrane domain of the human calcium sensing receptor and localization of an allosteric binding site, *Journal of Biological Chemistry*, 279, pp. 7254–7263.

Mirzabekov, A. (1994). DNA sequencing by hybridization - a megasequencing method and a diagnostic tool?, *Trends in Biotechnology*, 12, pp. 27–32.

Misteli, T. (2001). Protein dynamics: Implications for nuclear architecture and gene expression. *Science*, 291, pp. 843–847.

Miyasaka, T., Koyama, K. and Itoh, I. (1991). Quantum Conversion and Image Detection by a Bacteriorhodopsin–Based Artificial Photoreceptor, *Science*, 255, pp. 342–344.

Modiano, J. F., Mayor, J., Ball, C., Fuentes, M. K. and Linthicum, D. S. (2000). CDK4 expression and activity are required for cytokine responsiveness in T cells, *Journal of Immunology*, 165, pp. 6693–6702.

Molenda, J., Ojczyk, W., Marzec, M., Marzec, J., Przewoznik, J., Dziembaj, R. and Molenda, M. (2003). Electrochemical and chemical deintercalation of $LiMn_2O_4$, *Solid State Ionics*, 157, pp. 73–79.

Molenda, J., Swierczek, K., Kucza, W., Marzec, J. and Stokłosa, A. (1999). Electrical properties of $LiMn_2O_4$–delta at temperatures 220–1100K, *Solid State Ionics*, 123, pp. 155–163.

Mombelli, A., Van Oosten, M. A. C., Schurch, E. and Lang, N. P. (1987). The microbiota associated with successful or failing osseointegrated titanium implants, *Oral Microbiology and Immunology*, 2, pp. 145–151.

Monconduit, L., Jumas, J. C., Alcantara, R., Tirado, J. L. and Perez Vicente, C. (2002). Evaluation of discharge and cycling properties of skutterudite-type $Co_{1-2y}Fe_yNi_ySb_3$ compounds in lithium, cells, *Journal of Power Sources*, 107, pp. 74–79.

Morpurgo, A. F., Marcus, C. M. and Robinson, D. B. (1999). Controlled fabrication of metallic electrodes with atomic separation, *Applied Physics Letters*, 74, pp. 2084–2086.

Mozzarelli, A. and Rossi, G. L. (1996). Protein function in the crystal, *Annual Review of Biophysics and Biomolecular Structure*, 25, pp. 343–365.

Mullen, K., Ben–Jacob, E., Jaklevic, R. C. and Shuss Z. (1988). I-V characteristics of coupled ultrasmall-capacitance normal tunnel junctions, *Physics Review B*, 37, pp. 98–105.

Müller-Buschbaum, P., Casagrande, M., Gutmann, J., Kuhlmann, T., Stamm, M., von Krosigk, G., Lode, U., Cunis, S. and Gehrke, R. (1998). Determination of micrometer length scales with an X-ray reflection ultra small-angle scattering set-up, *Europhys Letters*, 42, pp. 517–519.

Müller-Buschbaum, P., Gutmann, J. S., Stamm, M., Cubitt, R., Cunis, S., Von Krosigk, G., Gehrke, R. and Petry, W. (2000). Dewetting of thin polymer–blend films examined with GISAXS, *Physica B*, 283, 53–59.

Müller-Buschbaum, P., Roth, S. V., Burghammer, M., Diethert, A., Panagiotou, P. and Riekel, C. (2003). Multiple-scaled polymer surfaces investigated with micro-focus grazing incidence small-angle X-ray scattering, *Europhysics Letters*, 61, pp. 639–645.

Nagira, M., Imai, T., Ishikawa, I., Uwabe, K. I. and Yoshie, O. (1994). Mouse homologue of C33 antigen (CD82), a member of the transmembrane 4 superfamily: Complementary DNA, genomic structure, and expression. *Cell Immunol.*, 157, pp. 144–157.

Nakano, H., Nonaka, T., Okuda, C. and Ukyo, Y. (2003). In situ XAFS study of LiNi0.5Mn0.5O2 cathode for Li rechargeable batteries, *Journal of the Ceramic Society of Japan*, 111, pp. 33–36.

Nanjundiah, C., McDevit, F. and Koch, V. R. (1997). Differential capacitance measurements in solvent–free ionic liquids at Hg and C interfaces, *Journal of Electrochemistry Society*, 144, pp. 3392–3397.

Narizzano, R. and Nicolini, C. (2005). Mechanism of conjugated polymer organization on SWNT surfaces, *Macromolecular Rapid Communications*, 26, pp. 381–385.

Narizzano, R., Erokhin, V. and Nicolini, C. (2005). A heterostructure composed of conjugated polymer and copper sulfide nanoparticles, *The Journal of Physical Chemistry B,* 109, pp. 15798–15802.

Nathans, J. (1992). Rhodopsin: structure, function, and genetics. *Biochemistry*, 31, pp. 4923–4931.

Nazar, L. F. and Crosnier, O. (2004). Anodes and composite anodes: an overview. In: *Lithium Batteries, Science and Technology,* G.A. Nazri, G. Pistoia (Eds.), Kluwer Academic Publishers, Boston, NY, London, pp. 112–143.

Nazeeruddin, M. K., Kay, A., Rodicio, J., Humphybaker, R., Muller, E., Liska, P., Vlachopoulos, N. and Gratzel, M. (1993). Conversion of light to electricity by cis-x2bis(2,2'-bipyridyl-4,4'-dicarboxylate)ruthenium(ii) charge–transfer sensitizers (x=cl-,br-,I-, cn-, and scn-) on nanocrystalline TiO_2 electrodes, *Journal of the American Chemical Society*, 115, pp. 6382–6390.

Neuman, K. and Block, S. (2004). Optical tweezers, *Review of Scientific Instruments*, 75, pp. 2787–2809.

Ngo, H. L., LeCompte, K., Hargens, L. and McEwen, A. B. (2000). Thermal properties of imidazolium ionic liquids, *Thermochimica Acta*, 357, pp. 97–102.

Nicastro, G., De Chiara, C., Pedone, E., Tato, M., Rossi, M. and Bartolucci, S. (2000). NMR solution structure of a novel thioredoxin from Bacillus acidocaldarius possible determinants of protein stability. *Eur. J. Biochem.*, 267, pp. 403–413.

Nicolini, C. (1983). Chromatin structure: from nuclei to genes, *Anticancer Research*, 3, pp. 63–86.

Nicolini, C. (1986). Nuclear structure: from pores to the high-order gene structure, *Cell Biophysics*, 9, pp. 67–90.

Nicolini, C. (1995). From neural chip and engineered biomolecules to bioelectronic devices: an overview, *Biosensors & Bioelectronics*, 10, pp. 105–127.

Nicolini, C. (1996). Heat-proof enzymes by Langmuir–Blodgett technique, *Annals New York Academy of Science*, 799, pp. 297–311.

Nicolini, C. (1996a). *Molecular Bioelectronics*, New York: World Scientific, pp. 1–266.

Nicolini, C. (1996b). Supramolecular architecture and molecular bioeletronics, *Thin Solid Films*, 284–285, pp. 1–5.

Nicolini, C. (1996c). *Molecular Manufacturing*, EL.B.A. Forum Series Vol. 2, Plenum Press, New York.

Nicolini, C. (1997). Protein monolayer engineering: principles and application to biocatalysis. *Trends in Biotechnology*, 15, pp. 395–401.

Nicolini, C. (1998a). *Biophysics of electron transfer and molecular bioelectronics*, EL.B.A. Forum Series Vol. 3, Plenum Press, New York, pp. 1–196.

Nicolini, C. (1998b). Engineering of enzyme monolayer for industrial biocatalysis, *Annals of the New York Academy of Sciences*, 864, pp. 435–441.

Nicolini, C. (2006). Nanogenomics for medicine, *Nanomedicine*, 1, pp. 147–151.

Nicolini, C. and Beltrame, F. (1982). Coupling of chromatin structure to cell geometry during the cell cycle. Transformed versus reverse-transformed CHO, *Cell Biology International Reports*, 6, pp. 63–71.

Nicolini, C. and Kendall, F. (1977). Differential light-scattering in native chromatin: corrections and inferences combining melting and dye-binding studies. A two-order superhelical model, *Physiological Chemistry & Physics*, 9, pp. 265–283.

Nicolini, C. and Pechkova, E. (2004). Nanocrystallography: an emerging technology for structural proteomics, *Expert Review of Proteomics*, 1, pp. 253–256.

Nicolini, C. and Pechkova, E. (2006). Structure and growth of ultrasmall protein microcrystals by synchrotron radiation: I µGISAXS and microdiffraction of P450scc, *Journal of Cellular Biochemistry*, 97, pp.544–552.

Nicolini, C. and Pechkova, E. (2006a). Nanostructured biofilms and biocrystals, *Journal of Nanoscience and Nanotechnology*, 6, pp. 2209–2236.

Nicolini, C. and Rigo, A. (1992). *Biofisica e Tecnologie Biomediche*, Zanichelli Editore S.p.A., pp. 1–721.

Nicolini, C., Ajiro K., Borun T. W. and Baserga R. (1975). Chromatin changes during the HeLa cell cycle, *Journal of Biological Chemistry*, 250, pp. 3381–3385.

Nicolini, C., Kendall F. and Baserga R. (1976). DNA structure in sheared and unsheared chromatin, *Science*, 192, pp. 796–798.

Nicolini, C., Kendall, F., Baserga, R., Dessaive, C., Clarkson, B. and Fried, J. (1977). The G0-G1 transition of WI38 cells, *Experimental Cell Research*, 106, pp. 111–118.

Nicolini, C., Linden, W. A., Zietz, S. and Wu, C. T. (1977a). Identification of non-proliferating cells in melanoma B16 tumor, *Nature*, 270, pp. 607–609.

Nicolini, C., Carlo, P., Martelli, A., Finollo, R., Bignone, F. A., Patrone, E., Trafiletti, V. and Brambilla G. (1982). Viscoelastic properties of native DNA from intact nuclei of mammalian cells. Higher-order Dna packing and cell function, *Journal of Molecular Biology* 161, 155–175,

Nicolini, C., Trafiletti, V., Cavazza, B., Cuniberti, C., Patrone, E., Carlo, P. and Brambilla, G. (1983). Quaternary and quinternary structure of native chromatin DNA in liver nuclei: differential scanning calorimetry, *Science,* 219, pp. 176–178

Nicolini, C., Cavazza, B., Trefiletti, V., Pioli, F., Beltrame, F., Brambilla, G., Maraldi, N., Patrone, E., (1983a). Higher-order structure of chromatin from resting cells. II. High–resolution computer analysis of native chromatin fibres and freeze–etching of nuclei from rat liver cells. *Journal of Cell Science,* 62, pp. 103–115.

Nicolini, C., Vernazza, G., Chiabrera, A., Maraldi, I. N. and Capitani, S. (1984). Nuclear pores and interphase chromatin: high-resolution image analysis and freeze etching, *Journal of Cell Science,* 72, pp. 75–87.

Nicolini, C., Diaspro, A., Bertolotto, M., Facci, P. and Vergani, L. (1991). Changes in DNA superhelical density monitored by polarized light scattering, *Biochemical and Biophysical Research Communications,* 177, pp. 1313–1318.

Nicolini, C., Adami, M., Antolini, F., Beltram, F., Sartore, M. and Vakula, S. (1992). Biosensors: a step to bioelectonics, *Physics World*, 5, pp. 30–34.

Nicolini, C., Erokhin, V., Antolini, F., Catasti, P. and Facci, P. (1993). Thermal stability of protein secondary structure in Langmuir-Blodgett films, *Biochem Biophys Acta*, 1158, pp. 273–278.

Nicolini, C., Adami, M., Dubrovsky, T., Erokhin, V., Facci, P., Paschkevitch, P. and Sartore, M. (1995). High-sensitive biosensor based on LB technology and on nanogravimetry, *Sensors and Actuators–B Chemical*, 24, pp. 121–128.

Nicolini, C., Lanzi, M., Accossato, P., Fanigliulo, A., Mattioli, F. and Martelli, A. (1995a). A silicon–based biosensor for real-time toxicity testing in normal versus cancer liver cells, *Biosensors & Bioelectronics*, 10, pp. 723–733.

Nicolini, C., Sartore, M., Zunino, M. and Adami, M. (1995b). A new instrument for the simultaneous determination of pH and redox potential, *Review of Scientific Instrument*, 66, pp. 4341–4346.

Nicolini, C., Erokhin, V., Facci, P., Rossi, A., Guerzoni, S. and Paschkevitch, P. (1997). DNA Based multiquartz sensor, *Biosensors & Bioelectronics*, 12, pp. 613–618.

Nicolini, C., Erokhin, V., Paddeu, S. and Sartore, M. (1998). Towards light–addressable transducer bacteriorhodopsin based, *Nanotechnology*, 9, pp. 223–227.

Nicolini, C., Erokhin, V., Paddeu, S., Paternolli, C. and Ram, M. K. (1999). Toward bacteriorhodopsin based photocells, *Biosensors and Bioelectronics*, 14, pp. 427–433.

Nicolini, C., Erokhin, V., Ghisellini, P., Paternolli, C., Ram M. K. and Sivozhelezov V. (2001). P450scc engineering and nanostructuring for cholesterol sensing, *Langmuir*, 17, pp. 3719–3726.

Nicolini, C., Erokhin, V. and Ram M. K., (2001a). Supramolecular layer engineering for industrial nanotechnology. In *Nano–surface chemistry*, Rosoff, M. (ed.), Marcel Dekker New York pp. 141–212.

Nicolini, C., Malvezzi, A. M., Tomaselli, A., Sposito, D., Tropiano, G. and Borgogno, E. (2002). DNASER I: Layout and Data Analysis, *IEEE Transactions on Nanobioscience*, 1, pp. 61–72.

Nicolini, C., Narizzano, R. and Bavastrello, V. (2005). New materials by organic nanotechnology and their applications. In: *Recent Research Development in Materials Science* (edited S. Pandalai) Research SignPost, Kerala, India, 6, pp. 17–40.

Nicolini, C., Adami, M. and Paternolli, C. (2006). Nanostructured organic matrices and intelligent sensors. In: *Smart Biosensor Technology*, (G. Knopf & Bassi A.S. eds), CRC Press, pp. 231–244.

Nicolini, C., Spera, R., Stura, E., Fiordoro, S. and Giacomelli, L. (2006a). Gene expression in the cell cycle of human T lymphocytes: II. Experimental determination by DNASER technology, *Journal of Cellular Biochemistry*, 97, pp. 1151–1159.

Odom, T. W., Huang, J. L. and Lieber, C. M. (2002). Single-walled carbon nanotubes: from fundamental studies to new device concepts, *Annals New York Academy of Science*, 960, pp. 203–215.

Oesterhelt, D., Brauchle, C. and Hampp, N. (1991). Bacteriorhodopsin - a biological-material for information-processing, *Quarterly Reviews of Biophics*, 24, pp. 425–478.

Ohmori, Y., Hironaka, Y., Yoshida, M., Fujii, A., Tada, N. and Yoshino, K. (1997). Enhancement of electroluminescence intensity in poly(3–alkylthiophene) with different alkyl side–chain length by doping of fluorescent dye, *Polyer for Advanced Technologies*, 8, pp. 403–407.

Ohta, A., Koshina, H., Okuno, H. and Murai, H. (1995). Relationship between carbonaceous materials and electrolyte in secondary lithium-ion batteries, *Journal of Power Sources*, 54, pp. 6–10.

Ohta, S., Shiomi, Y., Sugimoto, K., Obuse, C. and Tsurimoto, T. (2002). A proteomics approach to identify proliferating cell nuclear antigen (PCNA)-binding proteins in human cell lysates. Identification of the human CHL12/RFCs2–5 complex as a novel PCNA-binding protein, *Journal of Biological Chemistry*, 277, pp. 40362–40367.

Okada, S., Sawa, S., Egashira, M., Yamaki, J., Tabuchi, M., Kageyama, H., Konishi, T. and Yoshino, A. (2001). Cathode properties of phospho–olivine LiMPO$_4$ for lithium secondary batteries, *Journal of Power Sources*, 97–98, pp. 430–432.

Okada, S., Sawa, S., Uebou, Y., Egashira, M., Yamaki, J., Tabushi, M., Kobayashi, H., Fukumi, K. and Kageyama, H. (2003). Charge–discharge mechanism of LiCoPO$_4$ cathode for rechargeable lithium batteries, *Electrochemistry*, 71 1136–1138.

Okada, T., Fujiyoshi, Y., Silow, M., Navarro, J., Landau, E. M. and Shichida, Y. (2002). Functional role of internal water molecules in rhodopsin revealed by X–ray crystallography, *Proceedings of the National Academy of Science USA*, 99, pp. 5982–5987.

Olashaw, N. and Pledger, W. J. (2002). Paradigms of growth control: relation to Cdk activation, *Sci. STKE*, RE7.

Omura, T. and Sato, R. (1964). The carbon monoxide binding pigment of liver microsomes. II. Solubilization, purification and properties, *Journal of Biological Chemistry*, 239, pp. 2370–2378.

Oosterwegel, M. A., Greenwald, R. J., Mandelbrot, D. A., Lorsbach, R. B. and Sharpe, A. H. (1999a). CTLA-4 and T cell activation, *Current Opinion in Immunology*, 11, pp. 294–300.

Oosterwegel, M. A., Mandelbrot, D. A., Boyd, S. D., Lorsbach, R. B., Jarrett, D. Y., Abbas, A. K. and Sharpe, A. H. (1999b). The role of CTLA-4 in regulating Th2 differentiation. *Journal of Immunology*, 163, pp. 2634–2639.

Ortiz de Montellano, P. R. (1986). In: *Cytochrome P450: Structure Mechanism and Biochemistry*, ed. Ortiz de Montellano P.R., Plenum Press, New York and London.

Ortiz de Montellano, P. R. (1995), In Ortiz de Montellano, ed., *Cytochrome P450 Structure, Mechanism and Biochemistry*, 2nd ed., Plenum Press, New York, pp. 201.

Oster, S. K., Ho ,C. S., Socie, E. L. and Penn, L. Z. (2002). The myc oncogene: Marvelousl Y complex, *Advanced Cancer Research*, 84, pp. 81–154.

Owaku, K., Shinohara, H., Ikariyama, Y. and Aizawa M. (1989). Preparation and characterization of protein Langmuir-Blodgett films, *Thin Solid Films*, 180, pp. 61–64.

Ozaki, M., Peebles, D., Weinberger, B. R., Heeger, A. J. and MacDiarmid, A. G. (1980). Semiconductor properties of polyacetylene p–(CH)x: n–CdS heterojunctions, *Journal of Applied Physics*, 51, pp. 4252–4256.

Paddeu, S., Antolini, F., Dubrovsky, T. and Nicolini, C. (1995b). Langmuir-Blodgett film of glutathione S-transferase immobilised on silanized surfaces, *Thin Solid Films*, 268, pp. 108–113.

Paddeu, S., Fanigliulo, A., Lanzi, M., Dubrovsky, T. and Nicolini, C. (1995a). LB-based PAB immunosystem: activity of immobilized urease monolayer, *Sensors and Actuators B: Chemical*, 25, pp. 876–882.

Paddeu, S., Erokhin, V. and Nicolini, C. (1996). Kinetics study of glutathione S-transferase Langmuir–Blodgett films, *Thin Solid Films*, 284–285, pp. 854–858.

Paddeu, S., Ram, M. K. and Nicolini, C. (1997). Investigation of ultrathin films of processable poly(o–anisidineconducting polymer obtained by the Langmuir-Blodgett technique, *The Journal of Physical Chemistry B*, 101, pp. 4759–4766.

Paddeu, S., Ram, M. K., Carrara, S. and Nicolini, C. (1998). Langmuir-Schaefer films of poly(o –anisidine) conducting polymer for sensors and displays, *Nanotechnology*, 9, pp. 228–236.

Padhi, A. K., Nanjundaswamy, K. S. and Goodenough, J. B. (1997). Phospho-olivines as positive–electrode materials for rechargeable lithium batteries, *Journal of the Electrochemical Society*, 144, pp. 1188–1194.

Pandey, S. S., Ram, M. K., Srivastava, V. K. and Malhotra, B. D. (1997). Electrical properties of metal (indium)/polyaniline Schottky devices, *Journal of Applied Polymer Science*, 65, pp. 2745–2748.

Papageorgiou, N., Athanassov, Y., Armand, M. and Bonhote, P. (1996). The performance and stability of ambient temperature molten salts for solar cell applications, *Journal of Electrochemistry Society*, 143, pp. 3099–3108.

Parmar, A., Kumar, H., Marwahe, S. S. and Kennedy, J. F. (2000). Advances in enzymatic transformation of penicillins to 6–aminopenicillanic acid (6–APA), *Biotechnology Advances*, 18, pp. 289–301.

Pasquadi, M., Passerini, S. and Pistoia, G. (2004). In: G.A. Nazri, G. Pistoia (Eds.), *Lithium Batteries, Science and Technology*, Kluwer Academic Publishers, Boston, NY, London, p. 347 (Chapter 11).

Pastorino, L., Disawal, S., Nicolini, C., Lvov, Y. M. and Erokhin, V. V. (2003). Complex catalytic colloids on the basis of firefly luciferase, *Biotechnology and Bioengineering*, 84, pp. 286–291.

Pastorino, L., Pioli, F., Zilli, M., Converti, A. and Nicolini, C. (2004). Lipase–catalyzed degradation of poly(epsilon–caprolactone), *Enzyme and Microbial Technology*, 35, pp. 321–326.

Paternolli, C., Ghisellini, P. and Nicolini, C. (2002a). Development of immobilization techniques of cytochrome P450-GST fusion protein, *Colloid and Surface B*, 23, pp. 305–311.

Paternolli, C., Ghisellini, P. and Nicolini, C. (2002b). Pollutant sensing layer based on cytochrome P450, *Materials Science and Engineering C*, 22, pp. 155–159.

Paternolli, C., Antonini, M., Ghisellini, P. and Nicolini, C. (2004). Recombinant cytochrome P450 immobilization for biosensor applications, *Langmuir*, 20, pp. 11706–11712.

Paternolli, C., Ghisellini, P. and Nicolini C. (2007). Nanostructuring of heme-proteins for biodevice applications, *IET Nanobiotechnology*, 1, pp. 22-26.

Paternolli, C., Neebe, M., Stura, E., Barbieri, F., Ghisellini, P., Hampp, N. and Nicolini, C. (2008). Photoreversibility and photostability in films of octopus rhodopsin isolated from octopus photoreceptor membranes, *Journal of Biomedical Materials Research Part A*, in press.

Patro, S. Y. and Przybycien, T. M. (1996). Simulations of reversible protein aggregate and crystal structure, *Biophysical Journal*, 70, pp. 2888–2902.

Paul, E. M., Ricco, A. J. and Wrighton, M. S. (1985). Resistance of polyaniline films as a function of electrochemical potential and the fabrication of polyaniline–based microelectronic devices, *Journal of Physical Chemistry*, 89, pp. 1441–1447.

Pearson, D. L., Schumm, J. S. and Tour, J. M. (1994). Iterative divergent convergent approach to conjugated oligomers by a doubling of molecular length at each iteration - a rapid route to potential molecular wires, *Macromolecules*, 27, pp. 2348–2350.

Pease, A. C., Solas, D., Sullivan, E. J., Cronin, M. T., Holmes, C. P. and Fodor, S. P. (1994). Light-generated oligonucleotide arrays for rapid DNA sequence analysis, *Proc. Natl. Acad. Sci.*, 91, pp. 5022–5026.

Pebay-Peyroula, E., Rummel, G., Rosenbusch, J. P. and Landau, E. M. (1997). X-ray structure of bacteriorhodopsin at 2,5 Å from microcrystals grown in lipidic cubic phases, *Science* 227, pp. 1676–1681.

Pechkova, E. and Nicolini, C. (2001). Accelerated protein crystal growth onto the protein thin film, *Journal of Crystal Growth* 231, pp. 599–602.

Pechkova, E. and Nicolini, C. (2002). Protein nucleation and crystallization by homologous protein thin film template, *Journal of Cellular Biochemistry*, 85, pp. 243–251.

Pechkova, E. and Nicolini, C. (2002a). From art to science in protein crystallization by means of thin film technology, *Nanotechnology*, 13, pp. 460–464.

Pechkova, E. and Nicolini, C. (2003). *Proteomics and Nanocrystallography*, New York: Kluwer-Plenum, pp. 1–210.

Pechkova, E. and Nicolini, C. (2004). Atomic structure of a CK2α human kinase by microfocus diffraction of extra-small microcrystals grown with nanobiofilm template, *Journal of Cellular Biochemistry,* 91, pp. 1010–1020.

Pechkova, E. and Nicolini, C. (2004a). Protein nanocrystallography: a new approach to structural proteomics, *Trends in Biotechnology*, 22, pp. 117–122.

Pechkova, E. and Nicolini, C. (2004b). From art to science in protein crystallography by means of nanotechnology – one year later. In: *Trends in Nanotechnology Research*, Nova Science Publishers, pp. 31–50.

Pechkova, E. and Nicolini, C. (2005). Synchrotron radiation and nanobiosciences – introductory overview, *Journal of Synchrotron Radiation,* 12, pp. 711.

Pechkova, E. and Nicolini, C. (2006). Structure and growth of ultrasmall protein microcrystals by synchrotron radiation: II. μGISAXS and microscopy of lysozyme, *Journal of Cellular Biochemistry*, 97, pp. 553–560.

Pechkova, E., Zanotti, G. and Nicolini, C. (2003). Three-dimensional atomic structure of a catalytic subunit mutant of human protein kinase CK2, *Acta Crystallographica D*, 59, pp. 2133–2139.

Pechkova, E., Tropiano, G., Riekel, C. and Nicolini, C. (2004) Radiation stability of protein crystals grown by nanostructured templates: synchrotron microfocus analysis, *Spectrochimica Acta*, 59, pp. 1687–1693.

Pechkova, E., Fiordoro, S., Fontani, D. and Nicolini, C. (2005). Investigating crystal-growth mechanisms with and without LB template: protein transfer from LB to crystal, *Acta Crystallographica D*, 61, pp. 809–812.

Pechkova, E., Roth, S. V., Burghammer, M., Fontani, D., Riekel, C. and Nicolini C., (2005a). μGISAXS and protein nanotemplate crystallization methods and instrumentation, *Journal of Synchrotron Radiation*, 12, pp. 713–716.

Pechkova, E., Vasile, F., Spera, R., Fiordoro, S. and Nicolini, C. (2005b). Protein nanocrystallography growth mechanism and atomic structure of crystals induced by nanotemplates, *Journal of Synchrotron Radiation*, 12, pp. 772–778.

Pechkova, E., Sivozhelezov, V., Tropiano, G., Fiordoro, S. and Nicolini C. (2005c) Comparison of lysozyme structures derived from thin film-based and classical crystals, *Acta Crystallographica D*, 61, 803–808.

Pechkova, E., Innocenzi, P., Malfatti, L., Kidchob, T., Gaspa, L. and Nicolini, C. (2007a). Thermal stability of lysozyme Langmuir-Schaefer films by FTIR spectroscopy, *Langmuir*, 23, pp. 1147–1151.

Pechkova, E., Sartore, M., Giacomelli, L. and Nicolini, C. (2007b). Atomic force microscopy of protein films and crystals, *Review of Scientific Instruments*, 78, pp. 093704_1–093704–7.

Pechkova, E., Sivozhelezov, V. and Nicolini, C. (2007c). Protein thermal stability: the role of protein structure and aqueous environment, *Archives of Biochemistry and Biophysics*, 466, pp. 40–48.

Penn, S. G., He, L. and Natan, M. J. (2003). Nanoparticles for bioanalysis, *Current Opinion in Chemical Biology*, 7, pp. 609–615.

Pepe, I. M. and Nicolini, C. (1996). Langmuir-Blodgett films of photosensitive proteins, *Journal of Photochemistry and Photobiology B–Biology*, 33, pp. 191–200.

Pepe, I. M., Cugnoli, C. and Schwemer, J. (1990). Rhodopsin reconstitution in bleached rod outer segment membranes in the presence of a retinal–binding protein from the honeybee. *FEBS Letters*, 268, pp. 177–179.

Pepe, I. M., Ram, M. K., Paddeu, S., and Nicolini, C. (1998). Langmuir–Blodgett films of rhodopsin: an infrared spectroscopic study, *Thin Solid Films*, 327, pp. 118–122.

Pernecky, S. J. and Coon, M. J. (1996). N-terminal modifications that alter P450 membrane targeting and function, *Methods in Enzymology*, 272, pp. 25–34.

Persson, L. G., Berglundh, T., Sennerby, L. and Lindhe, J. (2001). Re–osseointegration after treatment of peri–implantitis at different implant surfaces. An experimental study in the dog, *Clinical Oral Implants Research*, 12, pp. 595–603.

Petrigliano, A., Tronin, A. and Nicolini, C. (1996). Deposition and enzymatic activity of Langmuir–Blodgett films of alkaline phosphatase, *Thin Solid Films*, 284–285, pp. 752–756.

Pickett, C. B. and Lu, A. Y. H. (1989), Glutathione S-transferases: gene structure, regulation and biological function, *Annual Review of Biochemistry*, 58, pp. 743–764.

Piras, L., Adami, M., Fenu, S., Dovis, M. and Nicolini, C. (1996). Immunoenzymatic application of a redox potential biosensor, *Analytica Chimica Acta*, 335, pp. 127–135.

Pirmohamed, M., Williams, D., Madden, S., Templeton, E. and Park, B. K. (1995). Metabolism and bioactivation of clozapine by human liver in vitro, *J Pharmacol Exp Ther*, 272, pp. 984–990.

Pogozheva, I. D., Lomize, A. L. and Mosberg, H. I., (1997). The transmembrane 7-alpha–bundle of rhodopsin: distance geometry calculations with hydrogen bonding constraints, *Biophysical Journal*, 72, pp. 1963–1985.

Prigodin, V. N., Efetov, K. B. and Iida, S. (1993). Statistics of conductance fluctuations in quantum dots, *Physical Review Letters*, 71, pp. 1230–1233.

Pu, W., He, X., Ren, J., Wan, C. and Jiang, C. (2005). Electrodeposition of Sn–Cu alloy anodes for lithium batteries, *Electrochimica Acta*, 50, pp. 4140–4145.

Pulsinelli, E., Vasile, F., Vergani, L., Parodi, S. and Nicolini, C. (2003). Structural investigation of proapoptotic peptide by CD and NMR spectroscopy, *Protein and Peptide Letters*, 10, pp. 541–549.

Rahden-Staron, I., Czeczot, H. and Szumilo, M. (2001). Induction of rat liver cytochrome P450 isoenzymes CYP1A and CYP2B by different fungicides, nitrofurans, and quercetin, *Mutation Research - Genetic Toxicology and Environmental Mutagenesis*, 498, pp. 57–66.

Raiteri, R., Grattarola, M., Butt, H. J. and Skladal, P. (2001). Micromechanical cantilever-based biosensors, *Sensors and Actuators B: Chemical*, 79, pp. 115–126.

Ram, M. K. and Nicolini, C. (2000). Thin conducting polymeric films and molecular electronics. In: *Recent Research Development in Physical Chemistry*, Kerala, India, Transworld Publishing 4, pp. 219–258.

Ram, M. K., Carrara, S., Paddeu, S., Maccioni, E. and Nicolini, C. (1997). Effect of annealing on physical properties of conducting poly(ortho-anisidineLangmuir-Blodgett films, *Thin Solid Films*, 302, pp. 89–97.

Ram, M. K., Maccioni, E. and Nicolini, C. (1997a). The electrochromic response of polyaniline and its copolymeric systems, *Thin Solid Films*, 303, pp. 27–33.

Ram, M. K., Mascetti, G, Paddeu, S, Maccioni, E, and Nicolini, C (1997b). Optical, structural and fluorescence microscopic studies on reduced from of polyaniline: the leucomeraldine, *Synthetic Metals*, 89, pp. 63–69.

Ram, M. K., Paddeu, S., Carrara, S., Maccioni, E. and Nicolini, C. (1997c). Poly(o–anisidineLangmuir–Schaefer films: fabrication and characterization, *Langmuir*, 13, pp. 2760–2765.

Ram, M. K., Adami, M., Sartore, M., Paddeu, S. and Nicolini, C. (1999). Comparative studies on Langmuir–Schaefer films of polyanilines, *Synthetic Metals*, 100, pp. 249–259.

Ram, M. K., Salerno, M., Adami, M., Faraci, P. and Nicolini, C. (1999a). Physical properties of polyaniline films: Assembled by the layer-by-layer technique, *Langmuir*, 15, pp. 1252–1259.

Ram, M. K., Adami, M., Faraci, P. and Nicolini, C. (2000). Physical insight in the in-situ self-assembled films of polypyrrole, *Polymer*, 41, pp. 7499–7505.

Ram, M. K., Adami, M., Paddeu, S. and Nicolini, C. (2000a). Nanoassembly of glucose oxidase on the in situ self-assembled electrochemical characterizations, *Nanotechnology*, 11, pp. 112–119.

Ram, M. K., Bertoncello, P., Ding, H., Paddeu, S. and Nicolini, C. (2001). Cholesterol biosensors prepared by layer by layer technique, *Biosensors & Bioelectronics*, 16, pp. 849–856.

Ram, M. K., Bertoncello, P. and Nicolini, C. (2001a). Langmuir-Schaefer films of processable poly(o-ethoxyaniline) conducting polymer: fabrication and characterization as sensor for heavy metals, *Electroanalysis*, 13, pp. 574–581.

Ramachandran, N., Hainsworth, E., Bhullar, B., Eisenstein, S., Rosen, B., Lau, A. Y., Walter, J. C. amd LaBaer, J. (2004). Self-assembling protein microarrays. *Science*, 305, pp. 86–90.

Ravelli, R. B. G., Leiros, H. K. S., Pan, B., Caffrey, M. and McSweeney, S. (2003). Specific radiation damage can be used to solve macromolecular crystal structures, *Structure*, 11, pp. 217–224.

Reed, M. A. (1999). Molecular-scale electronics, *Proceedings of the IEEE*, 87, pp. 652–658.

Reed, M. A., Zhou, C., Muller, C. J., Burgin, T. P. and Tour, J. M. (1997). Conductance of a molecular junction, *Science*, 278, pp. 252–254.

Richard, M. N. and Dahn, J. R. (1999). Accelerating rate calorimetry study on the thermal stability of lithium intercalated graphite in electrolyte. I. Experimental, *Journal of Electrochemistry Society*, 146, pp. 2068–2077.

Richmond, T. J., Finch, J. T., Rushton, B., Rhodes, D. and Klug, A. (1984). The structure of the nucleosome core particle at 7 Å resolution, *Nature*, 311, pp. 532–537.

Riekel, C. (2000). New avenues in X-ray microbeam experiments, *Reports on Progress in Physics*, 63, pp. 233–262.

Riekel, C., Burghammer, M. and Muller, M. (2000). Microbeam small-angle scattering experiments and their combination with microdiffraction, *Journal of Applied Crystallography*, 33, pp. 421–423.

Riley, J. D., Bavastrello, V., Covani, U., Barone, A. and Nicolini, C., (2005). An in-vitro study of the sterilization of titanium dental implants using low intensity UV–radiation, *Dental Materials,* 21, pp. 756–760.

Rodrigo, J. G., García-Martín, A., Sáenz, J. J. and Vieira, S. (2002). Quantum Conductance in Semimetallic Bismuth Nanocontacts, *Physics Review Letters*, 88, pp. 246801–246805.

Rom, I., Wachtler, M., Papst, I., Schmied, M., Besenhard, J. O., Hofer, F. and Winter, M. (2001). Electron microscopical characterization of Sn/SnSb composite electrodes for lithium–ion batteries, *Solid State Ionics*, 143, pp. 329–336.

Rossi Albertini, V., Appetecchi, G. B., Caminiti, R., Cillocco, F., Croce, F. and Sadun, C. (1997). Crystallization kinetics of PEO-alkaline perchlorate solutions observed by energy dispersive x-ray diffraction, *Journal of Macromolecular Science Part B*, 36, pp.629–641.

Roth, E. P. (1999). Abstract 388, In: The Electrochemical Society Meeting Abstracts, vol. 99–2, Honolulu, HI, October 17–22.

Roth, S. V., Burghammer, M., Riekel, C., Müller-Buschbaum, P., Diethert, A., Panagiotou, P. and Walter, H. (2003). Self-assembled gradient nanoparticle-polymer multilayers investigated by an advanced characterisation method: Microbeam grazing incidence X-ray scattering, *Applied Physics Letters*, 82, pp. 1935–1937.

Royant, A., Edman, K., Ursby, T., Pebay-Peyroula, E., Landau, E. M. and Neutze, R. (2000). Helix deformation is coupled to vectorial proton transport in the photocycle of bacteriorhodopsin, *Nature*, 406, pp. 645–648.

Royant, A., Nollert P., Edman K., Neutze, R., Landau, E. M., Pebay-Peyroula, E. and Navarro, J. (2001). X-ray structure of sensory rhodopsin II at 2.1 Å resolution, *Proceeding National Academy of Sciences USA*, 98, pp. 10131–10136.

Rupp, B. (2003). Maximum-likelihood crystallization, *Journal Structural Biology*, 142, pp. 162–169.

Salaneck, W. R. and Brédas, J. L. (1994). Conjugated polymers, *Solid State Communicatons*, 92, pp. 31–36.

Salerno, M., Sartore, M. and Nicolini, C. (1999). Towards a neural networks based on AFM, *Probe Microscopy*, 1, pp. 333–334.

Sambrook, J., Fritsch, E. and Maniatis, T. (1989). *Molecular Cloning*, Cold Spring Harbor Laboratory Press.

Sanchez–González, J., Ruiz–García, J. and Gálvez-Ruiz, M. J. (2003). Langmuir-Blodgett films of biopolymers: a method to obtain protein multilayers, *Journal of Colloid and Interface Science*, 267, pp. 286–293

Santner, H. J., Korepp, C., Winter, M., Besenhard, J. O. B. and Moller, K. C. (2004). In-situ FTIR investigations on the reduction of vinylene electrolyte additives suitable for use in lithium-ion batteries, *Analytical and Bioanalytical Chemistry*, 379, pp. 266–271.

Sarakonsri, T., Johnson, C. S., Hackney, S. A. and Thackeray, M. M. (2005). Solution route synthesis of InSb, Cu_6Sn_5 and Cu_2Sb electrodes for lithium batteries, *Journal of Power Sources*, 153, pp. 319–327.

Saridakis, E. and Chayen, N. E. (2003). Systematic improvement of protein crystals by determining the supersolubility curves of phase diagrams, *Biophysical Journal*, 84, pp. 1218–1222.

Saridakis, E. E. G., Stewart, P. D. S., Lloyd, L. F. and Blow, D. M. (1994). Phase-diagram and dilution experiments in the crystallization of carboxypeptidase-g(2), *Acta Crystallogr. D*, 50, pp. 293–297.

Sarkar, N., Ram, M. K., Sarkar, A., Narizzano, R., Paddeu, S., and Nicolini, C. (2000). Nanoassemblies of sulfonated polyaniline multilayers, *Nanotechnology*, 11, pp. 30–36.

Sartore, M., Adami, M. and Nicolini, C. (1992a). Computer simulation and optimization of a light addressable potentiometric sensor, *Biosensors & Bioelectronics*, 7, pp. 57–68.

Sartore, M., Adami, M., Baldini, E., Rossi, A. and Nicolini, C. (1992b). New measuring principles for LAPS devices, *Sensors and Actuators B*, 9, pp. 25–36.

Sartore, M., Adami, M., Nicolini, C., Bousse, L., Mostarshed, S. and Hafeman, D. (1992c). Minority carrier diffusion lenght effects on light-addressable potentiometric sensor (LAPSdevices). *Sensors and Actuators A*, 32, pp. 431–439.

Sartore, M., Pace, R., Faraci, P., Nardelli, D., Adami, M., Ram, M. K. and Nicolini, C. (2000). Controlled-atmosphere chamber for atomic force microscopy investigations, *Rev Sci Instrum*, 71, pp. 1–5.

Sauerbrey, G. Z. (1964). Messung von platten schwingungen sehr kleiner amplitude durch lichtstrommodulation, *Z. Phys.*, 178, pp. 457-462.

Sauter, C., Lorber, B., Kern, D., Cavarelli, J., Moras, D. and Giegé, R. (1999). Crystallogenesis studies on yeast aspartyl-tRNA synthetase. Use of phase diagram to improve crystal quality, *Acta Crystallogr D*, 55, pp. 149–156.

Scheffzek, K., Stephan, I., Jensen, O. N., Illenberger, D. and Gierschik, P. (2000). The Rac–RhoGDI complex and the structural basis for the regulation of Rho proteins by RhoGDI, *Nature Structural Biology*, 7, pp. 122–126.

Schlittler, R. R., Seo, J. W., Gimzewski, J. K., Durkan, C., Saifullah, M. S. M. and Welland, M. E. (2001). Single crystals of single-walled carbon nanotubes formed by self-assembly, *Science*, 292, pp. 1136–1139.

Scrosati, B. (2002). in: W.A. Van Schalkwijk, B. Scrosati (Eds.), Advances in Lithium–Ion Batteries, Kluwer Academic/Plenum Publishers, NY, Boston, London, p. 251 (Chapter 8).

See, E. G., Joannopoulos, J. D., Meade, R. D. andWinn, J. N. (1995). *Photonic crystals: molding the flow of light*, (Princeton, NJ: Princeton University Press).

Sennett, R. S. and Scott, G. D. (1950). The structure of evaporated metal films and their optical properties, *Journal of Optical Society American*, 40, pp. 203–211.

Sette, F., Ruocco, G., Krisch, M., Masciovecchio, C., Verbeni, R. and Bergmann, U. (1996). Transition from normal to fast sound in liquid water, *Physical Review Letters*, 77, pp. 83–86.

Shao–Horn, Y. and Middaugh, R. L. (2001). Redox reactions of cobalt, aluminum and titanium substituted lithium manganese spinel compounds in lithium cells, *Solid State Ionics*, 139, pp. 13–25.

Shen, Y., Safinya, C. R., Liang, K. S., Ruppert, A. F. and Rothshild, K. J. (1993). Stabilization of the membrane-protein bacteriorhodopsin to 140-degrees–c in 2-dimensional films, *Nature*, 336, pp. 48–50.

Shewale, J. G. and Sivaraman, H. (1989). Penicillin acylases: Enzyme. production and its application in the manufacture of 6-APA, *Process Biochemistry*, 8, pp. 146–154.

Shi, H .Z., Lan, T. T. and Pinnavaia, J., (1996). Interfacial effects on the reinforcement properties of polymer–organoclay nanocomposites, *Chemistry of Materials*, 8, pp. 1584–1584.

Shibata, A., Kohara, J., Ueno, S., Uchida, I., Mashimo, T. and Yamashita, T. (1994). Monovalent anions stabilize the structure of bacteriorhodopsin at the air-water interface, *Thin Solid Films*, 244, pp. 736–739.

Shibli, J. A., Martins, M. C., Nociti, F. H., Garcia, V. G. and Marcantonio, E. (2003). Treatment of ligature–induced peri–implantitis by lethal photosensitization and guided bone regeneration: A preliminary histologic study in dogs, *Journal of Periodontology*, 74, pp. 338–345.

Shimakawa, Y., Numata, T. and Tabuchi, J. (1997). Verwey-type transition and magnetic properties of the LiMn2O4 spinels, *Journal of Solid State Chemistry*, 131, pp. 138–143.

Shönenberger, C., van Houten, H. and Donkersloot, H. C. (1992a). Single-electron tunneling observed at room–temperatue by scanning tunneling microscopy, *Europhysics Letters*, 20, pp. 249–254.

Shönenberger, C., van Houten, H., Donkersloot, H. C., van der Putten, A. M. T. and Fokkink, L. G. J. (1992b). Single-electron tunneling up to room-temperature, *Physica Scripta*, T45, pp. 289–291.

Shumyantseva, V. V., Bulko, T. V, Usanov, S. A., Schmid, R. D. and Nicolini, C. (2001). Construction and characterization of bioelectrocatalytic sensors based on cytochromes P450, *Journal of Inorganic Biochemistry*, 87, pp. 185–190.

Shumyantseva, V., De Luca, G., Bulko, T., Carrara, S., Nicolini, C., Usanov, S. A. and Archakov, A. (2004). Cholesterol amperometric biosensor based on cytochrome P450scc, *Biosensors & Bioelectronics*, 19, pp. 971–976.

Shumyantseva, V. V., Carrara, S., Bavastrello, V., Riley, D. J., Bulko, T. V., Skryabin, K. G., Archakov, A. I. and Nicolini, C. (2005). Direct electron transfer between cytochrome P450scc and gold nanoparticles on screen-printed rhodium-graphite electrodes, *Biosensors & Bioelectronics*, 21, pp. 217–222.

Shumyantseva, V. V., Ivanov, Y. D., Bistolas, N., Scheller, F. W., Archakov, A. I. and Wollenberger, U. (2004). Direct electron transfer of cytochrome P4502B4 at electrodes modified with nonionic detergent and colloidal clay nanoparticles, *Analytical Chemistry*, 76, pp. 6046–6052.

Simonyan, V. V., Diep, P., Johnson, J. K., (1999). Molecular simulation of hydrogen adsorption in charged single–walled carbon nanotubes, *Journal of Chemical Physics*, 111, pp. 9778–9783.

Siodmiak J., Gadomski A., Pechkova E., Nicolini C., Computer model of a lysozyme crystal growth with/without nanotemplate – a comparison, *International Journal of Modern Physics C* 17, 1359–1366, 2006.

Sivozhelezov, V. and Nicolini, C. (2005). Homology modeling of cytochrome P450scc and the mutations for optimal amperometric sensor, *Journal of Theoretical Biology*, 234, pp. 479–485.

Sivozhelezov, V. and Nicolini, C. (2006). Theoretical framework for octopus rhodopsin crystallization, *Journal of Theoretical Biology*, 240, pp. 260–269.

Sivozhelezov, V. and Nicolini, C. (2007). Prospects for octopus rhodopsin utilization in optical and quantum computation, *Physics of Particles and Nuclei Letters*, 4, pp. 189–196.

Sivozhelezov, V., Giacomelli, L., Tripathi, S. and Nicolini, C. (2006a). Gene expression in the cell cycle of human T lymphocytes: Predicted gene and protein networks, *Journal of Cellular Biochemistry*, 97, pp. 1137–1150.

Sivozhelezov, V., Pechkova, E. and Nicolini, C. (2006b). Mapping electrostatic potential of a protein on its hydrophobic surface: Implications for crystallization of cytochrome P450scc, *Journal of Theoretical Biology*, 241, pp. 73–80.

Sivozhelezov, V., Braud, C., Giacomelli, L., Pechkova, E., Giral, M., Soulillou, J. P., Brouard, S. and Nicolini C. (2008). Immunosuppressive drug-free operational immune tolerance in human kidney transplants recipients: II. Nonstatistical gene microarray analysis, *Journal of Cellular Biochemistry*, 103, pp. 1693–1706.

Smotkin, E. S., Lee, C., Bard, A. J., Campion, A., Fox, M. A., Mallouk, T. E., Webber, S. E. and White, J. M. (1988). Size quantization effects in cadmium sulfide layers formed by a Langmuir-Blodgett technique, *Chemical Physics Letters*, 152, pp. 265–268.

Soo, P. P., Huang, B. Y., Jang, Y. I., Chiang, Y. M., Sadoway, D. R. and Mayes, A. M. (1999). Rubbery block copolymer electrolytes for solid–state rechargeable lithium batteries, *Journal of Electrochemistry Society*, 146, pp.32.

Sotiropoulou, S., Gavalas, V., Vamvakaki, V. and Chaniotakis, N. A. (2003). Novel carbon materials in biosensor systems, *Biosensors and Bioelectronics*, 18, pp. 211–215.

Sotton, A.P. (1996). Deformation mechanisms, electronic conductance and friction of metallic nanocontacts, *Current Opinion in Solid State & Materials Science*, 1, pp. 827–833.

Southern, E. M. (1996). DNA chips: analysing sequence by hybridization to oligonucleotides on a large scale, *Trends Genetics*, 12, pp. 110–115.

Spera, R. and Nicolini, C. (2007). cAMP induced alterations of chinese hamster ovary cells monitored by mass spectrometry, *Journal of Cellular Biochemistry*, 102, pp. 473–482.

Spera, R. and Nicolini, C. (2008). Nappa microarrays and mass spectrometry: new trends and challenges, *Essential in Nanoscience Booklet Series* (Taylor & Francis Group/CRC Press).

Spera R., Bruzzese, D., Vasile, F. and Nicolini C. (2007). Correlation of changes of CHO-K1 cells metabolism to changes in protein expression in c-AMP differerentiation, *International Journal of Mass Spectrometry*, submitted.

Steuerman, D. W., Star, A., Narizzano, R., Choi, H., Ries, R. S., Nicolini, C., Stoddart, F. and Heat, J. R. (2002). Interaction between conjugated polymers and single-walled carbon nanotubes, *The Journal of Physical Chemistry B*, 106, pp. 3124–3130.

Stone, A. D., Jalabert, R. A. and Alhassid, Y. (1992). In *"Proceedings of the 14th Taniguchi Symposium"*, Springer-Verlag, Berlin, pp. 39.

Storrs, M., Merhl, D. J. and Walkup, J. F. (1996). Programmable spatial filtering with bacteriorhodopsin, *Applied Optics*, 35, pp. 4632–4636.

Strathmann, M. and Simon, M. I. (1990). G protein diversity: A distinct class of alpha subunits is present in vertebrates and invertebrates. *Proc. Natl. Acad. Sci. USA*, 87, pp. 9113–9117.

Straume, M. (2004). DNA microarray time series analysis: automated statistical assessment of circadian rhythms in gene expression patterning, *Methods Enzymology*, 383, pp. 149–166.

Stura, E., Erokhin, V. and Nicolini, C. (2002). Hybrid organic–inorganic electrolytic capacitors, *IEEE Transaction on Nanosciences*, 1, pp. 141–145.

Stura, E., Carrara, S., Bavastrello, V., Bertolotti, F. and Nicolini, C. (2004). Hydrogen storage as stabilization for wind power: completely clean system for insulated power generation. In: Hage When, Why – International Conference. *Chemical Engineering Transactions*, 4, pp. 317–323, (ed. Sauro Pierucci), AIDIC, Italy,

Stura, E. and Nicolini, C. (2006). New nanomaterials for light weight lithium batteries, *Analytica Chimica Acta*, 568, pp. 57–64.

Stura, E. Bruzzese, D., Grasso,, Valerio, Perlo, P. Nicolini, C. (2007), Anodic porous alumina as mechanical stability enhancer for Ldl–cholesterol sensitive electrodes, *Biosensors & Bioelectronics*, 23, pp. 655–660.

Su, W. P., Schrieffer, J. R. and Heeger, A. J. (1979). Solitions in polyacetylene, *Physical Review Letters*, 42, pp. 1698-1701.

Sugiyama, Y., Inoue, T., Ikematsu, M., Iseki M. and Sekiguchi, T. (1997). Determination of the amount of native structural bacteriorhodopsin in purple membrane Langmuir–Blodgett films by a spectroscopic surface denaturation quantifying technique, *Biochem Biophys Acta – Biomembranes*, 1326, 138–148.

Sukhorukov, G. B., Erokhin, V. V. and Tronin, A. Y. (1993). Formation and investigation of Langmuir films of nucleic acid octadecylamine complexes, *Biofizika*, 38, pp. 257–262.

Sumetskii, M. (1993). Resistance resonances for resonant-tunneling structures of quantum dots, *Phys Rev B*, 48, pp. 4586–4591.

Sun, J., Forsyth, M. and MacFarlane, D. R. (1998). Room-temperature Molten salts based on the quaternary ammonium ion, *The Journal of Physical Chemistry B*, 102, pp. 8858–8864.

Sutherland, R. L. and Musgrove, E. A. (2004). Cyclins and breast cancer, *J. Mammary Gland. Biol. Neoplasia*, 9, pp. 95–104.

Suzuki, Y., Saito, J., Horii, Y. and Kasagi, N. (2003). Development of micro catalytic combustor with Pt/Al_2O_3 thin films *The Int. Symp. on Micro-Mechanical Engineering*, ISMME 2003 (Tsuchiura, December 2003), pp 171–176.

Swierczek, K., Marzec, J., Marzec, M. and Molenda, J. (2003). Crystallographic and electronic properties of Li1–delta Mn2–delta O4 spinels prepared by HT synthesis, *Solid State Ionics*, 157, pp. 89–93.

Swift, J. Q., Jenny, J. E. and Hargreaves, K. M. (1995). Heat–generation in hydroxyapatite–coated implants as a result of CO_2–laser application, *Oral Surgery Oral Medicine Oral Pathology Oral Radiology and Endodontics*, 79, pp. 410–415.

Syljuasen, R. G., Sorensen, C. S., Hansen, L. T., Fugger, K., Lundin, C., Johansson, F., Helleday, T., Sehested, M., Lukas, J. and Bartek, J. (2005). Inhibition of human Chk1 causes increased initiation of DNA replication, phosphorylation of ATR targets, and DNA breakage, *Molecular Cellular Biology*, 25, pp. 3553–3562.

Szilagyi, A. and Zavodsky, P. (2000). Structural differences between mesophilic, moderately thermophilic and extremely thermophilic protein subunits: results of a comprehensive survey, *Structure*, 8, pp. 493–504.

Takahashi, M., Tobishima, S., Takei, K. and Sakurai, Y. (2002). Reaction behavior of $LiFePO_4$ as a cathode material for rechargeable lithium batteries, *Solid State Ionics*, 148, pp. 283–289.

Tarascon, J. M., McKinnon, W. R., Coowar, F., Bowmer, T. N., Amatucci, G. and Guyomard, D. (1994). Aynthesis conditions and oxygen stoichiometry effects on li insertion into the spinel LiMn2O4, *Journal of the Electrochemistry Society*, 141, pp. 1421–1431.

Tarascon, J. M., Wang, E., Shokoohi, F. K., McKinnon, W. R. and Colson, S. (1991). The spinel phase of limn2o4 as a cathode in secondary lithium cells, *Journal of the Electrochemistry Society*, 138, pp. 2859–2864.

Templin, M. F., Stoll, d., Schrenk, M., Traub, P. C., Vohringer, C. F. and Joos, T. O. (2002). Protein microarray technology, *Drug Discovery Today*, 7, pp. 815–822.

Thackeray, M. M., Vaughey, J. T. and Fransson, L. M. L. (2002). Recent developments in anode materials for lithium batteries, *Journal of the Minerals & Materials Society*, 54, pp. 20–23.

Thayer A. M., (2003). Industrial review, *Chemical Engineering News*, 81, pp. 18–26.

Thirunakaran, R., Babu, B. R., Kalaiselvi, N., Periasamy, P., Kumar, T. P., Renganathan, N. G., Raghavan, M. and Muniyandi, N. (2001). Electrochemical behaviour of LiMyMn2–yO4 (M = Cu, Cr; $0 <= y <= 0$ center dot 4), *Bulletin of Materials Science*, 24, pp. 51–55.

Thom, R., Cummin. I, Dixon, D. P., Edwards, R., Cole, D. J. and Lapthorn, A. J. (2002). Structure of a tau class glutathione S-transferase from wheat active in herbicide detoxification, *Biochemistry*, 41, pp. 7008–7020.

Thompson, G. E. andWood, G. C. (1983). *Treatise on materials science and technology*, Vol. 23, ed J. C. Scully (New York: Academic) p 205.

Thorsen, T., Maerkl, S. J. and Quake, S. R. (2002). Microfluidic Large-Scale Integration, *Science*, 298, pp. 580–584.

Tiede, D. (1985). Incorporation of membrane proteins into interfacial films: model membranes for electrical and structural characterization, *Biochim Biophys Acta – Rewiew on Bioenergetics*, 811, pp. 357–379.

Tiefenauer, L. X., Kossek, S., Pedeste, C. and Thiébaud, P. (1997). Towards amperometric immunosensor devices, *Biosensors and Bioelectronics*, 12, pp. 213–223.

Tjio, J.H. and Puck, T.T. (1958). Genetics of somatic mammalian cells: II. Chromosomal constitution of cells in tissue culture, *J. Exp. Med.*, 108, pp. 259–268.

Toeroe, I., Thore, S., Mayer, C., Basquin, J., Seraphin, B. and Suck, D. (2001). RNA binding in an Sm core domain: X-ray structure and functional analysis of an archaeal Sm protein complex. *The EMBO Journal*, 20, pp. 2293–2303

Tonge, J. S. and Shriver, D. F. (1987). Increased dimensional stability in ionically conducting polyphosphazenes systems, *Journal of Electrochemistry Society*, 134, pp. 269–270.

Torgler, R., Jakob, S., Ontsouka, E., Nachbur, U., Mueller, C., Green, D. R. and Brunner, T. (2004). Regulation of activation–induced Fas (CD95/Apo–1) ligand expression in T cells by the cyclin B1/Cdk1 complex, *Journal of Biological Chemistry*, 279, pp. 37334–37342.

Tostmann, H., Kropf, A. J., Johnson, C. S., Vaughey, J. T. and Thackeray, M. M. (2002). In situ X-ray absorption studies of electrochemically induced phase changes in lithium-doped InSb, *Physical Review B*, 66, art. No. 014106.

Trabanino, R. J., Hall, S. E., Vaidehi, N., Floriano, W. B., Kam, V. W. and Goddard, W. A. (2004). 3rd First principles predictions of the structure and function of g-protein-coupled receptors: validation for bovine rhodopsin, *Biophysical Journal*, 86, pp. 1904–1921.

Tredgold, R. H. and Smith, G. W. (1983). Surface potential studies on Langmuir-Blodgett multilayers and adsorbed monolayers, *Thin Solid Films*, 99, pp. 215–220.

Troitsky, V. I, Sartore, M., Berzina, T. S., Nardelli, D. and Nicolini, C. (1996b). Instrument for depositing Langmuir-Blodgett films composed of alternating monolayer using a protective layer of water, *Review of Scientific Instruments*, 67, pp. 4216–4223.

Troitsky, V. I., Berzina, T. S., Petrigliano, A. and Nicolini, C. (1996a). Deposition of alternating LB monolayers with a new technique, *Thin Solid Films*, 285, pp. 122–126.

Troitsky, V., Berzina, T. S., Pastorino, L., Bernasconi, L. and Nicolini C. (2003). A new approach to the deposition of nanostrructured biocatalytic films, *Nanotechnology*, 14, pp. 597–602.

Troitsky, V., Ghisellini, P., Pechkova, E. and Nicolini, C. (2002). DNASER II. Novel surface patterning for biomolecular microarray, *IEEE Transactions on Nanobioscience*, 1, pp. 73–77.

Tronin, A., Dubrovsky, T. and Nicolini, C. (1995). Comparative-study of langmuir monolayers of immunoglobulins-G formed at the air-water interface and covalently immobilized on solid supports, *Langmuir*, 11, pp. 385–389.

Tronin, A., Dubrovsky, T., De Nitti, C., Gussoni, A., Erokhin, V. and Nicolini, C. (1994). Langmuir-Blodgett films of immunoglobulines IgG - ellipsometric study of the deposition process and of immunological activity, *Thin Solid Films*, 238, pp. 127–132.

Tronin, A., Dubrovsky, T., Radicchi, G. and Nicolini, C. (1996). Optimisation of IgG Langmuir film deposition for application as sensing elements, *Sensors and Actuators B*, 34, pp. 276–282.

Tsuda, M., Arai, H., Takahashi, M., Ohtsuka, H., Sakurai, Y., Sumitomo, K. and Kageshima, H. (2005). Electrode performance of layered $LiNi_{0.5}Ti_{0.5}O_2$ prepared by ion exchange, *Journal of Power Sources*, 144, pp. 183–190.

Tuerk, C., Gold, L. (1990). Systematic evolution of ligands by exponential enrichment: RNA ligands to bacteriophage T4 DNA polymerase. *Science*, 249, pp. 505–510.

Ulman, A. (1991). *An introduction to ultrathin organic films from Langmuir-Blodgett to self-assembly*, Academic Press, Inc., New York.

Ulrich, R., Zwanziger, J. W., De Paul, S. M., Reiche, A., Leuninger, H., Spiess, H. W. and Wiesner, U. (2002). Solid hybrid polymer electrolyte networks: nano-structurable materials for lithium batteries, *Advanced Materials*, 14, pp. 1134–1137.

Usha, R., Johnson, J., Moras, D., Thierry, J. C., Fourme, R. and Kahn, R. (1984). Macromolecular crystallography with synchrotron radiation: collection and processing of data from crystals with a very large unit cell, *Journal of Applied Crystallography*, 17, pp. 147-153.

Valentini, L., Bavastrello, V., Armentano, I., D'Angelo, F., Pennelli, G., Nicolini, C. and Kenny, J. M. (2004b). Synthesis and electrical properties of CdS Langmuir-Blodgett multilayers nanoparticles on self–assembled carbon nanotubes, *Chemical Physics Letters*, 392, pp. 214–219.

Valentini, L., Bavastrello, V., Stura, E., Armentano, I., Nicolini C. and Kenny, J. M. (2004a). Sensors for inorganic vapor detection based on carbon nanotubes and poly(o–anisidine) nanocomposite material, *Chemical Physics Letters*, 383, pp. 617–622.

Vasile, F., Pechkova, E. and Nicolini C. (2008). Solution structure of the β-subunit of the translation initiation factor Aif2 from Archaebacteria Sulfolobus Solfataricus, *Proteins Structure, Function and Bioinformatics,* 70, pp. 1112–1115.

Vaughey, J. T., Fransson, L. M. L., Swinger, H. A., Edstrom, K. and Thackeray, M. M. (2003). Alternative anode materials for lithium–ion batteries: a study of Ag_3Sb, *Journal of Power Sources*, 119, pp. 64–68.

Vaz, A. D. N., McGinnity, G. N. and Coon, M. J. (1998). Epoxidation of olefins by cytochrome P450: Evidence from site-specific mutagenesis for hydroperoxo–iron as an electrophilic oxidant, *PNAS USA*, 95, pp. 3555–3560.

Vendrell, J. A., Magnino, F., Danis, E., Duchesne, M. J., Pinloche, S., Pons, M., Birnbaum, D., Nguyen, C., Theillet, C. and Cohen, P. A. (2004). Estrogen regulation in human breast cancer cells of new downstream gene targets involved in estrogen metabolism, cell proliferation and cell transformation, *J. Mol. Endocrinol.*, 32, pp. 397–414.

Vergani, L., Gavazzo, P., Facci, P., Diaspro, A., Mascetti, G., Arena, N., Gaspa, L. and Nicolini, C. (1992). Fluorescence cytometry of microtubules and nuclear DNA during cell-cycle and reverse-transformation, *Journal of Cellular Biochemistry*, 50, pp. 201–209.

Vergani, L., Mascetti, G. and Nicolini, C. (2001). Changes of nuclear structure induced by increasing temperatures. *Journal of Biomolecular Structure & Dynamics*, 18, pp. 535-544.

Versluijs, J. J., Bari, M., Ott, F., Coey, J. M. D. and Revcolevschi, A. (2000). Non-linear I-V curves in nanocontacts between crystals of $(La_{0.7}Sr_{0.3})MnO_3$, *Journal of Magnetism and Magnetic Materials*, 211, pp. 212–216.

Vestergaard, B., Bjerum, N. J., Petrushina, I., Hjuler, H. A., Gerg, R. W. and Begtrup, M. (1993). Molten triazolium chloride systems as new aluminum battery electrolytes, *Journal of Electrochemistry Society*, 140, pp. 3108–3113.

Vianello, F., Cambria, A., Ragusa, S., Cambria, M. T., Zennaro, L. and Rigo, A. (2004). A high sensitività amperometric biosensor using a monomolecular layer of laccase as biorecognition element, *Biosensors & Bioelectronics*, 20, pp. 315–321.

Vincent, C. A. and Scrosati, B. (1993). *Modern Batteries. An Introduction to Electrochemical Power Sources*, second ed., Arnold, London.

Vo–Dinh, T., Cullum, B. M. and Stokes, D. L. (2001). Nanosensors and biochips: frontiers in biomolecular diagnostics, *Sensors and Actuators B: Chemical*, 74, pp. 2–11.

von Mering, C., Jensen, L. J., Snel, B., Hooper, S. D., Krupp, M., Foglierini, M., Jouffre, N., Huynen, M. A. and Bork, P. (2005). STRING: known and predicted protein–protein associations, integrated and transferred across organisms, *Nucleic Acids Research*, 33, D433–D437.

von Sacken, U., Nodwell, E., Sundher, A. and Dahn, J. R. (1994). Comparative thermal stability of carbon intercalation anodes and lithium metal anodes for rechargeable lithium batteries, *Solid State Ionics*, 69, pp. 284–290.

Wachtler, M., Besenhard, J. O. and Winter, M. (2001). Tin and tin–based intermetallics as new anode materials for lithium-ion cells, *Journal of Power Sources*, 94, pp. 189–193.

Wada A., Mathew, P. A., Barnes, H. J., Sanders, D., Estabrook, R. W. and Waterman, M. R. (1991). Expression of functional bovine cholesterol side-chain cleavage cytochrome P450 (P450scc) in Escherichia coli, *Archives of Biochemistry and Biophysics*, 290, pp. 376–381.

Walsh, M. A., Dementieva, I., Evans, G., Sanishvili, R. and Joachimiak, A. (1999). Taking MAD to the extreme: ultrafast protein structure determination, *Acta Crystallographica Section D*, 55, pp. 1168–1173

Wang, Z., and Pinnavaia, T. J. (1998b). Hybrid organic–inorganic nanocomposites: exfoliation of magadiite nanolayers in an elastomeric epoxy polymer, *Chemistry of Materials*, 10 pp. 1820–1826.

Wang, Z., Sun, Y., Chen, L. and Huang, X. (2004). Electrochemical characterization of positive electrode material LiNi1/3Co1/3Mn1/3O2 and compatibility with electrolyte for lithium-ion batteries, *Journal of the Electrochemistry Society*, 151, pp. A914-921.

Watanabe, M., Nagano, S., Sanui, K., and Ogata, N. (1986). Ionic conductivity of network polymers from poly(ethylene oxide) containing lithium perchlorate, *Polymer Journal*, 18, pp. 809–817.

Waterman, M. R. and Simpson, E. R. (1985). Regulation of the biosynthesis of cytochromes P-450 involved in steroid hormone synthesis. *Mol. Cell. Endocrinol.*, 39, pp. 81–89.

Weik, M., Ravelli, R. B. G., Kryger, G., McSweeney, S., Raves, M. L., Harel, M., Gros, P., Silman, I., Kroon, J. and Sussman, J. L. (2000). Specific chemical and structural damage to proteins produced by synchrotron radiation, *Proc. Natl. Acad. Sci. USA*, 97, pp. 623–628.

Weston, J. E. and Steele, B.C.H. (1982). Effects of inert fillers on the mechanical and electrochemical properties of lithium salt-poly(ethylene oxide) polymer electrolytes, *Solid State Ionics*, 7, pp.75–79.

Wiechmann, M., Enders, O. Zeilinger, C. and Kolb, H.-A. (2001). Analysis of protein crystal growth at molecular resolution by atomic force microscopy. *Ultramicroscopy*, 86, pp. 159–166.

Wieczorek, W., Florjanczyk, Z. and Stevens, J. R. (1995). Composite polyether based solid electrolytes, *Electrochimica Acta*, 40, pp. 2251–2258.

Wieczorek, W., Such, K., Wycislik, H. and Plocharski, J. (1989). Modifications of crystalline structure of peo polymer electrolytes with ceramic additives, *Solid State Ionics*, 36, pp.255–257.

Wilkins, R., Ben–Jacob, E. and Jaklevic, R. C. (1989). Scanning–tunneling–microscope observations of Coulomb blockade and oxide polarization in small metal droplets, *Physics Review Letters*, 63, pp. 801–804.

Willbrand, K., Radvanyi, F., Nadal J. P., Thiery, J. P. amd Fink, T. M. (2005). Identifying genes from up-down properties of microarray expression series, *Bioinformatics*, 21, pp. 3859–3864.

Williams, G. M. (1977). Detection of chemical carcinogens by unscheduled DNA synthesis in rat liver primary cultures, *Cancer Research*, 37, pp. 1845–1851.

Williams, K. A., Veenhuizen, P. T., de la Torre, B. G., Eritja, R. and Dekker, C. (2002) Nanotechnology: carbon nanotubes with DNA recognition. *Nature*, *420*, pp. 761–761.

Williams, P. A., Cosme, J., Sridhar, V., Johnson, E. F. and McRee, D. E. (2000). Mammalian microsomal cytochrome P450 monooxygenase: structural adaptations for membrane binding and functional diversity, *Mol. Cell.*, 5, pp. 121–131.

Wilson, R. J., Meijer, G., Bethune, D. S., Johnson, R. D., Chamblis, D. D., De Vries, M. S., Hunziker, H. E. and Wendt, H. R. (1990). Imaging C_{60} clusters on a surface using a scanning tunnelling microscope, *Nature*, 348, pp. 621–622.

Winter, M. and Besenhard, J. O. (1999). Electrochemical lithiation of tin and tin-based intermetallics and composites, *Electrochimica Acta*, 45, pp. 31–50.

Winter, M. and Brodd, J. R. (2004). What are batteries, fuel cells, and supercapacitors?, *Chemical Reviews*, 104, pp. 4245–4269.

Wise, K. J., Gillespie, N. B., Stuart, J. A., Krebs, M. P. and Birge, R. R. (2002). Optimization of bacteriorhodopsin for bioelectronic devices, *Trends in Biotechnology*, 20, pp. 387–394.

Wolde, P. R. and Frenkel, D. (1997). Enhancement of protein crystal nucleation by critical density fluctuation, *Science*, 277, pp. 1975–1978.

Wrighton, S.A. and Stevens, J.C., (1992). The hepatic cytochromes P450 involved in drug metabolism, *Critical Reviews in Toxicology*, 22, pp. 1–21.

Wu, L. Q. and Payne, G. F. (2004). Biofabrication: using biological materials and biocatalysts to construct nanostructured assemblies, *Trends in Biotechnology*, 22, pp. 593–599.

Wu, R., Schumm, J. S., Pearson, D. L. and Tour, J. M. (1996). Convergent synthetic routes to orthogonally fused conjugated oligomers directed toward molecular scale electronic device applications, *Journal of Organic Chemistry*, 61, pp. 6906–6921.

Wu, T. X., Liu, G. M., Zhao, J. C., Hidaka, H. and Serpone, N. (1998). Photoassisted degradation of dye pollutants. V. Self- photosensitized oxidative transformation of Rhodamine B under visible light irradiation in aqueous TiO_2 dispersions. *Journal of Physical Chemistry B*, 102, pp. 5845–5851.

Xia, D. W. and Smid, J. (1984). Solid polymer electrolyte complexes of polymethacrylates carrying pendant oligo-oxyethylene(glyme) chains, *Journal of Polymer Science: Polymer Letters,* 22, pp. 617–621.

Xie, J., Zhao, X. B., Cao, G. S., Zhao, M. J. and Su, S. F. (2005). Solvothermal synthesis and electrochemical performances of nanosized $CoSb_3$ as anode materials for Li–ion batteries, *Journal of Power Sources*, 140, pp. 350–354.

Xu, K., Zhang, S., Jow, T. R., Xu, W. and Angell, C. A. (2002). Lithium bis(oxalato)borate stabilizes graphite anode in propylene carbonate, *Electrochemical and Solid State Letters*, 5, pp. A259–A262.

Yablonovitch, E., (1987). Inhibited spontaneous emission in solid-state physics and electronics, *Physics Review Letters*, 58, pp. 2059–2062.

Yamada, A., Chung, S. C. and Hinokuma, K. (2001). Optimized LiFePO$_4$ for lithium battery cathodes, *Journal of the Electrochemical Society*, 148, pp. A224–A229.

Ye, Y., Ahn, C., Witham, C., Fultz, B., Liu, J., Rinzler, A. G., Colbert, D., Smith, K A., and Smalley, R. E. (1999). Hydrogen adsorption and cohesive energy of single-walled carbon nanotubes, *Applied Physics Letters*, 74, pp. 2307–2309.

Yin, J., Wada, M., Tanase, S. and Sakai, T. (2004). Electrode properties and lithiation/delithiation reactions of Ag–Sb–Sn nanocomposite anodes in Li-ion batteries, *Journal of the Electrochemical Society*, 151, pp. A867–A872.

Yoneda, Y. (1963). Anomalous surface reflection of X-rays, *Physical Review*, 161, pp. 2010–2013.

Yoon, W. S., Grey, C. P., Balasubramanian, M., Yang, X.Q. and McBreen, J. (2003). In situ X–ray absorption spectroscopic study on LiNi$_{0.5}$Mn$_{0.5}$O$_2$ cathode material during electrochemical cycling, *Chemistry of Materials*, 15, pp. 3161–3169.

Yurke, B., Turberfield, A. J., Mills, A P., Simmel, F. C. and Neumann, J. L. (2000). A DNA-fuelled molecular machine made of DNA, *Nature*, 406, pp. 605–608

Zeisel, D. and Hampp, N. (1992). Spectral relationship of light-induced refractive-index and absorption changes in bacteriorhodopsin films containing wildtype brwt and the variant BRD96N, *Journal of Physical Chemistry*, 96, pp. 7788–7792.

Zeisel, D. and Hampp, N. (1996). Bacteriorhodopsin applications in optical information processing. In: *Molecular Manufacturing*, EL.B.A. Forum Series, (C. Nicolini, ed.) New York: Plenum, Vol. 2, pp. 175–188.

Zemansky, M. W. and Dittman, R. H. (1997). *Heat and Thermodynamics*, 7th ed., McGraw–Hill, Boston.

Zhang, F., Basinski, M. B., Beals, J. M., Briggs, S. L., Churgay, L. M., Clawson, D. K., DiMarchi, R. D., Furman, T. C., Hale, J. E., Hsiung, H. M., Schoner, B. E., Smith, D. P., Zhang, X. Y., Wery, J. P. and Schevitz, R. W. (1997a). Crystal structure of the obese protein leptin–E100, *Nature*, 387, pp. 206–209.

Zhang, S. S., Xu, K. and Jow, T. R. (2003). Tris(2,2,2-trifluoroethyl) phosphite as a co–solvent for nonflammable electrolytes in Li-ion batteries, *Journal of Power Sources*, 113, pp. 166–172.

Zhang, S., Marini, D. M., Hwang, W. and Santoso, S. (2002). Design of nanostructured biological materials through self-assembly of peptides and proteins, *Current Opinion in Chemical Biology*, 6, pp. 865–871.

Zhang, Y. and Urquidi-Macdonald, M. (2005). Hydrophobic ionic liquids based on the 1-butyl-3-methylimidazolium cation for lithium/seawater batteries, *Journal of Power Sources*, 144, pp. 191–196.

Zhang, Z., Fouchard, D. and Rea, J. R. (1998). Differential scanning calorimetry material studies: implications for the safety of lithium-ion cells, *Journal of Power Sources*, 70, pp. 16–20.

Zhao, X. and Zhang, S. (2004). Fabrication of molecular materials using peptide construction motifs, *Trends in Biotechnolgy*, 22, pp. 470–476.

Zhou, M., Morais–Cabral, j. H., Mann, S. and MacKinnon, R. (2000). Potassium channel receptor site for the inacivation gate and quaternary amine inhibitors, *Nature*, 411, pp. 657–661.

Zietz, S., Belmont, A. and Nicolini, C. (1983). Differential scatering of circularly polarized light as a unique probe of polynucleosome superstructures. A simulation by multiple scattering of dipoles, *Cell Biophyics*, 5, pp. 163–187.

Zouni, A., Witt, H. T., Kern, J., Fromme, P., Krauss, N., Saenger, W. and Orth, P. (2001). Crystal strucure of photosystem II from Synechococcus elongatus at 3.8 angstrom resolution, *Nature*, 409, pp. 739–743.

Zylberajch, C., Ruaudel-Teixier, A. and Barraud, A. (1989). Properties of inserted mercury sulphide single layers in a Langmuir-Blodgett matrix, *Thin Solid Films*, 179, pp. 9-14.

Index

adrenodoxine, 10
AFM, *see* atomic force microscopy
alcohol dehydrogenase, 311
alkaline phosphatase, 302; 311
amperometric sensor, 226-227
anodic porous alumina (APA), 48-53; 81, 230-232; 307
APA, *see* anodic porous alumina
archaea, 16
atomic force microscopy (AFM), 54; 58-65; 88; 115-116; 118; 127; 230
ATP, 179

bacteriorhodopsin (bR), 6-7; 13; 43-45; 93-95; 97; 100; 110; 220-222; 232-236; 261-263; 279-281
BAM, *see* Brewster angle microscopy
batteries, 251; 275; 282-290; 292-296
bioactuators, 312
biocatalysis, 310; 311
bioinformatics, 67; 90-91; 93; 158; 160; 164-166; 172; 174
biomaterials, 1; 306
biosensor, 183; 186; 190; 223-227; 229; 232; 237
Bone Morphogenetic Proteins (BMP), 14
bovine rhodopsin, 13; 93; 95; 97
bovR, *see* bovine rhodopsin
Bragg reflections, 108; 111; 235

Brewster angle microscopy (BAM), 36; 70-71

cadmium arachidate, 34-35; 66; 267-269; 274
cadmium sulphide, 267
cAMP, 101; 103-104; 175-180
cantilever, 183
capacitor, 246; 250-253; 290
carbon nanotubes (NT), 148-151; 184; 241; 299; 301
cardiomyocytes, 148-149; 151
cardiovascular cells, 148
CCD, 36; 159
CD, *see* circular dichroism
CdA, 36
CdS, 18; 35; 66; 248; 255; 267-270; 274; 276
CHO-K1, 2-3; 5; 101-104; 175-178
cholesterol, 72-73; 223-226; 229; 230-232
chromatin, 197-199
chromatography, 5; 11
circular dichroism, 44-45; 61-62; 69-70
CK2α, *see* human kinase CK2α
clustering algorithms, 91
conducting polymer, 18-19; 25; 27-31; 38; 67; 258; 265; 277
contact mode, 58-59
Coulomb Blockade, 268-271

crystal growth, 113; 115
crystallization, 59; 93; 95-98; 100;
    113-116; 118; 121; 123-124;
    126-127; 129-132; 134; 136-137;
    141-145; 211; 217
crystallography, 118; 131-132; 136;
    157; 197-198; 202; 216; 283
CV, see cyclic voltammetry; 72
cyclic voltammetry (CV), 72-73;
    225-226; 231
cytochrome C, 74-75

dental implants, 152
differential scanning calorimetry
    (DSC), 213
diffraction patterns, 96; 109-110; 119;
    130; 144
diffractometry, 108
DNA microarray, 15; 51; 81-83; 93;
    158-159; 174
DNASER, 82-83; 88; 159

ELEctrochemical NAnodeposition
    (ELENA), 187; 189
electron spray ion (ESI), 101-104; 175;
    177-179
ELENA, see ELEctrochemical
    NAnodeposition
ellipsometry; 71
ESI, see electron spray ion

FIB, see focused ion beam
focused ion beam (FIB), 189
Fourier transform infrared spectroscopy
    (FTIR), 74-76
FTIR, see Fourier transform infrared
    spectroscopy
fuel cells, 275; 285; 296
fullerene, 17; 18; 277; 300

G0, 159-160
G1, 159; 161-164; 166; 174

G2, 159-163; 165
gel electrophoresis, 12
gene expression, 158-160; 166; 168;
    171; 173; 198
glutathione-S-transferase (GST), 11;
    13; 223; 301-303; 305; 311
gold nanoparticles, 228-230; 232
GST, see glutathione-S-transferase

H9c2, 2-3; 148-151
HeLa, 2
histone, 199-200
homology modeling, 64; 93-94; 97-98
HPLC, 10; 101-104; 175; 177-180
human kinase CK2α, 77; 109; 120;
    125; 129-131
human T lymphocytes, 4; 84; 91-93;
    158-167; 173
hydrogen storage, 296-298

infrared spectroscopy (IR), 73; 285
IR, see infrared spectroscopy

kidney transplant, 84
Kiessig fringes, 108; 111

Label Free, 65
Langmuir-Blodgett (LB), 2; 13; 34-35;
    37-47; 54-55; 57; 59-61; 63; 66; 69;
    72-73; 75; 77-78; 97; 100; 107-108;
    111; 116 124-125; 127; 135; 145;
    203; 213-218; 223-226; 228; 233,
    235; 237; 247-249; 255-256; 259;
    264; 266-269; 277-279; 301- 308;
    310; 312
Langmuir-Schaefer (LS), 24-25; 29;
    36; 39-41; 47; 70-71; 73; 76; 137;
    141;
layer-by-layer (LBL), 37-38; 223; 237;
    259
LB, see Langmuir-Blodgett
LBL, see layer-by-layer

LDL-cholesterol, 53
leader genes, 91-93; 158-163; 166-169; 171-173
LED, see light emitting devices
light emitting devices (LED), 19; 25; 256-261
lipase, 306-307
LS, see Langmuir-Schaefer
lysozyme, 13; 59-61; 63; 76; 102; 124-125; 127; 133; 135; 139; 142; 145-147; 217

MALDI TOF, 87; 89; 102; 105-106; 177; 181
mass spectrometry (MS), 5-6; 87-90; 93; 101-106; 174-175; 177-181
metalloproteins, 6; 8; 10; 13
micro Grazing Incidence Small Angle X-ray Scattering (μGISAXS), 54; 120; 131; 135-137; 140-143; 145-147
microarray, 50-51; 65; 81-82; 84-89; 91-93; 105-106; 158; 160; 167; 169-173; 180-182
microcrystals; 63; 96; 98; 109-111; 116; 119-120; 130; 136; 139; 143
microfocus; 98; 109-110; 119-120; 127; 129-131; 136; 137; 145
molecular modelling, 78; 90; 93
MS, see mass spectrometry
multi-walled nanotubes (MWNT), 26; 243; 299

nanoarrays, 132
nanobiofilm, 127; 130; 132; 143; 147
nanobiotechnology, 10; 90; 93; 100; 220; 229; 255; 275; 282
nanocomposites, 17; 25; 27
nanocontacts, 184-190
nanocrystallography; 77; 109; 118-120; 131; 136

nanocrystals; 112; 114; 117-118; 138
nanoelectronics; 220; 267
nanogenomics, 4; 158
nanogravimetry, 118; 127; 203
nanomaterials, 1-2; 6; 37
nanomechanics, 183
nanomedicine, 148
nanooptics, 183
nanoparticles, 17-18; 34-36; 38; 74; 118; 132-134; 137; 186; 228-229; 241; 247-249; 255; 267-269; 271-272; 277; 283-284; 288; 294-295
nanoscale, 219; 249; 255; 282; 293
nanosensors, 220; 237
nanostructures, 1; 41; 132-133
nanotechnology, 1; 12; 17-19; 21; 23-25; 27; 30-31; 118; 128; 132; 136; 184; 191; 219; 258; 270; 276; 280; 295; 305
nanotemplate, 59; 98; 113-114; 118; 124-125; 132; 135-137; 140-143; 145-147; 217-218
nanotubes (NT), 3; 17; 25-32; 34; 148-149; 151; 241-243; 284; 297-300
NAPPA, see Nucleic Acid Programmable Protein Array
NMR, see nuclear magnetic resonance
non-contact mode, 58
NT, see nanotubes
nuclear magnetic resonance (NMR), 12-14; 67-68; 202; 216
Nucleic Acid Programmable Protein Array (NAPPA), 65; 87-90; 105-106; 174; 181
nucleosome, 197-200

octopus rhodopsin (octR), 13; 93-95; 97; 100
octR, see octopus rhodopsin

optical tweezer, 193-195
opto-electronics, 7; 18
osteogenesis, 91; 173-174

P450, 56-58; 69; 72-73; 98; 223-228; 231
P4501A2, 12; 13; 223-228
P4502B4, 11-13; 96; 99; 223-226
P450scc, 8-10; 12-13; 51-53; 63; 69; 72-73; 77; 93; 96; 98; 100; 107-108; 116: 120; 125; 131; 135-137; 140-144; 223-226; 228-232
PANI, 256; 259; 265-266
PAOA, 265-267
PAOT, 265
penicillin G-acylase, 307-309; 311
PGA, see penicillin G-acylase
photocells, 232; 278
photosynthetic reaction centers (RC), 6; 13; 43; 46 57; 278-279
photovoltaic cells, 256; 275; 277; 281; 288
POAS, 241-242; 256-257; 266; 271
POAT, 266
polyaniline, 27; 29-30; 37
potentiometric stripping analysis (PSA), 238-240
PPV; 259-260; 295
PPY; 265
protective plate; 38; 41-43
protein microcrystals; 120; 126; 128; 131; 135; 140-141; 143; 146-147
protein-chip; 81
proteomics; 77; 81; 109-110; 118; 120; 127-128; 173
PSA, see potentiometric stripping analysis
purple membrane, 94; 97; 220; 280

QC, see quantum computing
quantum computing (QC), 100
quartz crystal balance; 77

radiation damage; 110-111; 115; 120; 128
Raman spectroscopy; 192; 285
RC, see photosynthetic reaction centers
reflectometry, 108
rhodamine B, 153-155
rhodopsin, 93-95; 97; 100; 261; 263-264

scanning probe microscopes (SPM), 183
Schottky diode, 256
SEM, 230; 255
semiconductor, 255-256; 260; 267; 272-273
sensors, 26; 29; 30
SFM, 137
STM, 17; 46; 65; 66; 189; 254; 257
surface potential, 54-57
synchrotron radiation, 108; 111; 112; 116; 118-120; 127; 129; 131; 135-136; 140-141; 143; 146-147; 195; 198; 217

tapping mode; 58; 59; 60; 62; 63
thermal stability; 43-45; 47; 202; 204-206; 208; 210-215; 217-219; 264; 291- 292; 301-306
thin films; 118; 124-125; 136-137; 139; 145; 220; 232-233; 241; 243; 259; 261; 263; 265; 277
thioredoxin; 13; 203; 208; 212-213; 215-218

urease; 303-305; 311

X-ray; 17; 54; 61; 63; 66; 107; 108-112; 115-116 119-121; 128; 131-132; 136-137; 139; 143-145; 157; 191-192; 195-198; 216; 235; 284

Yoneda, 113;-114; 135; 139-140; 142-143; 144-145; 147

µGISAXS, *see* micro Grazing Incidence Small Angle X-ray Scattering